Synchrotron Radiation in Structural Biology

BASIC LIFE SCIENCES

Ernest H. Y. Chu, Series Editor

The University of Michigan Medical School
Ann Arbor, Michigan

Alexander Hollaender, Founding Editor

Recent volumes in the series:

A Continuation Order Plan is available for this series. A continuation order will bring delivery of
each new volume immediately upon publication. Volumes are billed only upon actual shipment.
For further information please contact the publisher.

Synchrotron Radiation in Structural Biology

Edited by

**Robert M. Sweet
and Avril D. Woodhead**

*Brookhaven National Laboratory
Upton, New York*

With the assistance of

**Benno P. Schoenborn, John C. Sutherland,
Stephen W. White, Malcolm S. Capel,
and Elizabeth C. Theil**

Technical Editor

Katherine Vivirito

*Brookhaven National Laboratory
Upton, New York*

Springer Science+Business Media, LLC

Library of Congress Cataloging in Publication Data

Brookhaven Symposium in Biology (35th: 1988: Brookhaven National Laboratory)
    Synchrotron radiation in structural biology / edited by Robert M. Sweet and Avril
D. Woodhead, with the assistance of Benno P. Schoeborn . . . [et al.]; technical editor,
Katherine Vivirito.
        p.   cm. — (Basic Life Sciences; v. 51)
    "Proceedings of Brookhaven Symposium in Biology, no. 35 (BNL-52209) . . . held
May 22–25, 1988, at Brookhaven National Laboratory, Upton, New York" — T.p. ver-
so.
    Bibliography: p.
    Includes index.
    ISBN 978-1-4684-8043-6   ISBN 978-1-4684-8041-2 (eBook)
    DOI 10.1007/978-1-4684-8041-2
    1. Synchrotron radiation — Congresses. 2. Biology — Technique — Congresses. I.
Sweet, Robert M. II. Woodhead, Avril D. III. Brookhaven National Laboratory. IV.
Title. V. Series.
    QH324.9.S95B76   1989   89-16163
    574.19′28 — dc20   CIP

Proceedings of Brookhaven Symposium in Biology No. 35 (BNL-52209),
on Synchroton Radiation in Structural Biology, held May 22–25,
1988, at Brookhaven National Laboratory, Upton, New York

© Springer Science+Business Media New York 1989
Originally published by Plenum Pres, New York in 1989
Softcover reprint of the hardcover 1st edition 1989

PREFACE

The development of synchrotron radiation (SR) as a research tool was driven largely by the needs of materials scientists and solid-state physicists. However, the availability of SR has extended significantly the capability of scientists who study biological structure with radiation. This volume contains some of the results reported at a symposium held at Brookhaven National Laboratory in May 1988 to discuss the application of synchrotron radiation to structural biology.

We are grateful for financial support from the U.S. Department of Energy, the National Institutes of Health, Genentech, Inc., Blake Industries, Inc., Evans and Sutherland Co., The Upjohn Company, Eli Lilly and Company, Enraf-Nonius Service Corp., and Associated Universities, Inc. We warmly thank Ms. Nancy Siemon for her tireless efforts with correspondence and the manuscripts for this symposium volume.

Symposium Committee:

Robert M. Sweet, Chair
Malcolm S. Capel
Benno P. Schoenborn
John C. Sutherland
Elizabeth C. Theil
Stephen W. White
Avril D. Woodhead

Helen Z. Kondratuk, Coordinator

CONTENTS

AN INTRODUCTION TO THE SYMPOSIUM

Robert M. Sweet

Biology Department
Brookhaven National Laboratory
Upton, NY 11973

The symposium represented by the contents of this volume was held at Brookhaven National Laboratory during 22-25 May 1988 to discuss applications of synchrotron radiation to structural biology. The program included descriptions of the facilities available for biological studies at seven synchrotron radiation sources around the world. There are a dozen beamlines effectively dedicated to this research.

There were four invited lectures for each of several topics: X-Ray Crystallography, Small-Angle X-Ray Scattering, VUV Spectroscopy and Photobiology, Extended X-Ray Absorption Fire Structure Spectroscopy, Advances in Instrumentation, and New Techniques. In the Symposium Lecture, Hugh Huxley reviewed the history of the development of x-ray instrumentation for research on biological structure.

Mention of just a few details included in chapters in this volume will demonstrate the scope and power of synchrotron studies:

Workers have displayed diffraction from muscle fibers that reveals a 25,000 Å lowest-order maximum. This demonstration extends use of the technique to the scale accessible by light microscopes.

Analysis of protein crystal structure was accomplished with crystals only 50 microns thick.

Determination of atomic-resolution structures of virus particles has become almost routine with the use of synchrotron sources.

The tunability of the synchrotron x-ray beam allows the solution of crystal structures with only the anomalous scattering from a small number of selenium atoms, used for phasing and introduced into the structure by biological methods.

The "white" or broad bandwidth nature of SR allows use of Laue photography in crystallography. With this technique one may measure diffraction data several orders of magnitude more quickly than with monochromatic methods. This technique presents the possibility of using crystallographic methods to monitor dynamic processes in crystals.

A particular theme that was related to these sorts of research was repeated during the meeting. There are two sorts of advance that can be made in experimentation: advances in equipment and advances in methods. The very existence of synchrotron radiation sources represents a triumph in equipment development, the full impact of which can only be imagined. The development of x-ray storage phosphors by the Fujifilm and Eastman Kodak companies will provide a new generation of precise x-ray detectors for diffraction science.

On the other hand, development of clever methods can amplify the use of equipment manyfold. One such method is that of quick freezing of protein crystals, making them almost completely resistant to radiation damage. Another is the use of Laue photography for diffraction intensity measurement. This method requires no new apparatus, but demands a precise analysis of the physics of diffraction and careful experimentation. Finally, the solving of a protein crystal structure by multi-wavelength anomalous diffraction represents the culmination of years of effort by many workers. Again in this case, the apparatus was not particularly unusual; the use made of it was.

A final message from the symposium is that new advances in instrumentation are absolutely required. The most glaring need is for x-ray detectors. Existing synchrotron sources produce sufficient x-rays so that complete data sets could be measured from most protein crystals in a fraction of an hour, if there were x-ray area detectors coupled to sufficient computing power. The film methods that are in use must be supplanted by true electronic detectors. Only then will we be able to realize the full potential of synchrotron radiation for the study of biological structures.

DEVELOPMENTS IN X-RAY TECHNOLOGY AND

THEIR CONTRIBUTION TO STRUCTURAL BIOLOGY

Hugh E. Huxley

Brandeis University
Rosenstiel Basic Medical Sciences
Research Center and Department of Biology
415 South Street
Waltham, MA 02254-9110

INTRODUCTION

I want to take a broad view of this subject and trace some of the strands of technical development in X-ray diffraction which led, over a long period of time, to the present widespread use of synchrotron radiation. I will emphasize the earlier days when relatively few people were working in the field, and I will focus on the way that the technical developments actually took place, rather than on their contributions to structural biology since many of the present community may be unfamiliar with these early origins.

There is another qualification that I would like to make: I realize that it is generally considered a bad sign when someone who is supposed to be an active scientist starts talking about history. However, I thought that in an after-dinner talk I might be given a little more latitude. Also, this talk will be a very anecdotal account of what I happened to see because I was in a particular place at a particular time. Other people may know about other different pieces of this puzzle and might arrange them differently.

On the front cover of the program for this Symposium there is a figure with a time scale in milliseconds (msecs) on the horizontal axis. The vertical axis is the total number of photons in the particularly interesting part of the muscle pattern--the (143 Å) meridional reflection. The total number of counts in each of the time intervals is about a thousand; so statistics are not too bad. The data points show the behavior of the reflection, in 1 msec time channels, when we apply a very rapid 1% length change to an otherwise isometrically contracting muscle (by hooking it up in essence, to a loud-speaker solenoid with feedback, which can produce the length change in a well-controlled manner). We published the results of this experiment in 1983 with Bob Simmons and Joan Bordas. The work was done on the DORIS storage ring at the European Molecular Biology Outstation in Hamburg, and represented more or less the limit of what was technically feasible at that time. Rather than explain the detailed implication of the results, I would like to analyze the long and tortuous course of events that made it possible to carry out that particular experiment in Hamburg. These events were also connected with the development of protein crystallography.

I became acquainted with the realities of X-ray diffraction in 1948, almost exactly 40 years ago, just after the war, when I was a research student with John Kendrew and Max Perutz in the Cavendish Laboratory in Cambridge, where Sir Lawrence Bragg was professor of physics. To many people, forty years ago must sound not just like history, but positively prehistory. At that time, nevertheless, we really were already thinking about what are still current problems in the field of structural biology. Our immediate concern however, was how to get some useful information from all those patterns of spots given by hemoglobin and myoglobin crystals. These spots could be registered perfectly well, but could not be interpreted because Max Perutz had not yet discovered that heavy atom phasing would work on protein crystals.

As well as this problem, there were also concerns of a practical nature. First, how does one collect the X-ray data in a reasonable time? At that time, some exposures might be as long as a month. One also had to put up with the awful drudgery of measuring many reflections and calculating Fourier series by hand, which was the way things were done then. In my case, there was another problem; how could I extract myself from a field into which I had plunged, without too much thought, as sort of a refugee from nuclear physics and radar? The field was beginning to seem to me not only boring on a day-to-day basis, but to have very little future in it. I thought, and I was not alone in this, that protein crystallography was a hopeless task!

Quite early on in my research studentship, when I worked on lattice changes in hemoglobin crystals, Bragg suggested that it would be good experience for me to calculate a two-dimensional Patterson projection of one of the hemoglobin forms of moderately high resolution. This was a terrible chore with Beevers-Lipson strips and an adding machine cranked by hand. It took me about two weeks of solid work, and much loud complaining. One of the people I complained to was my best friend, John Bennett, who was an engineer-mathematician from Australia, and a fellow research student in Christ's College. (I think this little story illustrates how valuable it sometimes can be to have acquaintances outside your own immediate field, and to have things organized so that you meet other people and not just those in your own laboratory). Bennett sympathized with my complaints, and as it happened, was working on the original EDSAC computer in the Math Laboratory with Wilkes. He immediately realized that the calculation that was giving me so much trouble was programmable. At that time, as I remember, the computer had only about five hundred and twelve memory positions, so all the sines and cosines which we used had to be calculated by the program at each single step because there was nothing that would store them. Bennett produced a program which could do my two weeks work in about half an hour, although it did take about another half an hour to print out the results! However, by the time this system was operating properly, I had left the field of protein crystallography and had started working on muscle. In a reversal of perhaps a more normal procedure, I turned over the developments of the crystallographic computations to my Ph.D. supervisor, John Kendrew, who had a strong personal interest in computing Fouriers at that time. After much more work on that problem, he and Bennett got the system working really well with the Math Laboratory's computer so that it was all in place when the real work with isomorphous replacement began in the mid fifties. An article on the system appeared in Acta Crystallographica in 1952 (Bennett and Kendrew, 1952). The paper is a classic; it is the first one written about this type of computation done with a digital computer.

Later on, during the latter part of the fifties and in the sixties, in Cambridge and at the Royal Institution in London to which Bragg had gone, methods were worked out for recording the reflections on automatic diffractometers and also for scanning film automatically, so that progressively more and more of the chores were taken out of the initial stages of the crystallographic solution.

Having left protein crystallography, I started to learn some biology. At night when we watched over the x-ray tubes, to make sure they didn't burn the laboratory down, I read a lot of reprints and realized that people had no clear idea of the molecular mechanism of muscle, or even of the molecular structures that were present in muscle, although a tremendous amount of physiology had been done and a certain amount of biochemistry was known about the system. Some low-angle diffraction patterns had been collected by Schmitt and Bear, but with very long exposure times, of the order of a month or so in some cases, using pieces of dried muscle. I think this was done deliberately with the idea of getting more tissue in the beam--if one had 80% of water it unnecessarily diluted the patterns. Clearly, these patterns did not show a great deal about what might be happening in a live and fully hydrated muscle. The perpetual problem was then, as it is now, how does one reduce the exposure time? At that time, it was a question of reducing it to the point where one could keep an isolated muscle alive for the duration of the exposure! From the work I had done on protein crystals, I developed a strong feeling that muscles would probably give better patterns if they were fully hydrated, in their native state, and if they were alive as well, and so I set out to try to record such low-angle diagrams.

The limitations of the x-ray tubes were very apparent at that time, in 1949. The 17 procession pictures that John Kendrew was taking (on a rather inefficient sealed-off tube) needed weeks of exposure and so Max Perutz and John Kendrew, with Bragg's encouragement, hired Tony Broad, an engineer, to work on the design of a reliable rotating anode x-ray tube. There had been earlier designs, in fact there was one in the Royal Institution in London, but it was much too temperamental to be of use for the systematic collection of a large amount of data. However, at the time Broad had started working, it was clear that at least two or three years were needed to design good rotating seals, which was where the main problem was, and to design high current electron guns that would not be flashing over all the time. I could not wait that long because I had to produce my Ph.D. thesis. As luck would have it, John Kendrew heard from J. D. Bernal in London that W. Ehrenberg in his department had produced an x-ray tube with an extraordinary bright, though rather small, focal spot (Fig. 1). A. Franks, also in London, was using this x-ray tube in experiments in x-ray mirrors as a Ph.D. project. On looking into this, I learned that the tube had a very neat little electron gun which produced a focal spot about 50 microns in diameter. The small size of the spot meant that there was very efficient lateral heat transfer, because everywhere under the electron beam was very close to a part of the copper surface that did not have any beam impinging on it. It was possible to have a relatively high current density targeted into this copper anode before it began to melt, which was the basic limitation on all x-ray tube design. In this tube a current up to half a milliamp could be used. This does not sound like very much, but with a 50 micron spot, it gave a loading of nearly 10 kw per square mm, which was about 100 times better than the conventional sealed-off tubes available at that time.

The next problem was how to use this small source efficiently on a low-angle camera, because there were no focussing optics available. The solution for the muscle work was fairly straightforward, but a little tricky technically, and was to use a miniaturized low-angle slit camera. It had slits about five or ten microns in width, which matched the focal size, which was about five microns with the usual ten-to-one foreshortening of the 50 micron spot. The specimen-to-film distance was only about 3 cm but the resolution was quite high, about 1500 Å or more, so it gave a very respectable signal-to-noise ratio on the muscle reflections. These muscle reflections turned out to have spacings of about 400 Å in the wet state, and the

lines on the film were about 100 microns apart. The pattern, which was recorded on film, had to be examined in the microscope because it was so
small, but it was, in fact, a very efficient way of recording the data because virtually every single silver grain that could be developed could be
seen under the microscope. One could probably see the trace left by as few
as about 1000 photons in these very narrow lines which were only about 10
microns in width. The equatorial pattern (Fig. 2) from muscle took about
four hours to record and it indicated the presence of a double hexagonal
array of filaments. The meridional pattern (Fig. 3), including the 143 Å
reflection, took somewhat longer, about one or two days. It showed a number
of axial reflections from long repeat of a little over 400 Å (which we now
know corresponds to the helical array of myosin crossbridges) which, strik-
ingly, was unchanged by passive stretch. The most important fact, however,
was that the pattern could be recorded from a live functioning muscle.

These patterns were quite important from several points of view, but
particularly because they provided strong criteria by which the plausibility
of structures seen in the electron microscope or deduced in other ways could
be judged. When Jean Hanson and I looked at the electron microscope pictures of muscle and at light microscope pictures, we were confident that our
ideas about the nature of the structure were soundly based because they were
consistent with the x-ray patterns from the live muscle. The pictures suggested an overlapping filament structure which agreed with the double array
of actin and myosin filaments indicated by the equatorial x-ray diagrams.

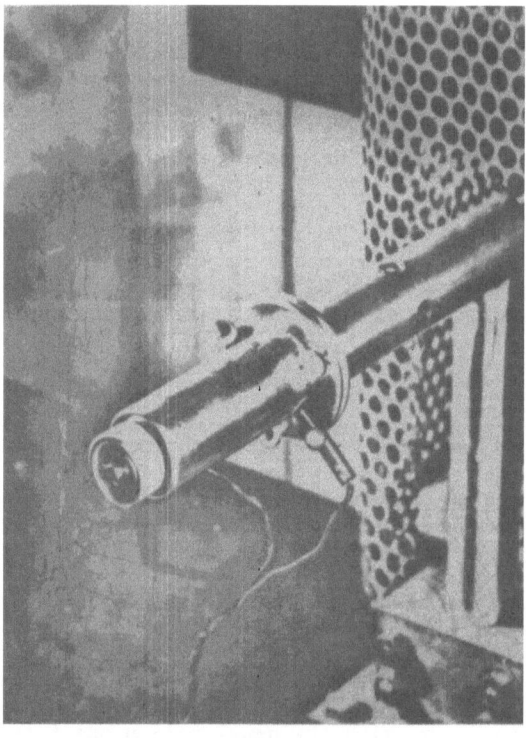

Fig. 1.   Ehrenberg's fine focus tube. The diameter of the tube is about an
          inch and the anode is at high voltage. The figure shows the
          adjustable cathode (for focussing), the windows and the protective
          screen around the high voltage lead.

The light microscope pictures also suggested a sliding filament system with
filaments remaining constant in length. We thought that this must be the
correct interpretation, because passive stretch of the muscle did not alter
the spacing of the meridional reflections. Clearly, there was some structure with a repeat spacing which did not change when the muscle length
changed. These early x-ray results gave us a great deal more confidence in
our interpretations than we otherwise would have had. In fact, I worked for
perhaps another ten years with electron microscopy of muscle and muscle filaments before the model became generally accepted, confident that I had a

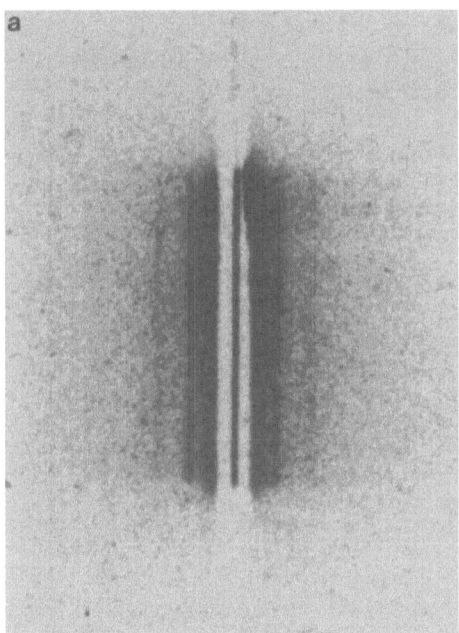 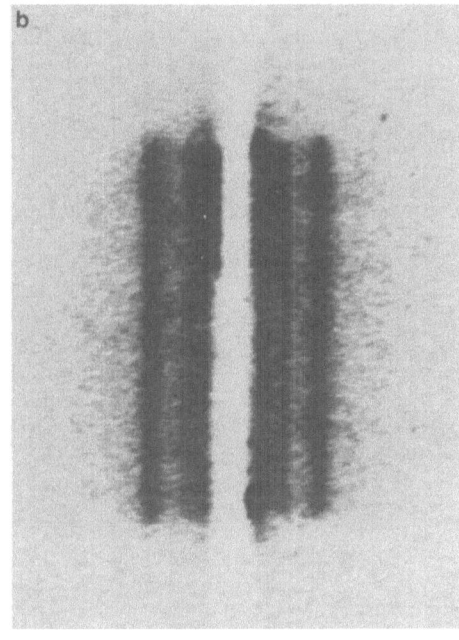

Fig. 2.   The pattern here is greatly enlarged, and shows the equatorial
          diagram from (a) relaxed muscle with the direct beam showing
          through the backstop, the 1,0 and the 1,1 equatorial reflection,
          and (b) muscle in rigor. These lines are a little more than 10
          microns in width on the original. The recording was made in about
          4 hours.

great deal of inside knowledge of what the structure was really like because
of the x-ray results. This course of events convinced me, and I think several other people, that the combination of x-ray diffraction and electron
microscopy was a very powerful tool. Also, these were days when biological
electron microscopy was only just at its beginning and people at first did
not have much confidence in the pictures that were being produced because,
in many cases, there was no way of telling whether they were artifacts or
not. However, in the case of muscles which were being processed in the same
way as many other tissues, the pictures that were produced clearly agreed
with the x-ray patterns, and this suggested that the preparative procedures
might indeed be reasonably reliable.

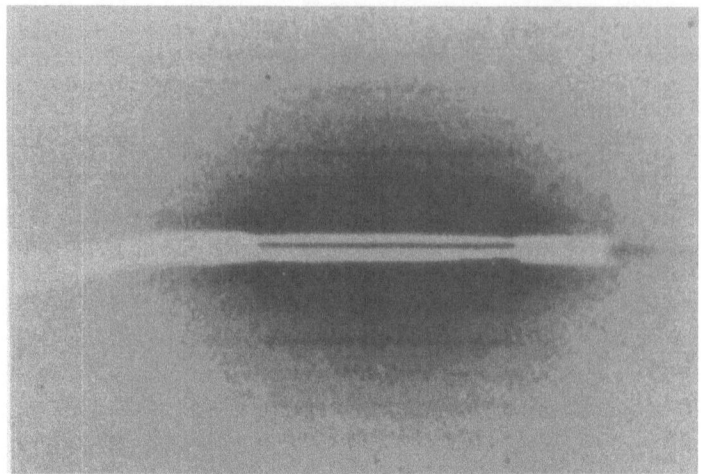

Fig. 3. Axial pattern from live relaxed frog sartorius muscle taken with early camera showing prominent 143 Å reflection and other layer lines now known to come from myosin filaments. The spacings stayed constant when the muscle was stretched.

The next phase of the these events occurred in the early and middle 1960s in Cambridge. We moved to a somewhat smaller version of the present MRC Molecular Biology Laboratory. Kendrew, Perutz, Crick, and Brenner moved from the hut outside Cavendish, and Sanger moved in from Biochemistry. Aaron Klug and his students moved in from Birkbeck, and I came from

Fig. 4. The figure shows one of the small mirrors we used in the first mirror-monochromator camera. It is essentially a Franks' mirror, about 5 cms long which was bent by being pressed onto steel pins at either end of the carriage. The degree of bending could be adjusted.

University College, London. Klug's students, Ken Holmes and Bill Longley, became interested in improving the x-ray equipment with which they had been collecting data on various kinds of viruses, in particular on the Tobacco Mosaic Virus in solution (TMV), and on crystals of the small spherical plant viruses: Turnip Yellow Mosaic Virus (TYMV), and Tomato Bushy Stunt Virus (BSV). These viruses had large unit cells, but gave only small crystals, and it took a long time to take pictures of them. At Birkbeck, Holmes and Longley successfully used sealed-off semi-fine focus stationary anode tubes from the French manufacturer Beaudoin (possibly a legacy from Rosalind Franklin).

When this group came to Cambridge, the x-ray protein data were being collected on Tony Broad's rotating anode tubes. These were being operated with relative large foci, about a millimeter after foreshortening, and very large currents and enormous power supplies were required. Holmes and Longley cleverly grafted the two tubes together with a hacksaw, putting a Beaudoin cathode that produced a small, high-brightness focus, onto the very satisfactory Broad rotating anode assembly. This combination was excellent, especially because of the high brightness and they could get a loading of about 10 kw per square mm. This was the prototype for the Elliott GX6, which many people may be familiar with and which has had a very long life.

In the meantime, I wanted to get back to x-ray work on muscle, now that I knew more about its structure, and of course I wanted to try to get patterns from contracting muscle, even though initially this seemed to be a long way off. The existence of the Holmes/Longley tubes was a big stimulus, and Holmes also was producing excellent x-ray diagrams from TMV, using a quartz crystal focussing monochromator, taking advantage of the very narrow foreshortened focus that one could get using the Baudoin cathodes on the rotating anode tubes. These gave an effective focal width of about 100 microns which matched the characteristics of the quartz monochromator fairly efficiently.

I wanted to look at muscle patterns with good resolution in two directions, so I thought it might be worthwhile to combine one of these focussing monochromators with a bent glass mirror of the type used in the Franks'low angle camera (Fig. 4). (Franks was the same man that we met in the 50s at Birkbeck). At that time we did not have very asymmetrically cut quartz crystals, so we put the mirror element first immediately in front of the X-ray tube, and then followed it with a monochromator (Fig. 5). Bernal had designed a camera in the 1920s which was arranged with a cylindrical film so that everything was in focus at about an 11 cm film distance. The resulting camera dimensions had become standardized and so the focal length of our monochromator was about 15 cm. With the monochromator stopped down, as was the normal practice at that time, this produced an excellent 2 D pattern in about a hour or so. However, quite unexpectedly, we found that we could open up the whole aperture of the monochromator, so that there were no slits at all in front of it in a horizontal direction, without getting too much background scattering. This change produced a dramatic increase in intensity and we were able to get usable pictures of muscle in about 10 minutes. This meant that we could now get the overall pattern from muscles during contraction, by taking a long series of about 600 one-second exposures. With an interval of a couple of minutes rest in between each contraction (interrupting the beam with a timed shutter) a well-dissected muscle would just about survive long enough.

We were able to do a great amount of work with this system, to characterize in detail the whole of the low-angle diffraction pattern, and to begin to see how it changed during contraction. From this work emerged the swinging crossbridge model with myosin heads able to pivot out with some linkage to the myosin backbone. Force was produced at the interacting site

9

Fig. 5. This figure is an enlarged view of the experimental set-up we were
using; the gap between the mirror and the monochromator was not as
large during actual operation as it is shown here. The mirror is
next to a prototype version of the GX6; the monochromator, specimen holder, and film are shown. Initially, the monochromator had
crystals permanently clamped to a fixed radius of curvature, but
then we found a much smoother and more usable bend could be
achieved in the same manner as in the mirrors by applying couples
to eith- end of the quartz crystal.

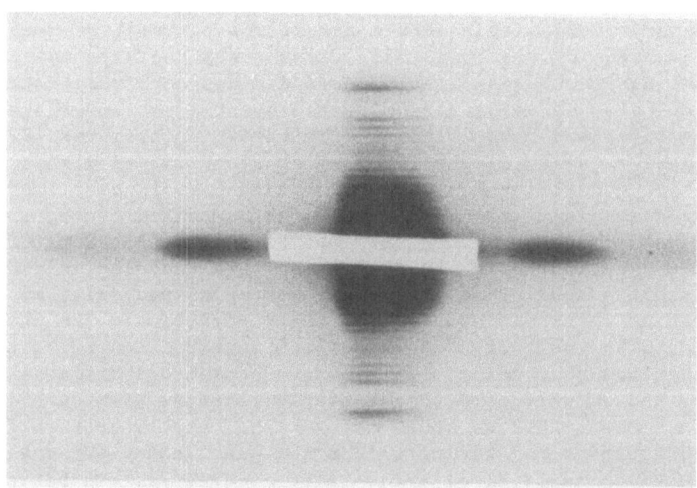

Fig. 6. Very low-angle pattern from muscle photographed on 2 meter long
monochromator camera. The figure shows the inner reflections
which are alternate orders of the sarcomere repeat of about 24,000
Å. This is from a stained muscle, where the inner part of the
pattern with the very long repeat shows much more strongly, but
similar patterns were obtained from live muscles.

between the myosin head and the actin alongside, either by a change in tilt of the myosin head or a change in its internal shape. We also constructed longer cameras and a two-meter long one, which on a rotating anode tube even gave excellent resolution of some of the orders of the sarcomere repeat (Fig. 6).

Although we took a great many pictures of contracting muscle, many of the changes we were looking at were on the borderline of detectability. This was also true of the patterns that Ken Holmes collected from insect flight muscle. And so we continued to try to think of ways to increase the x-ray intensity. The only way seemed to be to increase the velocity of the anode past the incident electron beam so that there was less time for any point on the anode to heat up. However, we believed that we could not increase the rotational speed of the anode and still stay with the same technology for the rotating vacuum seals, which were the weak point of the whole system. Therefore, it seemed that the only way to increase the x-ray intensity was to increase the anode diameter. We built various extraordinary prototypes of the so-called Big Wheel rotating anode tubes. Outside their housing, some of them looked more like a bicycle wheel than an anode, but they were the diameter we wanted; they were terrifying in action because the outer rim was traveling at almost the speed of sound. We enclosed the anode in a one-or two-inch thick outer steel casing, like an ultra centrifuge, which we hoped would contain the pieces if it exploded, which fortunately never happened. Later, Elliott Automation helped to sort out the design problems of the anode. Gerd Rosenbaum also was involved at this stage: Eventually the commercial big wheels were running and gave a useful gain of three or four times over their predecessors. One could now get up to a loading of about to 27 kw per sq mm on the anode, and with the best cameras we had then, we could get about 10^8 counts per second incident upon the muscle specimens with a useful low-angle resolution. However, it was clear that we would not advance much further down this road, because the x-ray output only improves approximately with the square root of the linear velocity of the outer surface of the anode, while the stress on the material of which the anode was constructed increased with the square of the anode velocity. So an adverse fourth power relationship worked against us, and it did not appear that the use of better alloys or carbon fibers would ever produce a dramatic improvement. Thus the prospect of taking advantage of all the information that might be in the muscle patterns looked rather bleak, unless some new approach could be found.

Ken Holmes became aware in about 1967 of the existence of synchrotron radiation in the useful x-ray range, emitted by various sorts of particle accelerators, particularly electron synchrotrons. At that time he calculated that with the machines available, the amount of radiation one could get would barely be competitive with our laboratory sources and would be much more difficult to get access to. However, in 1969, this situation changed and I have a very clear memory of the occasion. In October 1969 I chaired a committee that was trying to produce a scientific case for the establishment of a European laboratory of molecular biology. The general argument was that biology, and molecular biology in particular, would profit from the establishment of an international center for the development and application of advanced technology, particularly in structural biology which by now had become an outstandingly successful field.

There were several good reasons for wanting to establish the laboratory. First, to help science, particularly molecular biology, in Continental Europe, where even in the 1960s many of the existing universities were still rather old-fashioned and slow to change and admit new academic disciplines. Second, it would be a very good idea to have an international laboratory in Europe which could compete with laboratories in the United States, where many of the best postdoctoral researchers were going at that

time. The third reason was more vague but idealistic. It was thought that
a European laboratory would foster the cause of European unity and cooperation. However, these arguments had already been tried out on the Government
concerned and had not been strong enough to convince them that they should
go ahead and spend the necessary money. So someone--I think it was David
Phillips with John Kendrew's strong support--came up with the idea of argu-
ing the case for a European laboratory primarily on the basis of fostering
technological development. From this argument came the committee, and ul-
timately the laboratory.

Ken Holmes arrived at the first committee meeting in a state of considerable excitement. By this time he had moved to Heidelberg, where he was a
Max Planck professor, and a new research student from physics, Gerd
Rosenbaum, had arrived in his laboratory. In talking with him, Holmes found
out that the ring parameters at a relatively new electron synchrotron DESY
in Hamburg, (where Rosenbaum had been working) were much more favorable than
at earlier rings. Preliminary calculations and experiments showed that the
useful x-ray output of the ring would be about 100 times better than any
rotating anode tube that could ever be available. Moreover, in the future,
with the development of storage rings, it looked as though one would probably get at least another factor of 100 or more, so that output would then
be a factor of 10^4 better than our rotating anode tubes. The future seemed
to be very bright in all senses of the word. Rosenbaum, Holmes and Witz
published a classical paper in Nature in 1971 (Rosenbaum et al., 1971) using
the Hamburg synchrotron. The total flux was something on the order of a few
times 10^9 photons per second, in a narrow beam, considerably better than
anything possible with a rotating anode tube.

Another interesting fact about the use of the synchrotron, novel then
although it is very obvious now, was that a mirror-monochromator arrangement
had to be used for the camera. A monochromator was needed because the radiation coming out of the synchrotron was not monochromatic. A mirror was
needed because it was necessary to focus the radiation in a second direction. Thus our experience with mirror monochromators over the previous 5
years on rotating anode tubes turned out to be extremely useful, because a
lot of the instrumentation was just taken straight over and put on the synchrotron. This was all very exciting because it meant that we could look
forward, eventually, to doing all the rapid time-resolved experiments we had
always dreamed of doing on muscle but never dared to think would be possible. It was fortunate that we didn't realize how long it would take! It
also was clear that it would be a major undertaking to organize a facility
to use the radiation, with new buildings, new beamlines, elaborately controlled optics, rapid counters and visitors from all over the world. It was
a task that a European Molecular Biology Laboratory, focussing on structural
biology, would be particularly well placed to undertake. Hence, this provided very strong scientific arguments for the Laboratory, to which we added
several others. These were arguments about the importance of pursuing other
developments in techniques and instrumentation, in electron microscopy, in
crystallographic computing and in neutron diffraction. When all these arguments were put together, they were successful in persuading the governments
concerned to go ahead and establish the Laboratory. But I think that is was
the use of synchrotron radiation which tipped the balance because it was obvious that this development was scientifically feasible and likely to become
very important.

By about 1972, six or seven governments had joined the agreement to set
up the European Molecular Biology Laboratory (EMBL) in Heidelburg, with out-
stations at Hamburg and at Grenoble to look after the provision of photons
and neutrons for biologists. The initial report of the working group on
techniques and instrumentation had argued that one needed such a center to
develop these advanced techniques because it was difficult for most

biological laboratories at universities to do so. The group outlined a number of problems that might be tackled, such as x-ray analysis of large structures like viruses, ribosomes, and large multi-enzyme proteins, and electron microscope image analysis of biological macromolecular assemblies and the techniques that might be involved. It was pointed out that these intense sources should facilitate x-ray diffraction work on large molecules and small crystals, and also make dynamic experiments on crystalline enzymes possible. With synchrotron radiation it would also be possible to use x-ray microscopy in various forms on hydrated specimens to get a resolution far beyond that which could be obtained with a light microscope. Many of these predictions have turned out to be accurate.

By this time the German government had set up an initial small facility at DESY and research had begun. We also were beginning to work at the original synchrotron NINA at Daresbury. However, things took much longer to set up satisfactorily than we anticipated. There were three problems. One was that electron synchrotrons accelerated the electrons to maximum energy and then ejected them from the orbit because they were used for bombarding a stationary target with particle beams. Therefore there was a good x-ray output only for two or three msec out of each twenty msec cycle. Second, the position and size of the beams were very unstable. Third, the machines were not very reliable and they were owned and operated by physicists whose first objective in life was not to produce x-rays for biologists. So these were long, stressful, and somewhat unhappy years both in Hamburg and in Daresbury. However, sufficient results were obtained to keep up our spirits and to show others that the experiments were feasible and valuable, in principle. At the same time Wasi Faruqi, John Haselgrove and I went ahead with developing and gaining experience in our home laboratory with time-resolved electronics that would be necessary to exploit new sources when they became properly available. Andre Gabriel in Grenoble developed a one-dimensional positon-sensitive counter which had a limited maximum counting rate, but which was nevertheless extremely useful. By pushing everything to the limit I could get nearly 10^9 photons per second out of our best rotating anode tube, so I could begin to do some time-resolved muscle diffraction, with a resolution of about ten or twenty msecs, and with considerable difficulty. But when the next generation of synchrotron sources was developed, we did have a good idea how to do the experiments.

The first really good source was the storage ring DORIS in Hamburg, available in about 1977 or 1978. By then the EMBL had built an outstation for biology and, as far as I was concerned, that was the place where the real work on the synchrotron radiation began. The ring itself continued to be rather temperamental, and was still not a dedicated source. The beamlines and cameras were temperamental too. Gradually problems were resolved. But a perpetual headache remained; that was the detector problem. I emphasize this point, because, as many people are aware, it remains a problem. I was looking through some old papers last week, and I came across notes I had made for an instrumentation workshop, in 1978 in Hamburg. I had been invited as a user to suggest some technical requirements that would be appropriate. At that time we desperately wanted one- and two-dimensional position-sensitive counters which had total count rates between 2.10^5 and 10^6 per sec, preferably more. The counters that we were using, or that would very soon become available--one-dimensional counters--would not go much beyond about 2.3×10^5. We knew that there were potentially many more counts than that in the pattern from the muscle specimens. As I recall, we were assured that development of suitable instruments was in the pipeline and within a year or so satisfactory counters would be available. In practice it took until about 1984 before there was a really satisfactory one-dimensional position-sensitive counter, with a high counting rate that would go up to 10^6. And to the best of my knowledge, there is not yet a satisfactory two-dimensional counter which will reach 10^6 or more counts per sec. The lack of such instrumentation has slowed down and made much more laborious the whole exploitation of synchrotron radiation. I think it was a tragedy that more resources were not put into the development of counters very early, and that more people were not encouraged to work together and collaborate on this problem. I think that the European Molecular Biology Laboratory might have organized this cooperation better than it did. I think we should watch over problems like this in the future.

Nevertheless, thanks to the efforts of everyone at Hamburg we were able eventually to get a lot of work done, to get our beam of 10^{11} photons per sec, to get 1 msec time resolution, and do the experiment I began with and to show that there were a lot of interesting things to be found out about the structural dynamics of muscle. And as the reliability of the sources and the beamlines has improved, other users who were not as desperate as we were for as many photons as possible, but who were, nevertheless, very glad to have them, began to join in more and more. Soon the protein crystallography community became a major user of these new sources and it is difficult sometimes now for the people studying muscle to squeeze in between them! Thinking back to the beginning of this story, there seems a peculiar and somewhat ironic symmetry here!

What messages, if any, emerge from all this history? I think there is a general one: what leads to progress in science is often in the first instance not strokes of genius, but technical developments applied to good problems. This provides the stimulating new experimental data that provoke original thought, and lead eventually to the important new ideas. This confirms the concept that it is very useful to combine basic biological research and technical development in the same close environment. Secondly, there is a more specific message, that it is all too easy to build vastly expensive machines and then underuse them for years because of inadequate instrumentation. The two should be developed together, not sequentially, and there should be a major emphasis on trying to achieve this. The third message, and the best, is that it has been a great deal of fun working in this field, and I fully expect it to continue.

REFERENCES

Bennett, J. M., and Kendrew, J. C., 1952, The computation of Fourier syntheses with a digital electronic calculating machine, Acta Crystallographica, 5:109.

Rosenbaum, G., Holmes, K. C., and Witz, J., 1971, Synchrotron radiation as a source for x-ray diffraction, Nature, 230:434.

MacCHESS - A MACROMOLECULAR DIFFRACTION RESOURCE

AT THE CORNELL HIGH ENERGY SYNCHROTRON SOURCE

Wilfried Schildkamp[1,3], Keith Moffat[1], Boris Batterman[2,3],
Donald Bilderback[2,3], Tsu-Yi Teng[1], Alan LeGrand[1], and
Doletha Szebenyi[1]

Section of Biochemistry, Molecular and Cell Biology[1]
School of Applied and Engineering Physics[2]
Cornell High Energy Synchrotron Source[3]
Cornell University
Ithaca, NY 14853

INTRODUCTION

The Cornell High Energy Synchrotron Source (CHESS) operates within the
Wilson Synchrotron Laboratory of the Laboratory of Nuclear Studies at
Cornell University. CHESS runs three beamlines at the Cornell Electron-
Positron Storage Ring (CESR), which are generally operated in a parasitic
mode. CHESS is a national facility funded by the National Science Foundation. MacCHESS is a biomedical technology resource to promote macromolecular diffraction at CHESS, funded by the National Institutes of Health.

PURPOSE OF THE MACROMOLECULAR DIFFRACTION RESOURCE, MacCHESS

A major task for the Resource is to devise novel x-ray diffraction
techniques and apparatus suitable for biomolecular specimens. Projects that
have benefited from these developments include the study of macromolecules
and macromolecular assemblies such as enzymes, hormones, immunoglobulins,
membranes and membrane components, DNA and DNA-protein complexes, ribosomes, and plant, insect and mammalian viruses. Specimens are principally
in the form of single crystals, but are also of fibers, membranes and solutions. Techniques exploit the unique features of a synchrotron x-ray
source: its intensity, its polychromatic nature, and its time structure.
All developments are made accessible to scientists from universities, government, and industry who are interested in macromolecular diffraction as a
probe of structure and function by means of publications, oral presenta-
tions, conferences, and most importantly, by hands-on instruction during
visits by the user to CHESS.

USEFUL EXPERIMENTAL STATIONS AT CHESS

The most common technique to collect diffraction data from single crystals is the monochromatic oscillation technique. A versatile oscillation
camera with many useful attachments has been developed at MacCHESS. This
camera is based on an optical bench that is located on a computer-controlled

15

optical table, and can be used in all CHESS stations; however, its most frequent site of use is the A1 station. This station is a doubly focussed fixed wavelength station which receives its radiation from a 6-pole wiggler with a critical energy of 26 keV. Using a horizontally focussing Germanium (111) monochromator (Schildkamp and Bilderback, 1986), and a vertically focussing total reflection mirror (Bilderback et al., 1986), a flux of $1.5 \cdot 10^{12}$ photons/sec/mm^2 of 8-keV radiation can be achieved on the sample. The bandpass of 0.1% is adequate for protein crystallography, and the horizontal divergence of 2 mrad and the vertical divergence of .6 mrad generally provide enough collimation for high resolution data collection on film, with typical exposure times of 3 seconds to 5 minutes per degree oscillation. The camera has a variable crystal-to-film distance between 50 mm and 800 mm. Regularly used attachments are crystal coolers, monitoring ionization chambers, and helium pathways to reduce air scattering. Cassettes for two film sizes are available: 5 x 5 inches and 8 x 10 inches.

The development of the polychromatic Laue technique for data collection on single crystals and its associated software for data reduction is one of the major research areas of MacCHESS. Precise quantitation of Laue diffraction patterns yields structure factors of the same quality as monochromatic oscillation data (Temple and Moffat, 1987). However, the exposure times on a dipole beamline are much shorter, typically between 5 and 400 msec on direct exposure x-ray film. Hence, the data are less affected by the time-dependent component of radiation damage. It can be shown that with only a few Laue diffraction patterns a complete data set can be collected, in some cases from only one single crystal (Cruickshank et al., 1987). The experimental stations used for this type of studies are the B-cave and the C2 station. Laue diffraction can be done using the raw spectrum from the storage ring, but a bandpass filter is generally used. With a total reflection mirror as a low pass filter and an absorption foil as a high pass filter, a flux of approximately $1.5 \cdot 10^{13}$ photon/sec/mm^2 in a bandpass of 20% can be achieved on both dipole beamlines. Smaller bandpasses are of interest for Laue patterns from viruses, and these may be achieved by using layered synthetic microstructures (LSMs) (Moffat et al., 1986). Time-resolved crystallography has become possible for relatively slow reactions, such as thermal unfolding of proteins and certain chemically initiated enzyme reactions (Hajdu et al., 1987). This area is under active development.

For researchers with a need for higher fluxes, CHESS has the A3 station, which receives the raw spectrum from the 6-pole wiggler and yields about one order of magnitude more flux on the sample at 10 keV. However, the difficulties of working with a beam that generates such a substantial heat load are considerable. Heat loads of about 8 Watts/mm^2 can melt lead collimators and beamstops within seconds.

A far superior way to increase the speed of data collection is to use a faster detector. MacCHESS has recently received a KODAK storage phosphor system with a very high detection quantum efficiency and a dynamic range of 5 optical densities (Bilderback et al., 1988). Around 10 keV, this detector is about a factor of 10 faster than x-ray film. This system is presently being optimized for macromolecular crystallography and will soon be generally available to MacCHESS users.

Biological EXAFS is carried out in the C1 and C2 stations, with the latter one having the capability of receiving sagittally focussed radiation from the bending magnet. This station has been used, for instance, for time resolved EXAFS on Carboxymyoglobin, an experiment in which the flash of a copper vapor laser was synchronized to the time structure of the storage ring (Mills and Lewis, 1986).

The prospect of time-resolved Laue diffraction on an even shorter time scale is presently being explored at CHESS with the installation of an undulator with a fundamental peak tunable to about 9 keV. We will try to obtain a Laue pattern from the x-rays emitted by one single bunch of electrons circulating in the storage ring, which represents a time resolution of 120 picoseconds (Moffat et al., 1987).

FUTURE EXPANSION

The current CHESS laboratory was not designed with the safe handling of biohazards in mind. Experiments that might lead to aerosol production of pathogens have been banned by the CHESS safety committee, since the air-handling system of the entire building is tied together. Some experiments at biosafety level 2 were made possible by careful application of the federal guidelines in the light of the special conditions at a synchrotron radiation environment. Funding has been sought to build a new laboratory that will roughly double the capacity of CHESS, and that will contain a biosafety level 3 laboratory within an experimental station for synchrotron radiation.

At biosafety level 3, the experimental area must be hermetically sealed from the rest of Wilson laboratory. This requires a totally independent air-handling system with underpressurized rooms. Viable material is handled in a biological safety cabinet which is equipped with High Efficiency Particulate Air filters. A sterilizer is mandatory for the decontamination of disposable material, mainly glassware and syringes used in crystal mounting. After decontamination of the safety cabinet the facility can also be used for unhazardous experiments or for experiments with less stringent safety requirements.

The source for this station is a hybrid permanent magnet wiggler with a flat spectrum in the energy range of 7 to 25 keV. A new monochromator which will handle the heat load of about 5 kW has been designed (Schildkamp, 1988). It combines the good focussing properties of the present Al monochromator with a wide range of tunability and a relatively constant spectral brightness. We expect the brightness of this beamline to be higher by about one order of magnitude over the present Al beamline. Thus, the new beamline will be well suited for study of crystals of pathogenic viruses and other biohazards up to biosafety level 3.

The diffraction station is very large and has a movable helium beam pipe of several meters length inside it. With little effort this can be used to create a small angle scattering station by implementing a sample mounting stage within this beampipe. With a detector-to-specimen distance of 3000 mm, a small angle scattering resolution of approximately 3500 Å can be achieved.

The new laboratory will also contain beamlines that will allow focussed white radiation to be used, and hence will permit substantially shorter Laue exposures than those previously obtained using the B cave and C2 station.

ACCESS TO CHESS AND MacCHESS

The MacCHESS resource is operated in close concert with CHESS and access to beam time is handled by CHESS. Beam time is allocated on the basis of written, peer-reviewed proposals, in one of three modes. In a feasibility proposal up to 24 hours of beam time on any station is made available after review of a very brief proposal by the local Feasibility Study Committee. The turn-around time for feasibility studies varies between 4 and 8 weeks.

An express mode access was specifically designed to facilitate macro-molecular diffraction experiments. Up to 48 hours are made available on the Al station using standard crystallographic equipment after review of a slightly longer proposal by the local Express Mode Committee. The turn-around time for express mode proposals varies between 6 and 10 weeks due to the heavy oversubscription of the Al station.

Full proposals, which are allotted beam time up to the amount judged necessary, are sent out for written peer review. These reviews are in turn evaluated by the single CHESS Proposal Review Panel, which meets twice a year to consider all proposals from biology through physics. The proposals are given numerical ratings, and the beam time is allocated accordingly.

CONCLUSION

As a small laboratory, CHESS and MacCHESS have elected to concentrate on only certain areas of macromolecular diffraction: monochromatic oscillation, especially as applied to megastructures such as viruses and ribosomal subunits, Laue diffraction, the KODAK storage phosphor detector and its applications, and time-resolved crystallography. We anticipate that these areas will be further aided by the CHESS expansion noted above and that studies conducted now by MacCHESS staff and users will pave the way for research on the next generation of synchrotron sources, the Advanced Photon Source at Argonne National Laboratory and the European Synchrotron Radiation Facility at Grenoble.

REFERENCES

Bilderback, D., Henderson, C., and Prior, C., 1986, Elastically bent mirror for focusing synchrotron x-rays, Nucl. Instr. Methods, A246:194.
Bilderback, D., Moffat, K., Owen, J., Rubin, B., Schildkamp, W., Szebenyi, D., Smith Temple, B., Volz, K., and Whiting, B., 1988, Protein crystallographic data acquisition and preliminary analysis using KODAK storage phosphor plates, Nucl. Instr. Methods, A266:636.
Cruickshank, D., Helliwell, J., and Moffat, K., 1987, Multiplicity distribution of reflections in laue diffraction, Acta Cryst. A43:656.
Hajdu, J., Machin, P., Campbell, J., Greenhough, T., Clifton, I., Zurek, S., Gover, S., Johnson, L., and Elder, M., 1987, Milli-second x-ray diffraction and the first electron density map from laue photographs of a protein crystal, Nature, 329:178.
Mills, D., and Lewis, A., 1986, Time-resolved x-ray absorption spectroscopy of MbCO using synchrotron radiation, in: "Structural Biological Applications of X-Ray Absorption, Scattering, and Diffraction," H. D. Bartunik and B. Chance, eds., Academic Press, Orlando.
Moffat, K., Bilderback, D., Schildkamp, W., Szebenyi, D., and Loane, R., 1986, Laue diffraction: prospects for time-resolved macromolecular crystallography, in: "Structural Biological Applications of X-Ray Absorption, Scattering, and Diffraction," H. D. Bartunik and B. Chance, eds., Academic Press, Orlando.
Moffat, K., Bilderback, D., and Schildkamp, W., 1987, Protein crystal-lography with laue geometry on wigglers and undulators, in: "Workshop on PEP as a Synchrotron Radiation Source," R. Coisson and H. Winick, eds., SSRL, Stanford.
Schildkamp, W., and Bilderback, D., 1986, Helium cooling of x-ray optics during synchrotron heating, Nucl. Instr. Methods, A246:437.
Schildkamp, W., 1988, Design of a tunable and focusing single crystal mono-chromator for powerful x-ray xources, Nucl. Instr. Methods, A266:479.
Temple B., and Moffat, K., 1987, Laue film processing, in: "Computational Aspects of Protein Crystal Analysis," Helliwell, J., Machin, P., and Papiz, Z., eds., SERC, Daresbury, England.

FACILITIES AVAILABLE FOR BIOPHYSICS RESEARCH AT THE STANFORD

SYNCHROTRON RADIATION LABORATORY

R. Paul Phizackerley

Stanford Synchrotron Radiation Laboratory
Stanford Linear Accelerator Center
Stanford University
P.O. Box 4349, Bin 69
Stanford, CA 94305

OVERVIEW OF THE STANFORD SYNCHROTRON RADIATION LABORATORY

The Stanford Synchrotron Radiation Laboratory (SSRL) is a National
Facility that provides synchrotron radiation for research in the fields of
biology, chemistry, materials science, medicine, physics and other scientific fields. Of the more than 2000 research proposals that have been
submitted over the years from scientists throughout the United States and
the rest of the world, ~30% have been in the biological sciences.

SSRL originated in 1973 when a single port was installed on the
side of the SPEAR storage ring at the Stanford Linear Accelerator Center
(SLAC), making x-rays available parasitically during the operation of the
ring for high energy physics. Since those early days, the laboratory has
expanded rapidly and now has a total of twenty-two experimental stations on
nine beamlines on SPEAR and a further two beamlines, each with a single experimental station, on the much higher energy storage ring PEP. Combined,
these twenty-four experimental stations provide a broad range of photons
from 5 eV to 45 keV. The SPEAR and PEP storage rings were built and are
operated by SLAC for studies in high energy physics. Fig. 1 shows an aerial
view of the SLAC site which is located about one mile from the Stanford
University main campus. Both SPEAR and the PEP beamline buildings are indicated by arrows in the photograph. A schematic drawing of both storage
rings together with a SPEAR injector, which is currently being constructed,
is shown in Fig. 2.

SPEAR Storage Ring

SPEAR is generally run for about eight months per year and is dedicated
to synchrotron radiation research for half of this time. During dedicated
beamtime, SSRL can specify the running parameters of the ring. Although the
ring is capable of running at an electron energy of 3.5 GeV or more, it is
usually run at 3 GeV during dedicated beamtime which permits an electron
beam current of up to 100 mA to be initially stored, although between 70 and
80 mA is more typical. SPEAR has 280 possible electron "bunches" around its
circumference. However, it is normally run with much fewer bunches filled,
e.g., sixteen bunches equally spaced in four groups of four sequential
bunches. Occasionally SPEAR is run in timing mode to permit time-resolved

Fig. 1. Aerial photograph of the SLAC site. Both SPEAR and the PEP beamline buildings PB1B and PB5B are indicated by arrows. PEP interaction region 12 (IR12) in which SSRL plans to build future beamlines is also indicated. The circular dotted line shows the path of the PEP storage ring which is underground. The 2-mile long linear accelerator, currently used to inject both SPEAR and PEP, can be seen at the top. (Photograph by E. J. Faust).

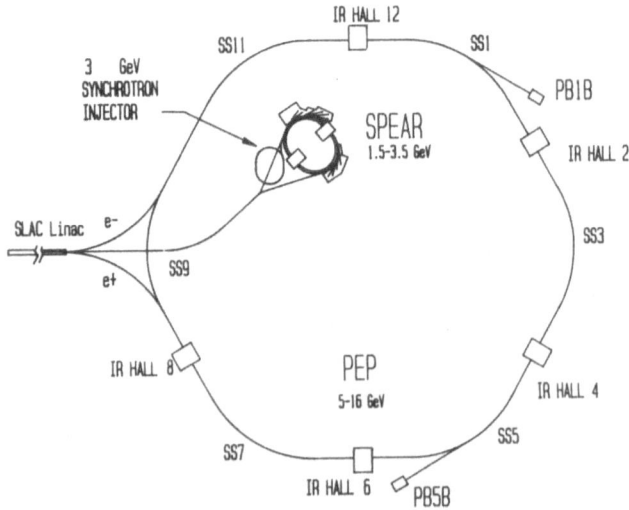

Fig. 2. Schematic layout of the linear accelerator, the large PEP storage ring (showing SSRL facilities PB1B, PB5B and IR Hall 12), the smaller SPEAR storage ring (showing both SSRL buildings with beamlines) and the SPEAR injector that is currently being built. (Drawing by R. M. Boyce).

experiments to be conducted. When operated in this mode, four equally spaced single bunches of electrons are stored producing a radiation pulse of the order of 300 psec every 195 nsec. In this mode the maximum total current is limited to about 60 mA and, therefore, the ring is often operated at a higher electron energy, such as 3.3 or 3.5 GeV.

Beam lifetime varies substantially depending on the condition of the ring. If the ring has been operating for several weeks, an average lifetime (i.e., the estimated time for the current to decay to 1/e of its current value) of up to 20 hours is frequently obtained. However, if the ring has recently been vented for modifications or repair, the beam lifetime is usually shorter than this. Shorter lifetimes also occur when the ring is operated in timing mode because of the higher electron current per bunch. Beam position monitors have been installed on each of the beamlines to determine the vertical height of the emitted photon beam. This information is used to accurately control the beam height by driving local electromagnetic trim coils in the ring. This system works very well and beam stability is normally excellent within a single fill. When SPEAR is used for colliding-beam high-energy physics experiments, the energy is usually only between 1.5 and 2.2 GeV with a beam current of less than 25 mA. Furthermore, beamline steering is normally quite limited. Consequently, parasitic synchrotron radiation experiments in the x-ray region are seldom conducted due to lack of flux and beam instabilities.

A 3-GeV injector, incorporating a 100-MeV linac and a 133 m circumference booster synchrotron is currently being constructed to permit the SPEAR storage ring to be injected independent of the existing linear

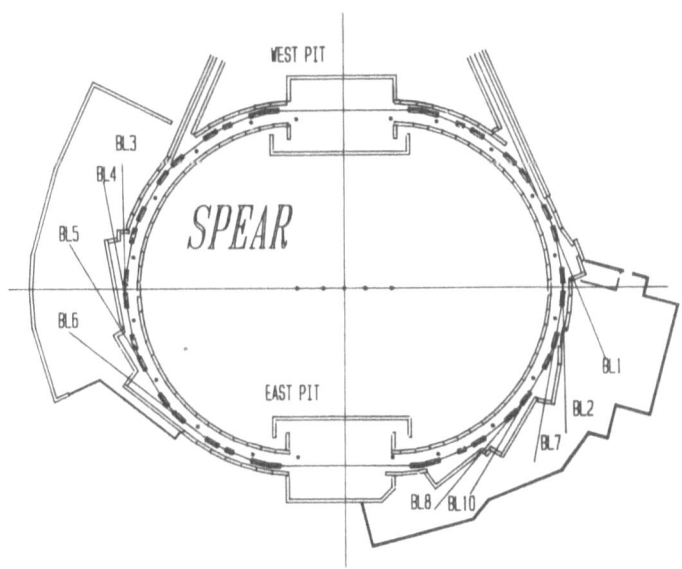

Fig. 3.   Schematic layout of the nine beamlines (BL) on the SPEAR storage ring that are housed in two separate buildings. The East and West experimental pits are used for high energy physics experiments only, with electrons circulating clockwise and positrons circulating anticlockwise. (Drawing by R. M. Boyce).

accelerator, which is heavily used for high energy physics experiments. This should lead to more time being available on SPEAR for synchrotron radiation research in future years. This new injector has been built in a way that will permit it to be upgraded at a future date to 5 GeV so that it could be used as a possible injector for the PEP storage ring.

The layout of the nine SPEAR beamlines, housed in two separate buildings, is shown in Fig. 3. Beamlines 1, 2, 3 and 8 are bending magnet lines. The remaining lines incorporate insertion devices. Beamlines 4 and 7 employ 8-pole wigglers; BL 10 a 31-pole wiggler; BL 6 a 54-pole wiggler and BL 5 has a set of four interchangeable undulators. Fig. 4 shows a set of curves of calculated photon flux as a function of photon energy for a bending magnet and two types of wiggler when SPEAR is operating at 3 GeV.

PEP Storage Ring

SSRL commissioned its first beamline on the 16-GeV PEP storage ring in November 1985. Currently there are two beamlines (PB1B and PB5B), each housed in a separate building. Both beamlines have similar beamline components. In each case the radiation source is a permanent magnet (SmCo) 2 m long 26-period undulator (λ = 7.7 cm) with a variable gap (0 to 2.1 kGauss). Radiation from the undulator is focussed onto the sample by a 1 m long Pt-coated quartz toroidal mirror which is 36 m from the source. This mirror,

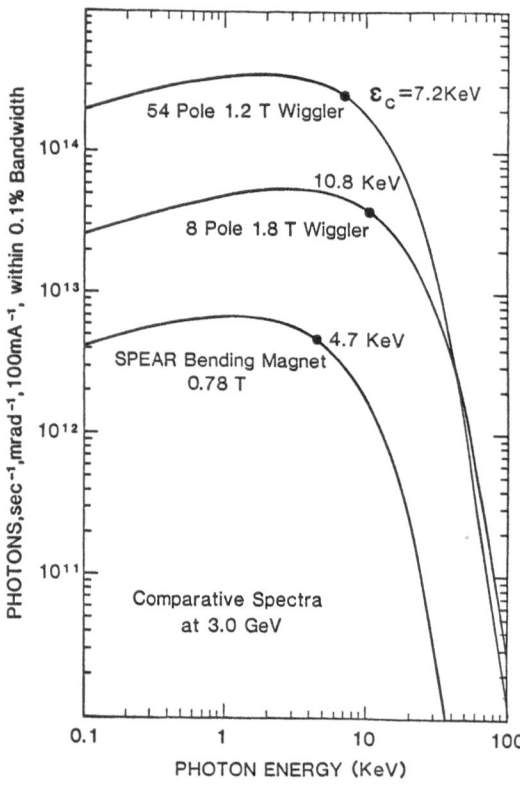

Fig. 4. Plot of calculated values of photon flux as a function of photon energy for the SPEAR bending magnets and two types of wiggler, when SPEAR is operated at 3 GeV. In each case the field strength in Tesla and the critical energy (ε_c) of the device is shown. (Plot provided by H. Winick).

which has a cutoff energy of 22 keV, can be lowered out of the beam for higher energy experiments up to ~45 keV. The lowest energy available is ~2.8 keV. The experimental enclosure for each line, positioned 60 m from the source, is a spacious room which can be operated in a "no-bars" mode. A water cooled 2-crystal monochromator is positioned 56 m from the source, just upstream of the experimental enclosure. A 6-circle Huber diffractometer is mounted on an alignment table in the PB1B experimental enclosure.

During a parasitic run, when the ring is being used for colliding-beam high-energy physics experiments, PEP is likely to operate at ~14.5 GeV with a peak current of about 12 mA and a lifetime of 3 hours. Under these conditions the beam brightness, using monochromatic radiation, is roughly an order of magnitude higher than can be produced by the 54-pole wiggler on SPEAR when operating in the dedicated mode. The peak of the x-ray flux distribution can be varied from ~12 keV to ~20 keV by changing the undulator gap. During dedicated beamtime on PEP, SSRL can specify the ring parameters. During these runs 8 GeV and ~20 mA are most likely to be encountered in the near term.

The vertical and horizontal beam size at the sample under parasitic beam conditions is 0.3 mm x 3 mm, respectively, or 6 mm x 8 mm if the focussing mirror is removed. Experiments to run PEP in a low emittance mode were conducted in December 1987, when an emittance of ~6 nm-rad at 7.1 GeV and up to 33 mA was achieved. Due to the large circumference of PEP (2200 m), the repetition time for a single bunch of electrons is 7 μsecs. The bunch length is approximately 60 psec long. This makes PEP well suited to time-resolved experiments, and experiments in protein crystallography using Laue photography (Amoros et al., 1975; Moffat et al., 1984; Moffat et al., 1986; Cruickshank et al., 1987; Hadju et al., 1987) are planned in the near future.

During the next few years, SSRL plans to reduce further the beam emittance and also hopes to build a beamline which will incorporate a 12 m undulator built from six 2-m sections that can be switched on or off independently. Furthermore, SSRL now has the possibility of developing a very long (117 m) straight section in the PEP interaction region 12 (IR12). A straight section of this length may be sufficient to accommodate a 'bypass' (Winick, 1987) and an accompanying beamline so that a flexible time structure can be generated for time-resolved experiments.

FACILITIES AND SUPPORT AVAILABLE FOR RESEARCH

The operation of SSRL is funded by the Department of Energy with additional funds from other agencies including the National Institutes of Health (NIH) and the National Science Federation (NSF) to support scientific research. Administratively, SSRL is divided into a number of different divisions. Biological research is supported by the Biotechnology division. In addition to DOE funds, the division is supported by an NIH research resources grant. This grant supports the development of instrumentation and methodology and scientific research in the fields of crystallography, x-ray absorption spectroscopy and small-angle scattering. It also supports a chemical/biochemical laboratory and computational resources. These facilities are briefly described below.

Area Detector Facility for Protein Crystallography

A facility incorporating a multi-wire proportional counter (MWPC) area detector has been built at SSRL specifically to investigate the method of multi-wavelength anomalous dispersion (MAD) phasing in macromolecular crystallography. This facility (Phizackerley, 1986), which was opened to users in the spring of 1982, was designed to take advantage of the tunability of the synchrotron x-ray source and is used to measure crystallographic data at x-ray wavelengths selected to enhance anomalous scattering effects in the sample crystal in order to derive crystallographic phases directly from the measured amplitudes. It is permanently in place at the end of the unfocussed bending magnet branch line 1-5, and has been designated beamline 1-5AD. The wavelength of the incident x-ray beam is tuned by a two-crystal monochromator and currently, the beamline does not include a mirror. Higher energy harmonics in the beam are eliminated by slightly detuning the monochromator by rotation of the water-cooled first crystal with respect to the second crystal. Although 1 milli radian of radiation is available at the experimental station, we have not yet installed focusing optics and, therefore, the intensity of the beam is no higher than is available from conventional x-ray sources.

Fig. 5. Photograph of the area detector facility experimental apparatus which includes a 5-circle diffractometer mounted on an alignment table, a MWPC area detector in the foreground that is mounted on a fully adjustable support, a fast x-ray shutter, an elemental filter holder, ion chambers, a fluorescence detector and a sample crystal cooling system. The monochromatic x-ray beam enters the experimental enclosure from the left. (Photograph by E. J. Faust).

The area detector filled with 90% Xe and 10% CO_2 at 1 atom was built in 1979 in collaboration with groups at the University of California, San Diego (UCSD) and Lawrence Berkeley Laboratory (Phizackerley et al., 1980). It is of the type that had been developed earlier for protein crystallography at UCSD (Cork, 1975). The detector has an active surface of 256 mm x 281 mm, with a spatial resolution of 2 mm and ~1 mm, respectively. It is mounted on an optical bench which is connected to the 2θ circle of a 5-circle single-crystal Huber diffractometer as shown in Fig. 5. This mounting arrangement permits the crystal-to-detector distance to be readily varied from 18 cm to 101 cm, providing an angular resolution which is sufficient to resolve unit cell dimensions of up to ~200Å. Two ion chambers mounted between collimators on the diffractometer input arm are used to measure the intensity of the incident x-ray beam and to measure x-ray absorption edges from elemental foils that can be placed between them for wavelength calibration. It is critical to the success of this experiment to measure the absorption spectra of the anomalously scattering element from the sample crystal itself, to determine the optimal x-ray wavelengths for data collection. To accomplish this, x-ray fluorescence spectra may be measured from the sample crystal using a scintillation counter which is mounted normal to the x-ray beam in the horizontal plane and can be positioned to within ~1 cm of the sample crystal. The complete system is controlled by a MicroVAX II GPX colour workstation which is also used to process the measured data. Raw diffraction images are stored on a Toshiba 5.25-inch diameter optical disk with a storage capacity of 2 x 0.4 GBytes per disk for archival purposes.

Much effort has been put into developing the MAD phasing technique using this system. Details of the MAD phasing method and some results are presented in these proceedings by W. A. Hendrickson. Crystals of the previously solved protein structure lamprey hemoglobin were used to test this new methodology (Hendrickson et al., in press). Recently, these methods have been applied to solve a number of unknown protein crystal structures, namely the 2.8Å structure of a basic 'blue'copper protein from cucumbers (Guss et al., 1988), the 5Å structure of Clostridium acidiurici ferredoxin (Murthy et al., unpublished data) and the 3.1Å structure of streptavidin/biotin complex (Pahler et al., unpublished results). The latter structure was solved using data from both SSRL and the Photon Factory in Japan.

In the case of the cucumber basic protein structure, it is worth noting that the signal from just a single Cu atom per molecule (M_r 10,000) was sufficient to solve the crystal structure using the MAD phasing technique at four wavelengths. This structure has now been refined to R=14.3%. It had resisted solution by conventional crystallographic techniques over a period of nine years because of the difficulty of producing isomorphous heavy atom derivatives. The anomalous dispersion signal in the streptavidin/biotin complex was from Se that had been chemically substituted for the sulphur in biotin. This structure has now been refined to R=16%.

Four-Circle Diffractometer

The bending magnet branch line 1-5 in fact has two experimental stations in tandem. The end station (BL 1-5AD) is occupied by the area detector system as mentioned above. The other experimental station, upstream of the area detector, is called BL 1-5 and is a general purpose station. A system incorporating an Enraf Nonius CAD4 diffractometer, mounted on its side because of the horizontal polarization of synchrotron radiation, can be temporarily installed at this station. The system was built in 1977 (Phillips et al., 1979) and was used for initial experiments on anomalous scattering phasing methods (Phillips, 1978) and for the investigation of anomalous scattering factors (Phillips et al., 1978; Templeton et al., 1982), dichroic effects (Templeton and Templeton, 1980; Templeton and

Templeton 1982; Templeton and Templeton, 1985) and beam polarization (Templeton and Templeton, 1984). At the time of writing, this diffracto-meter is being upgraded by Enraf Nonius Corp. to incorporate new micro-processor-based control and readout electronics and control software. After the upgrade, it will be interfaced to a MicroVAX II computer system.

Rotation Camera for Protein Crystallography

An Enraf Nonius rotation camera, for the photographic collection of crystallographic data from macromolecules, is permanently installed on the 8-pole wiggler side station 7-1 (Phizackerley et al., unpublished data). Although most branch lines at SSRL are general purpose, that is to say they are used for a number of different types of experiments from materials science to biology, BL 7-1 is dedicated to the rotation camera. This

Fig. 6. Photograph showing the rotation camera facility experimental apparatus. The focussing crystal monochromator and focussing mirror enclosures can be seen in the background. The rotation camera with four flat film cassettes mounted on a film carousel can be seen in the foreground. Ion chambers and a shutter/ elemental foil assembly are just upstream of the rotation camera. These components and the rotation camera are mounted on an alignment carriage. The cooling system temperature controller can be seen above the camera near the roof. (Photograph by E. J. Faust).

facility has been open to users since the spring of 1984. It has been in strong demand and so far, it has been used to collect data from ~50 different crystalline proteins. The experimental station has been equipped with an enclosure for radiation protection that is the size of a small room and can be operated in no-bars mode for easy access to the experimental apparatus (Fig. 6).

The high intensity beam available from the wiggler is focused by x-ray optical elements in both the horizontal and the vertical planes. This doubly-focused beam facilitates the collection of data several hundred times faster, and usually to a higher resolution, than with conventional x-ray sources. An asymmetrically cut and cylindrically bent triangular crystal monochromator is used to focus ~1 milli radian of radiation in the horizontal plane. The monochromator is followed by a platinum-coated fused-silica x-ray quality mirror which is inclined to the incident x-ray beam to reject the higher energy harmonic components in the beam. The mirror can be bent to focus the beam in the vertical plane. A remotely controlled high-resolution TV camera has been mounted inside the enclosure to simplify beam focusing and camera alignment.

The rotation camera, together with ion chambers to measure the intensity of the incident beam, a backstop and a film carousel capable of holding up to 8 flat 5 inch x 5 inch film cassettes or 4 high resolution angled film cassettes, have been mounted on an optical bench, which is in turn mounted on an alignment carriage. This carriage, which provides remotely controlled translational and rotational adjustments, is mounted on a second optical bench which can be rotated about a vertical axis positioned directly below the monochromator crystal to permit the x-ray wavelength to be changed over a limited range. A larger change of x-ray wavelength can be obtained by interchanging the monochromator crystal. This provides the opportunity to collect anomalous scattering data at any one of a large number of elemental absorption edges. However, the system has been used at either 1.5418Å (CuKα) or 1.08Å for extended periods of time to accommodate the maximum number of proposals. It has primarily been used for the rapid collection of high quality diffraction data to high resolution and for the investigation of smaller and also radiation-sensitive crystals, since the effects of radiation damage do not normally take place immediately.

The motions of the beamline optical elements and the rotation camera phi axis and film carousel are under computer control and software has been written for system alignment and data collection that is both flexible and straightforward to use. Furthermore, provision has been made to rotate the crystal phi axis at a speed determined by the incident x-ray flux, thereby making the intensity of the recorded diffraction pattern independent of beam intensity fluctuations. An elemental foil can be remotely interposed between two ion chambers for x-ray wavelength calibration. An FTS systems crystal cooling device provides temperature control at the sample position down to just below 0°C.

We are currently building a low temperature cryostat that will be capable of rapidly cooling the sample crystal down to near liquid nitrogen temperatures following the recent success of collecting crystallographic data from protein crystals at these temperatures (Hope, 1988). These experiments, and work carried out on the SSRL rotation camera facility on large ribosomal subunits (Hope, personal communication) using a low temperature cryostat developed by H. Hope at the University of California at Davis, have shown that radiation damage in the sample can be almost eliminated at these temperatures. After we at SSRL have gained experience with this technique, we also plan to use the cryostat on the area detector system mentioned earlier.

Fiducial marks can be recorded on the film packs after exposure using a specially adapted x-ray generator located near to the rotation camera. A large darkroom, also close to the camera, has been equipped with temperature controlled processing tanks with N_2 burst capability. No film densitometer is available at SSRL for film scanning.

It is hoped that in the near future, film will be replaced by an Imaging Plate Detector System (Miyahara et al., 1986; Amemiya et al., 1988; Whiting et al., 1988). It is reported that this system has a 100% detective quantum efficiency for 8 - 17 keV photons, the spatial resolution is less than 0.2 mm and the dynamic range is 10^5. In view of these excellent characteristics, we may either build a system of this type or buy one, if an acceptable system becomes commercially available in the near future.

The rotation camera facility is rigidly constructed and the user is protected against inadvertently missetting critical alignment motions by software protection. Consequently, in all but a few cases, users have been able to rely on the camera being optimized for the duration of their data collection. A rotation camera facility user guide (Merritt and Phizackerley, 1985) is available from SSRL.

First-Come First-Served Scheduling Procedure. A special scheduling procedure has been implemented for the rotation camera facility to provide easy access. Beamtime is obtained through the submission of a very short application form. This form is reviewed by the Biology subpanel of the Proposal Review Panel and given a rating. These ratings are only used to schedule the line in periods when it is oversubscribed. If the line is not oversubscribed, beamtime is assigned on a first-come first-served basis. Applications for the rotation camera may be submitted at any time. About two months before each dedicated run, past users are informed of the run and invited to submit proposals. Each application is good for one run within the year after its submission.

X-Ray Absorption Spectroscopy Apparatus

Over the years, there has been great activity in the field of x-ray absorption spectroscopy (XAS) at SSRL. There is no dedicated experimental station for XAS but experiments can be carried out on many of the bending magnet and wiggler/undulator beamlines (i.e., beamlines 1-5, 2-2, 4-1, 4-2, 6-2 and 7-3). Experimental equipment is available to perform both transmission and fluorescence XAS measurements at temperatures down to 4 K. Tables and plots of monochromator glitch spectra and EXAFS data collection and analysis software with documentation is available.

A set of three standard SSRL 6-inch long ion chambers, mounted on an optical bench with appropriate adjustments, complete with Huber X-Y slits and sample holders is available for transmission EXAFS measurements. Three Stern/Heald/Lytle ionization chambers, complete with soller slits and a set of filters, is available for fluorescent measurements. Two adjustable arrays, each with four NaI(Tl) scintillation detectors and readout electronics, are also available for fluorescence measurements. Lastly, a Ge solid state detector with ~200 eV resolution at liquid N_2 temperature, is also available. Currently, a joint project involving SSRL and the Physics Institute at USC (Warburton et al., 1986; Iwanczyk et al., 1988) is underway to develop an array of HgI_2 detectors that has been shown to provide a resolution of ~300 eV at room temperature.

Low temperature XAS measurements can be performed using an Oxford Instruments continuous flow liquid helium cryostat down to 4 K. It can be used on any of the beamlines mentioned above except BL 4-1. This cryostat allows one to study samples at ambient pressure and incorporates a set of large and thin (total thickness ~90μm) Al/mylar windows that permits efficient collection of fluorescent radiation even at fairly low x-ray energies. Transparent mylar windows are also available.

Using BL 6-2, with the 54-pole wiggler running as an undulator and with a new thinner set of beamline Be windows, fluorescence EXAFS and edge spectra can be recorded from dilute samples with absorbing elements that have edges in the 2-3 keV region. This region includes the P, S, and Cl K-edges and the L-edges of the second row transition metals (Hedman et al., 1986). Overall resolution in this region is very high (~0.5 eV), thus making it possible to obtain enhanced information about the electronic structure of the elements being studied.

Rapid Turn-Around Facility for XAS. A rapid turn-around XAS facility is currently being implemented at SSRL. This facility, which will use up to 30% of the time available on the experimental station BL 1-5, will provide a convenient way for users to perform straightforward XAS experiments. Basic experimental apparatus including ion chambers and fluorescence detectors will be provided. Experienced users can apply for up to two blocks of six 8-hour shifts per year on a letter-proposal basis which will be reviewed in-house for feasibility. New users, who will be apprenticed to an experienced user for training, can apply for two blocks of four shifts per year, that will be scheduled contiguously with the experienced user.

Small Angle X-ray Scattering Camera

A Small Angle X-ray Scattering (SAXS) camera system has been built at SSRL (Miake-Lye, 1983; Hubbard, 1987) for scattering studies of molecules in

Fig. 7.   Photograph of the SAXS camera. The x-ray beam runs from left to right. Two sets of X-Y slits, with an ion chamber between them, can be seen on the left. An adjustable detector support, shown without the detector attached, is on the right. A tilt/rotation sample stage is mounted between the downstream slits and the detector support. This support is connected to a Huber goniometer that is positioned directly below the sample stage so that it may be inclined to the direct x-ray beam. (Photograph by S. Wakatsuki).

solution and for diffraction studies from oriented membranes and from thin or mono layers. This camera system requires an end station, and has been used on both beamlines 2-2 and 4-2.

The camera, shown in Fig. 7, incorporates two sets of collimating X-Y slits upstream of the sample with an ion chamber interposed between them to measure the intensity of the collimated beam. Three types of sample holders are available. The first supports a capillary tube for solution scattering experiments, the second supports a disk for diffraction studies from oriented membranes, and the third incorporates a tilt-rotation stage which supports a helium filled enclosure for diffraction studies from thin and mono layers. Each of the above components are mounted on an optical bench that runs parallel to the x-ray beam, so that individual components can be moved with respect to each other. Each component can also be adjusted vertically and horizontally with respect to the optical bench, and the bench can be translated in a horizontal plane normal to the x-ray beam. A vertical linear position sensitive detector (LPSD) (Gabriel, 1977) is mounted on an arm which can rotate around a vertical axis directly below the sample position. The path length between the sample and the detector can be varied from 10 to 70 cm and is evacuated. The 90% Ar / 10% CO_2 filled LPSD has an active length of 72 mm which is divided into 256 pixels, and is typically run at a count rate of ~50 kHz. Stepping or DC motors are used to drive most of the camera adjustments.

Currently, the necessary hardware and software (on a MicroVAX II) is being developed to facilitate time-resolved x-ray diffraction studies down to the msec time scale (Wakatsuki and Spann, unpublished data). This system incorporates a powerful strobe lamp for sample excitation, optics to simultaneously perform time-resolved optical absorption spectroscopy, and a temperature-controlled sample enclosure.

Chemical and Biochemical Laboratory

The chemical and biochemical laboratory at SSRL includes two stereo microscopes (one with a total magnification sufficient to view micro-crystals down to 10μm in size), a Supper optical analyzer, wax melters, fiber optics light sources and adequate bench space for crystallographers to mount sample crystals. The laboratory is also equipped with an ultra-centrifuge (1000-65000rpm), a microfuge, a UV-VIS spectrometer, an inert atmosphere glovebox, ultrafiltration cells, a convection oven (225°C max.), a pH meter, analytical balances (0.001 - 400 gm), a water purifier, a refrigerated circulator (-30 to 100°C), a refrigerator/freezer, an ice machine, a dishwasher, a vortex mixer and an ultrasonic cleaning bath. There are also a number of common chemicals and a selection of basic laboratory glassware.

Computer Resources

The SSRL computer group is a part of the Biotechnology division. The group is responsible for all computer resources at SSRL. The central computer, which is used both for experimental data analysis and for administration purposes, is a DEC VAX 8810 with 64 MBytes of memory and 3.5 GBytes of disk storage. It is linked via Ethernet to various terminals through a series of DECServer terminal servers and also to an LPS40 and several LN03 laser printers placed strategically throughout the laboratory.

Until recently, each experimental station has been equipped with a DEC PDP-11/34 computer system, but these are currently being phased out and replaced with DEC MicroVAX II/GPX workstations. Each workstation typically has 5 MBytes of memory, a 70-MByte hard disk and a 90-MByte tape drive. The central VAX 8810 and all beamline computers are being run exclusively under the VMS operating system and provide FORTRAN 77 and 'C' compilers.

All VAX computers in the laboratory, including those at the remote PEP sites, are connected by DECnet via an Ethernet cable. The SSRL network is connected to, and forms part of, the worldwide HEPnet. Communication to the outside world is also possible using BITnet. A DECserver based dial-in telephone modem (1200 Baud) with 8 lines can be used to gain access to any SSRL VAX computer from outside the laboratory. A PDP-11/34 computer system, also linked by Ethernet, and equipped with a 800/1600 bpi TE16 9-track magnetic tape drive, three RL01 and three RL02 disk drives and an RX02 dual floppy disk drive is available as a data transfer system. It provides the capability of transferring data from one format and medium to another format and/or medium. A description of this facility is available from SSRL.

The computer group also provides user support and documentation for a number of software packages for CAMAC instrumentation control, graphics and experimental data acquisition and data analysis that were written in-house.

USER ACCESS TO THE FACILITY

In order to obtain access to beamtime and facilities at SSRL, a research proposal must be submitted which is then subjected to peer review. Based on scientific merit, the proposal is then assigned a rating by a Proposal Review Panel and beamtime is later.assigned based on that rating. However, it is sometimes possible to obtain a small amount of beamtime without review based on a letter of intent. Proposals can be submitted at any time; however, proposals should be received by 1 March to be rated by the Proposal Review Panel in June and by 1 September to be rated the following January. A detailed publication entitled "SSRL User Guide" is available from:

> User Research Administration,
> Stanford Synchrotron Radiation Laboratory,
> SLAC,
> P.O. Box 4349, Bin 69,
> Stanford, California 94309-0210

or

> Telephone (415) 926 2050
> FTS   462 2050

or

> FAX   (415) 926 4100
> BITNET   Carol@SSRL750

The user guide includes (i) the necessary application forms, (ii) description of the proposal submission, review and scheduling procedures, (iii) administrative procedures, (iv) list of available documentation, (v) information about support facilities and available equipment, (vi) maps of the area and local transport and motels, (vii) staff responsibilities and (viii) characteristics of each of the branch lines. There is no charge for beamtime or for the use of SSRL owned facilities.

ACKNOWLEDGMENTS

SSRL is supported by the Department of Energy, Office of Basic Energy Sciences and the National Institutes of Health, Biotechnology Resource Program, Division of Research Resources. Some of the equipment described above was developed with support from the National Science Foundation.

REFERENCES

Amemiya, Y., Matsushita, T., Nakagawa, A., Satow, Y., Miyahara, J., and Chikawa, J., 1988, Design and performance of an imaging plate system for x-ray diffraction study, Nucl. Instr. and Meth., A266:645.

Amoros, J. L., Buerger, M. J., and Canut de Amoros, M., 1975, "The Laue Method," Academic Press, New York.

Cork, C., Hamlin, R., Vernon, W., Xuong, Ng. H., and Perez-Mendez, V., 1975, A xenon-filled multiwire area detector for x-ray diffraction, Acta Cryst., A31:702.

Cruickshank, D. W. J., Helliwell, J. R., and Moffat, K., 1987, Multiplicity distribution of reflections in laue diffraction, Acta Cryst., A43:656.

Gabriel, A., 1977, Position sensitive x-ray detector, Rev. Sci. Instrum., 48:1303.

Guss, J. M., Merritt, E. A., Phizackerley, R. P., Hedman, B., Murata, M., Hodgson, K. O., and Freeman, H. C., 1988, Phase determination by multiple-wavelength x-ray diffraction: crystal structure of a basic "blue" copper protein from cucumbers, Science, 241:806.

Hajdu, J., Machin, P. A., Campbell, J. W., Greenough, T. J., Clifton, I. J., Zurek, S., Gover, S., Johnson, L. N., and Elder, M., 1987, Millisecond x-ray diffraction and the first electron density map from laue photographs of a protein crystal, Nature, 329:178.

Hedman, B., Frank, P., Penner-Hahn, J. E., Roe, A. L., Hodgson, K. O., Carlson, R. M. K., Brown, G., Cerino, J., Hettel, R., Troxel, T., Winick, H., and Yang, J., 1986, Sulfur K-edge x-ray absorption studies using the 54-pole wiggler at SSRL in undulator mode, Nucl. Instr. and Meth., A246:797.

Hendrickson, W. A., Pahler, A., Smith, J. L., Satow, Y., Merritt, E. A., and Phizackerley, R. P., 1988, Crystal structure of core streptavidin determined from multi-wavelength anomalous diffraction of synchrotron radiation, unpublished.

Hendrickson, W. A., Smith, J. L., Phizackerley, R. P., and Merritt, E. A., 1988, Crystallographic structure analysis of lamprey hemoglobin from anomalous dispersion of synchrotron radiation, Proteins, 4:77.

Hope, H., 1988, Cryocrystallography of biological macromolecules: a generally applicable method, Acta Cryst., B44:22.

Hope, H., and Yonath, A., 1987, Personal communication.

Hubbard, S. R., 1987, Small-angle x-ray scattering studies of calcium-binding proteins in solution, Ph.D. Thesis, Stanford Univ.

Iwanczyk, J. S., Warburton, W. K., Hedman, B., Hodgson, K. O., and Beyerle, A., 1988, The HgI_2 array detector development project, Nucl. Instr. and Meth., A266:619.

Merritt, E. A., and Phizackerley, R. P., 1985, Users' guide to the SSRL Rotation Camera Facility, SSRL Report, 127B-85-V1.

Miake-Lye, R. C., 1983, Anomalous X-ray scattering as a probe of biological structure, Ph.D. Thesis, Stanford Univ.

Miyahara, J., Takahashi, K., Amemiya, Y., Kamiya, N., and Satow, Y., 1986, A new type of x-ray area detector utilizing laser stimulated luminescence, Nucl. Instr. and Meth., A246:572.

Moffat, K., Bilderback, D., Schildkamp, W., and Volz, K., 1986, Laue diffraction from biological samples, Nucl. Instr. and Meth., A246:627.

Moffat, K., Szebenyi, D. M. E., and Bilderback, D. H., 1984, X-ray laue diffraction from protein crystals, Science, 223:1423.

Murthy, H. M. K., Hendrickson, W. A., Orme-Johnson, W. H., Merritt, E. A., and Phizackerley, R. P., 1988, Crystal structure of Clostridium acidi-urici ferredoxin at 5Å resolution based on measurements of anomalous x-ray scattering at multiple wavelengths, J. Biological Chem., in press.

Phillips, J. C., 1978, Crystal structure determination using synchrotron radiation, Ph.D. Thesis, Stanford Univ.

Phillips, J. C., Cerino, J. A., and Hodgson, K. O., 1979, A four-circle diffractometer on a focused, tuneable synchrotron radiation source: mechanical design, computer control and evaluation of system performance, J. Appl. Cryst., 12:592.

Phillips, J. C., Templeton, D. H., Templeton, L. K., and Hodgson, K. O., 1978, L_{III}-edge anomalous x-ray scattering by cesium measured with synchrotron radiation, Science, 201:257.

Phizackerley, R. P., Cork, C. W., Hamlin, R. C., Nielsen, C. P., Vernon, W., Xuong, Ng. H., and Perez-Mendez, V., 1980, Progress report on the development of an area detector data acquisition system for x-ray crystallography and other x-ray diffraction experiments, Nucl. Instr. and Meth., 172:393.

Phizackerley, R. P., Cork, C. W., and Merritt, E. A., 1986, An area detector data acquisition system for protein crystallography using multiple-energy anomalous dispersion techniques, Nucl. Instr. and Meth., A246:579.

Phizackerley, R. P., Cox, A. D., and Merritt, E. A., unpublished data.

Templeton, D. H., and Templeton, L. K., 1980, Polarized x-ray absorption and double refraction in vanadyl bisacetylacetonate, Acta Cryst., A36:237.

Templeton, D. H., and Templeton, L. K., 1982, X-ray dichroism and polarized anomalous scattering of the uranyl ion, Acta Cryst., A38:62.

Templeton, D. H., and Templeton, L. K., 1984, Anomalous scattering measured by single crystal diffraction, SSRL Activity Report 84/01:IX-27.

Templeton, D. H., and Templeton, L. K., 1985, X-ray dichroism and anomalous scattering of potassium tetrachloroplatinate(II), Acta Cryst., A41:365.

Templeton, L. K., Templeton, D. H., Phizackerley, R. P., and Hodgson, K. O., 1982, L_3-edge anomalous scattering by gadolinium and samarium measured at high resolution with synchrotron radiation, Acta Cryst., A38:74.

Wakatsuki, S., and Spann, U., unpublished data.

Warburton, W. K., Iwanczyk, J. S., Dabrowski, A. J., Hedman, B., Penner-Hahn, J. E., Roe, A. L., Hodgson, K. O., and Beyerle, A., 1986, Development of mercuric iodide detectors for XAS and XRD measurements, Nucl. Instr. and Meth., A246:558.

Whiting, B. R., Owen, J. F., and Rubin, B. H., 1988, Storage phosphor x-ray diffraction detectors, Nucl. Instr. and Meth. A266:628.

Winick, H., 1987, PEP bypasses, in: "Proceedings of SSRL Workshop on PEP as a Synchrotron Radiation Source," R. Coisson and H. Winick, eds., SSRL, Stanford.

SYNCHROTRON RADIATION FACILITIES FOR BIOLOGICAL USE AT THE INSTITUTE FOR SOLID STATE PHYSICS AND PHOTON FACTORY

Takashi Ito

University of Tokyo
Meguroku, Komaba 3-8-1, Tokyo 153
Japan

BEAMLINE AT THE INSTITUTE FOR SOLID STATE PHYSICS

Irradiation Experiments

Irradiation experiments at the Synchrotron Radiation Laboratory (SRL), Institute for Solid State Physics (ISSP) began over 10 years ago. The experiments started in the fall of 1977 with white synchrotron radiation from the electron storage ring called INS-SOR that was operated at 0.38 GeV. The first group of biological materials used was <u>Bacillus</u> <u>subtilis</u> spores and yeast cells. The synchrotron radiation was extracted through the MgF_2 window installed at the end of vacuum line of the storage ring, and was introduced into the vacuum irradiation chamber connected to it. The surprisingly short irradiation time required to get a significant biological effect prompted us to construct a monochromator system, especially in the vacuum-UV (VUV) region. In 1980 we installed a 2.2 m Wadsworth mount monochromator along with some accessories for the measurement of photon intensity. Surprised by a rapid deterioration of optical elements, so that there was insufficient time to perform our experiments, we inserted a pre-mirror chamber that houses two mirrors to remove unnecessary short wavelength radiation before reaching the monochromator. We found later that the most critical factor is the vacuum. After a period of trial and error, we were satisfied with a small-scale improvement for performing irradiation experiments in the wavelength range from 150 to 250 nm (Ito et al., 1984). Several important contributions in VUV photobiology were made with this arrangement.

To extend the usable wavelength to a range shorter than the limit set by the optics of the monochromator, we plan to remove the pre-mirror system and the MgF_2 window. This should widen the wavelength range and, at the same time, substantially improve the photon intensity. Fig. 1. shows the present irradiation system installed at the beamline 5 at the ISSP (INS-SOR) storage ring (Hieda et al., 1986). A set of differential pumping systems were installed upstream of the monochromator. The monochromator itself is evacuated by an ion pump, and the irradiation chamber is evacuated by a turbo-molecular pump. The vacuum, under routine conditions, is 10^{-9} Torr in the upstream vacuum line, and 10^{-7} Torr in the monochromator and sample chamber, respectively. A gate valve between the monochromator and the use sample chamber is equipped with the MgF_2 window as an option, by which the use of wavelength region above 130 nm is possible at 10^{-6} Torr in the

irradiation chamber without disturbing the vacuum of the monochromator and
the upstream components. The Al-coated and Au-coated elements of the
monochromator can be exchanged, depending on the range of wavelength needed.
This change is normally coupled with the insertion or removal of the MgF_2
window. The sample chamber also was improved during this period and the one
presently in use is the third version.

The photon intensity at the sample position was 1.3×10^{16} photons/m^2s
per 100 mA ring current and 5.5×10^{17} photons/m^2s per 100 mA ring current
for the wavelength range below and above 130 nm respectively (corresponding
to Au- or Al-coated optics). Under actual irradiation conditions, a few
hours and a fraction of an hour below and above 130 nm are required, respec-

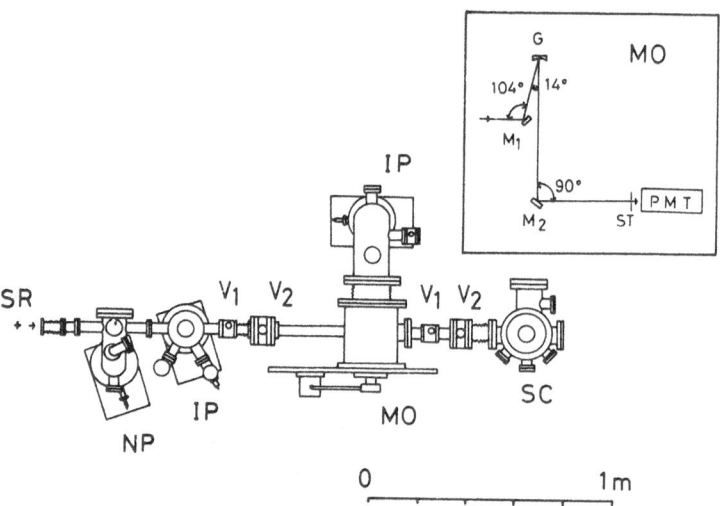

Fig. 1.   Schematic diagram of the irradiation system with a 2.2m modified
          Wadsworth monochromator (top view). M_1, collimating mirror; G,
          concave grating; M_2, deflecting mirror; ST, slit; PMT, photo-
          multiplier tube; NP, noble pump; IP, ion pump; V_1, fast close-
          ing valve; V_2, gate valve; MO, monochromator; SC, sample
          chamber. (Hieda et al., 1986).

tively, for the fluence of the order of 10^{20} photons/m^2. These requirements, however, also depend nearly proportionally on the ring current which
decays with time according to the ring parameters. The fluence mentioned
above (10^{20} photons/m^2) is sufficient to induce changes in a molecule with a
relatively small absorption cross-section. The action cross-section for the
inactivation of _Bacillus_ spores at 160 nm is about $10^{-19} m^2$/photon and that
of DNA single strand break is $10^{-18} m^2$/photon. The resolution of wavelength
is expected to be on the order of a few percent under usual irradiation conditions.

The selection of desired wavelength and the fluence can be controlled
by a desktop microcomputer. Steps for irradiation procedures are executed
automatically by a monitoring of the ring current.

Absorption and Photoacoustic Spectroscopy

Absorption spectroscopy can be performed at the same beamline using the same sample chamber with appropriate settings for the measurement. For measurements of transmission a slit is placed in front of or behind the sample film to define its area. Wavelength scanning and the recording of transmitted light are carried out by a computer-assisted device. The ring current is also monitored at every measuring point and is used to normalize the reading of photomultiplier output.

Devices for photoacoustic spectroscopy are set to the irradiation system as follows. The chopping device, a specially designed multi-fork type, is hung from the top flange of the irradiation chamber to provide a chopped light beam with a near sinusoidal time structure. This system has been operated at 60-100 Hz in the vacuum of 10^{-7} Torr. The gas-microphone photoacoustic cell is set by an adaptor flange on the air side of the vacuum irradiation chamber. The signal from the microphone is detected by a phase-sensitive-detection system as a function of wavelength, and it is fed into the same microcomputer system to process. The whole system has been successfully operated for the past several years (Inagaki et al., 1985).

General information on the SRL beamlines can be found in the Activity Report annually published by SRL of ISSP, University of Tokyo. A comprehensive account by Namioka (1986) on the VUV and soft X-ray beamlines at the synchrotron radiation facilities in Japan may also be useful to users.

BEAMLINE AT THE PHOTON FACTORY

Irradiation Experments

Several beamlines (BL-1B, BL-4A and BL-11B) are available for the irradiation experiments in the X-ray region at PF, where a storage ring is in operation at 2.5 GeV. An irradiation system with a channel-cut Si(111) crystal monochromator for the X-ray range from 4-16 KeV(0.08-0.31 nm) was constructed for radiobiological experiments (Fig. 2) (Kobayashi et al., 1987). Later, a channel cut InSb(111) monochromator for 1.7-3.1 KeV(0.4-

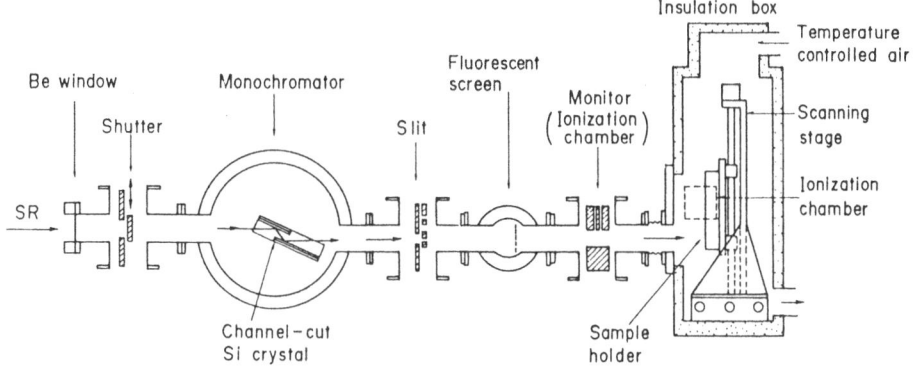

Fig. 2. Schematic diagram of the monochromatic X-ray irradiation system. (Kobayashi et al., 1987).

0.7 nm) was added. An irradiation chamber with a scanning stage for a wide, homogeneous beam area, suited for the irradiation under atmospheric conditions, was constructed for the same purpose. A similar but vacuum-type irradiation chamber also is in use for soft X-ray biological experiments.

Monochromators for soft X-rays for general use are available; they include a InSb(111) double-crystal monochromator for 1.8-3.3 KeV (0.38-0.70 nm) X-rays. A wiggler line with a Si(311) monochromator may also be used for hard X-rays (up to 43 KeV). A grazing monochromator for ultrasoft X-rays intended to irradiate biological samples at several hundred eV range (3-25 nm) is under construction. In principle, the same settings can be used for the transmission absorption measurements with appropriate preparation of samples. Under routine conditions, the intensity or exposure rate of monochromatized X-rays amounts to the order of 10^4 R per min at 0.11 nm at the sample position.

More details on the combinations of beamlines and monochromators and their characteristics can be found in the forthcoming Handbook on Synchrotron Radiation (Hieda and Ito, 1988).

Structural Experiments

One of the unique instrumentations at PF for structural studies is a beamline equipped with a demagnifying focusing mirror-monochromator optics for small-angle diffraction measurements of biological specimens (BL-15A) (Fig. 3). This provides good spatial resolution and high intensity of

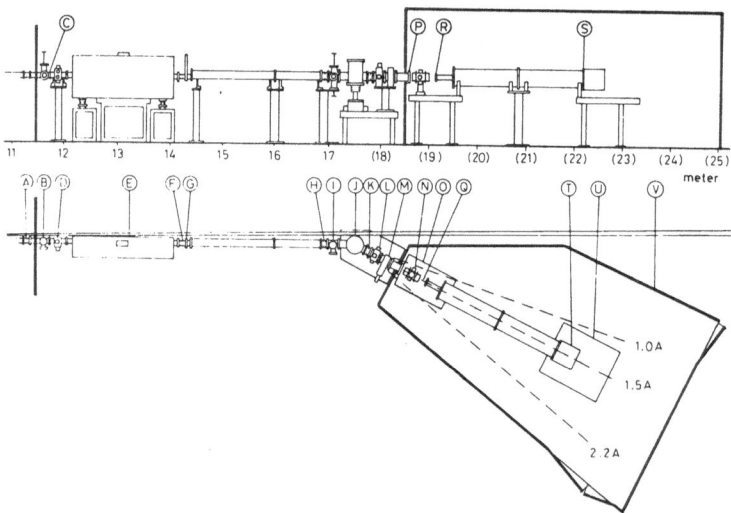

Fig. 3. Side view and plan view of the beamline for the diffractometer. A, bellows; B, attenuator and fluorescent screen; C, beryllium window; D, incidence slits; E, mirror chamber; F, bellows; G, vacuum gate valve; H, bellows; I, fluorescent screen and white beam stop; J, monochromator box; K, bellows; L, scatter suppressor slits; M, downstream beam shutter; N, specimen slits; O, table for specimen; P, beryllium window; Q, specimen position; R and S, kapton windows; T, vacuum chamber for film cassette; U, table for detector; V, safety hutch. (Reproduced by permission of Amemiya et al. (1983) and Nuclear Instruments and Methods. Copyright 1983 by the Elsevier Science Publishers.)

X-rays. A small-angle resolution of 1000 Å is obtainable at 1.5 Å wavelength with an angular resolution of 0.8 mrad and 0.6 mrad in the vertical
and horizontal directions, respectively. The intensity is 10^{11}-10^{12}
photon/s (Amemiya et al., 1983), which is so high that the system can be
used even for time-resolved diffraction studies (Tanaka et al., 1986). Combining the system with a special integrating X-ray area detector, Amemiya et
al. (1987) succeeded in recording a clear X-ray diffraction pattern of contracting frog skeletal muscle in as short a time as 10 s (Fig. 4).

With another type of small-angle X-ray scattering system (at BL-10C),
combined with a rapid mixing apparatus, the aggregation process of large
molecules was successfully measured (Ueki et al., 1985). This opens the
possibility of studying kinetics of transient phenomena in solutions and
biological systems.

Also noteworthy is a beamline (BL-6A$_2$) constructed for macro-molecular
crystallography with a Weissenberg camera (Sakabe, 1983; Kamiya et al.,
1984/1985). An over-bent triangle crystal is used as the monochromator. It
is particularly useful for the phasing of large molecules with anomalous
dispersion of heavy atoms at an absorption edge. A four-circle diffractometer with a data acquisition system is available for analysis of crystal
structures (BL-14A).

For further information on the beamlines and optics as well as current
projects for users see the recent Photon Factory Activity Report V (1987),
obtainable from the Technical Information Office, National Laboratory for
High Energy Physics, 1-1, Oho, Tsukuba-shi, Ibaraki-ken, 305, Japan.

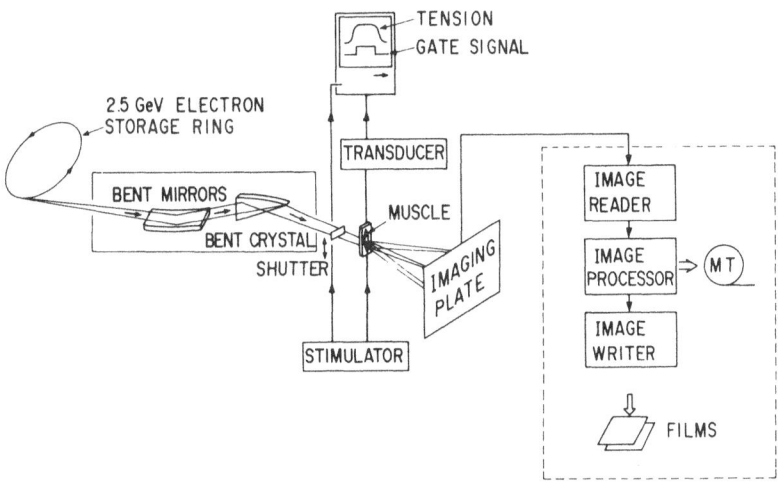

Fig. 4. Experimental configuration for recording two-dimensional x-ray
         diffraction patterns from a contracting muscle using a scanning
         laser-stimulated photoluminescence system and synchrotron radi-
         ation. (Reproduced by permission of Amemiya et al. (1987) and
         Science. Copyright 1987 by the AAAS.)

ACKNOWLEDGMENT

Drs. Y. Amemiya and K. Kobayashi kindly provided materials on PF beamlines.

REFERENCES

Amemiya, Y., Wakabayashi, K., Hamanaka, T., Wakabayashi, T., Matsushita, T., and Hashizume, H., 1983, Design of a small-angle X-ray diffractometer using synchrotron radiation at the Photon Factory, Nucl. Instrum. Methods, 208:471.

Amemiya, Y., Wakabayashi, K., Tanaka, H., Ueno, Y., and Miyahara, J., 1987, Laser-stimulated luminescence used to measure X-ray diffraction of a contracting striated muscle, Science, 237:164.

Hieda, K., and Ito, T., 1988, Radiobiological experiments in X-ray region with synchrotron radiation, in: "Handbook on Synchrotron Radiation," Vol. 4, S. Ebashi, E. Rubenstein, M. Koch, eds., Elsevier Science Publishers B. V., Amsterdam (in preparation).

Hieda, K., Maezawa, H., Ito, A., Kobayashi, K., Furusawa, Y., and Ito, T., 1986, Choice of coatings for the optical elements in the irradiation system of vacuum-ultraviolet radiation above 50 nm, Photochem. Photobiol., 44:417.

Inagaki, T., Ito, A., Motosuga, M., Hieda, K., Kobayashi, K., Maezawa, H., and Ito, T., 1985, Vacuum-ultraviolet photoacoustic spectroscopy of biological materials using synchrotron radiation as a light source, Photochem. Photobiol., 41:527.

Ito, T., Kada, T., Okada, S., Hieda, K., Kobayashi, K., Maezawa, H., and Ito, A., 1984, Synchrotron system for monochromatic UV irradiation (>140 nm) of biological material, Radiat. Res., 98:65.

Kamiya, N., Higashi, T., Sakabe, K., and Sakabe, N., 1984/1985, Anomalous dispersion photograph of myoglobin, Photon Factory Activity Report, VI:16.

Kobayashi, K., Hieda, K., Maezawa, H., Ando, M., and Ito, T., 1987, Monochromatic X-ray irradiation system (0.08-0.4 nm) for radiation biology studies using synchrotron radiation at the Photon Factory, J. Radiat. Res., 28:243.

Namioka, T., 1986, Vacuum ultraviolet/soft-X-ray spectroscopic facilities at the synchrotron radiation laboratories in Japan, Photochem. Photobiol., 44:245.

Sakabe, N., 1983, A focusing Weissenberg camera with multi-layer-line screens for macromolecular crystallography, J. Appl. Cryst., 16:542.

Tanaka, H., Kobayashi, T., Amemiya, Y., and Wakabayashi, K., 1986, Time-resolved X-ray diffraction studies of frog skeletal muscle isometrically twitched by two successive stimuli using synchrotron radiation, Biophys. Chem., 25:161.

Ueki, T., Hiragi, Y., Kataoka, M., Inoko, Y., Amemiya, Y., Izumi, Y., Tagawa, H., and Muroga, Y., 1985, Aggregation of bovine sirum albumin upon cleavage of its disulfied bonds, studied by the time-resolved small-angle X-ray scattering technique with synchrotron radiation, Biophys. Chem., 23:115.

BIOLOGICAL STRUCTURAL STUDIES

AT THE LURE SYNCHROTRON RADIATION FACILITY

R. Kahn

LURE, B^t 209d, Orsay Campus
91405 Orsay
France

LURE: SOME DATES AND SOME FIGURES

Laboratoire d'Utilisation de Rayonnement Electromagnetique (LURE) started in the early 1970s through the common will of a group of physicists to use the synchrotron radiation emitted by the Orsay storage rings, Anneau de Collision d'Orsay (ACO) in the UV range, followed in 1976 by Dispositif de Collision sous Igloo (DCI) in the X-ray range. Since 1985, all these machines, storage rings, and linac, are dedicated to the production of synchrotron radiation. In 1987, the first beam was stored in the new ring Super-ACO, an 800-Mev machine designed as a light source (Marin, 1988), and a superconducting wiggler was installed on DCI (Bazin et al., 1988). The performance of these instruments was up to expectations, and the first experimental set-ups are being installed progressively.

Every year on DCI 120 shifts of 16 to 20 hours are available. For Super-ACO 150 shifts of 16 hours are planned for the fully operational ring (1989). In 1987, 426 projects involving more than 800 scientists were granted beamtime at LURE (Activity Report, 1985-87).

Super-ACO

This machine, which replaces ACO, is becoming accessible to users working with radiation ranging from visible light to soft X-rays ($\sim$ 3 Å). Its main characteristics are reported in Table 1.

The main biological applications in this range are related to pulse-fluorescence spectroscopy.

Four stations, installed on the beam ports SA1-SB1 and SA4-SB4, equipped with quartz windows of large aperture (100 mrad) or with a narrower (40 mrad) sapphire window transparent down to 160 nm, are dedicated to fluorescence decay and anistropy decay measurements, using either photon counting or phase fluorimetry.

Before the installation of the 500-MHz cavity, planned for 1989, a picosecond laser is available for experiments requiring high time resolution. A high pressure cell (up to 4 Kbars) should become operational in 1989.

Table 1. Characteristics of Super-ACO

Positron filled		
Energy		800 MeV
Emittances	ε_x	$3.7 \ 10^{-8}$ m.rad
	ε_z	$3.7 \ 10^{-9}$ m.rad
$\lambda_{o \ c}$ from bending magnets		18.5 Å
Maximum intensity		500 mA (currently 360 mA)
Beam lifetime		10 h (currently 7 h)
Pulse separation		240 ns in single bunch mode
		60 ns in symetrical four bunch mode
Pulse width		180 ps (40 mA/bunch)
		23 ps (planned for end 1989)
Number of beam ports:		
from bending magnets		13
from insertion devices		6

The application of maximal entropy methods to the analysis of pulse fluorimetry data has turned the latter into a quantitative method of analysis of the rotational dynamics of biological macromolecules (Livesey and Brochon, 1987). A complete set of programs is now used to analyze the measurements of total intensity or anistropy decay, not as the sum of a few discrete exponentials, but in terms of distributions of lifetimes or of correlation times. This method, when applied to the fluorescence decay analysis of several proteins studied at LURE, gave much more detailed results than conventional ones. It renders feasible the analysis of dynamic structural heterogeneities and internal flexibilities in the molecules.

Three main lines of research are pursued by the local group: One is the dynamics of proteins and hormonal peptides in solution. Proteins are now available where the fluorescent residue(s) has(ve) been modified by site-directed mutagenesis, in number and/or in position. The physical or chemical environment or the mobility of each fluorophore in a protein can thus be specifically studied. This approach was recently applied to two enzymes: the ATCase from E. coli (Royer et al., 1987), and pancreatic phospholipase A_2.

The second line is peptide-lipid or protein-lipid interactions in model membranes (liposomes or inverse micelles). This is best illustrated by the recent study of the dynamics of Trp in ACTH and its derivatives as well as Glucagon in interaction with a membrane interface (Gallay et al., 1987).

The third line of research is time-resolved fluorescence spectroscopy of chlorophyll-protein complexes. After numerous studies of isolated bacterial centers, more complex systems are investigated from the isolated chlorophyll-protein complexes that form the peripheral antennae to the whole plant (Hodges and Moya, 1987).

DCI

This machine is used in the X-ray range. Its main characteristics are reported in Table 2.

The X-ray beam emitted by DCI exhibits an excellent time and spatial stability, with a lifetime of 40 to 60 hours and a position reproducibility from one injection to the next one (usually 24 hours), which in practice obviates any readjustment of the instruments. These qualities are of prime importance to precise measurements, as for instance in multi-wavelength data collection in protein crystallography.

Table 2. Characteristics of DCI

Positron filled	
Energy	1.85 GeV
Emittance	$7.0 \ 10^{-7}$ m.rad
Wiggler	superconducting (4.8 T), 5 poles
λ c from bending magnets	3.4 Å
λ c from the wiggler	1.09 Å
Maximum intensity	320 mA
Beam lifetime	40 h
Number of beam ports:	
from bending magnets	14
from the wiggler	6

To date, fourteen workstations are installed on the three bending magnet beamlines of DCI; five additional ports will be installed on the three wiggler beamlines. Six ports on the bending magnet are used either fully or partially for biological experiments in crystallography (D23, D41, D43) and solution X-ray scattering (D24); the experimental set-ups for diffuse scattering (D16) and for X-ray absorption spectroscopy (D21) are also used for biological applications, but these are a minority. Finally, an instrument for protein crystallography and for diffuse scattering is currently designed to be installed on the wiggler port W134.

Solution X-ray Scattering

The D24 instrument is being improved: an evacuated automatic sample changer is under test and a high counting rate 2D data acquisition system is set up.

Apart from rapid low-resolution studies, two lines of research are worth mentioning. The first one is concerned with structural modification; changes of the quarternary structure of the allosteric enzyme aspartate transcarbamylase from E. coli and of its mutants are studied in relation with the corresponding changes of activity (Cherfils et al., 1987); the assembly of a viral capsid (brome mosaic virus) from the purified protein has been investigated. The maximum entropy method was applied successfully to the latter to interpret the scattering data in terms of distribution of capsid fragments. The second theme is the study of the interactions in concentrated protein solutions where models developed for liquid physics have been successfully applied to several systems (crystallin, haemoglobin, ATcase) (Tardieu et al., 1986; Tardieu and Delaye, 1987).

Biological Crystallography

There are several subjects of research in biological crystallography. One is acceleration of the operation involved in the resolution of three-dimensional structures, in particular the diffraction data collection stage and the resolution of the phase problem. Others are dynamic and electrostatic properties of biological macromolecules, molecular graphics, and the resolution of new protein structures.

Instruments and Experimental Methods

Progress has been realized on the D23 diffractometer equipped with a bi-dimensional electronic detector. After mechanical improvements, the double crystal monochromator with sagittal focusing and a fixed exit gives a stable beam with good horizontal focussing. The detector can handle 350,000 counts per second without degradation of performance. Tests done on a lysozyme crystal show a good uniformity of the detector response (R_{sym} on intensities 4.7% without absorption correction, 2.7% for reflections collected in similar diffracting conditions but for different settings of the detector).

The adaptation to this instrument of the MADNES data acquisition and data reduction software package is under way (Kahn et al., 1987). This package was originally developed by J. Pflugrath and A. Messerschmidt and has been modified during a series of workshops sponsored by the European Economic Community (EEC) and organized by G. Bricogne. Both the off-line and the on-line version of this software will soon be operational for our instrument.

A new instrument, D41, brighter than its counterpart D43, is operational since the beginning of 1988. Each of these instruments uses a triangular shaped Ge crystal as a monochromator and is equipped with a rotation camera. They are used for macromolecular diffraction data collection.

The monochromator on the wiggler line is at the design stage. As with D41 and D43, it uses a triangular shaped crystal as a monochromator, but great care has been given to its cooling. A C/W multilayer reflector will be used to eliminate harmonics and to focus the beam vertically. A prototype of this reflector will be tested during 1988. Finally, an instrument using the Laue method and which will include a fast change for photographic or phosphor plates is planned.

The fast quenching technique for biological crystals using an immersion in liquid ethane followed by transfer to the goniometer head cooled by a gas flow is currently being developed in the laboratory. Tests done on tumor necrosis factor crystals, an unfavorable case due to the high solvent content, show that after freezing, the crystals diffract at a slightly worse resolution than before. In practice, at the time scale of the data collection, irradiation damage seems to be suppressed. Systematic work on the freezing parameters is under way.

Studies of the Phase Problem

New approaches to the direct solution of the phase problem using the entropy maximization has been generalized by G. Bricogne (in press); a unified formalism allows one to include and analyze all the available information on a crystal structure in terms of maximum likelihood.

The contrast variation method has been developed for X-rays, including experimental procedures and programs; contrast variation is used to evaluate the moduli of the Fourier transform of the molecular envelope so as to use direct phasing, in particular by entropy maximization. The first application was the successful determination of the molecular envelope of tryptophanyl tRNA synthetase using three different contrasts.

After the resolution at LURE of a protein using the multi-wavelength anomalous dispersion method, the work on this subject goes on (Fourme et al., 1986). Diffraction data at two wavelengths on an Yb-labelled industrial protein have recently been collected. A set of programs that compute the best electron density map from multi-wavelength data is available.

Dynamic Properties of Macromolecules

Diffuse scattering of lysozyme crystals has been measured. The data were interpreted in terms of correlated movements of molecules along two of the cell axes (Doucet and Benoit, 1987). This was the first application of a systematic analysis of diffuse scattering from a protein crystal. This kind of disorder was shown to be very common in protein crystals (for instance in deoxyhaemoglobin). Recent work suggests that in some cases diffuse scattering may also give structural information.

Theoretical work studying dynamic and electrostatic properties of biological macromolecules was undertaken. These studies give interesting results in fields as different as the study of psoralen, the theoretical interpretation of time-resolved fluorescence spectroscopy experiments, and the effect of dynamic fluctuations of a protein on its electrostatic properties.

Molecular Graphics

MANOSK, a molecular graphics program package and its link to various programs for crystallography and molecular dynamics, was developed for the E&S PS300 workstation. Besides their use for molecular modelling, these programs proved to be very useful for the evaluation and the interpretation of results of diffuse scattering modelling or from electrostatic potential computations; the interactive editing of molecular envelope from an electron density map is used in particular in the solvent-flattening methods.

Structures

Several structure determinations are under way at LURE: (i) The Tumor Necrosis Factor. Owing to the fragility of the crystals and their rapid decay, data have been collected on the native and two putative derivatives at 110 K. (ii) Work with nucleotides. (iii) Study of a suicide inhibitor-elastase complex, and (iv) Crystallization trials on several proteins are being performed.

All these programs rely heavily on the availability of adequate computing facilities. The computing center at LURE is organized around a VAX 780 (soon to be replaced by a VAX 6210) connected via DECnet to smaller VAX machines or stations, two of which run the instruments D23 and D24. Large size calculations are performed on a ST100 array-processor (100 Mflops).

Training and Relation With The European Synchrotron Radiation Facility (ESRF)

A graduate teaching program, shared between the Orsay and the Strasbourg Universities, train a dozen students in structural methods (crystallography, solution X-ray and neutron scattering, electron microscopy) in particular for protein engineering. It is carried out in collaboration with LURE.

Like all the European synchrotron radiation centers, LURE is involved in the development of instruments for the ESRF. Several projects, including the design of multi-layer reflectors, of a Reticon type solid detector, and a VME data acquisition system for protein crystallography, are carried out in collaboration with the ESRF.

Regarding fluorescence spectroscopy, the data analysis with the maximal entropy method will allow us to take full advantage of the 23 ps-wide pulse provided by the 500-Mhz cavity on Super-ACO planned for 1990. Crystallographers on DCI are mainly concerned with the wiggler line. The Laue method, which is not yet implemented at LURE, will require instrumental and methodological developments which will benefit from the experience gained at several other synchrotron radiation centers. As of next year, data should be routinely recorded using the electronic detector on the D23 diffractometer, with a strong emphasis on multi-wavelength anomalous dispersion experiments. Finally, the introduction of imaging plates is the object of collaborative effort.

REFERENCES

Brazin, C., Dubuisson, J. M., Jacquemin, J.P., Perot, J., and Raoux, D., 1988, A five-pole superconducting wiggler for the DCI ring at LURE, Nucl. Instr. Meth., A266:132.

Cherfils, J., Vachette, P., Tauc, P., and Janin, J., 1987, The pAR5 mutation and the allosteric mechanism of E. coli aspartate carbamoyltransferase, The EMBO Journal, 6:2843.

Doucet, J., and Benoit, J.-P., 1987, Analysis of the x-ray diffuse scattering from lysozyme crystals, Nature, 325:641.

Fourme, R., Kahn, R., and Chiadmi, M., 1986, Multi-wavelength anomalous diffraction in protein crystallography: experimental procedures, phase calculations and results, in: "International Conference on Biophysics and Synchrotron Radiation," A. Frascati, A. Bianconi, and A. Congiu-Castellano, eds., Springer Verlag.

Gallay, J., Vincent, M., Nicot, C., and Waks, M., 1987, Conformational aspects and rotational dynamics of ATCH (1-24) and glucagon in reverse micelles, Biochem., 26:5738.

Hodges, M., and Moya, I., 1987, Time-resolved chlorophyll fluorescence studies on photosynthetic mutants of Chlmydomanas reinhardtii: origin of the kinetic decay component, Photosyn. Res., 13:125.

Kahn, R., Fourme, R., Bosshard, R., Prange, T., and Lewit-Bentley, A., 1987, Area detector diffractometry with synchrotron: hardware and software for data acquisition at single or multiple wavelengths, in: "Proc. Computational Aspects of Protein Crystal Data Analysis," SERC Daresbury.

Livesey, A. K., and Brochon, J.-C., 1987, Analyzing the distribution of decay constants in pulse fluorimetry using the maximum entropy method, Biophys. J., 52:693.

LURE: activity report 1985-1987.

Marin, P., 1988, Storage rings at Orsay, Nucl. Instr. Meth., A266:18.

Royer, C., Tauc, P., Herve, G., and Brochon, J.-C., 1987, Ligand binding and protein dynamics: a fluorescence depolarization study of apartate transcarbamylase from E. coli., Biochem, 26:6472.

Tardieu, A., and Delaye, M., 1987, Colloidal dispersions of α-crystallin proteins. I. Small angle X-ray analysis of the dispersion structure, J. de Phys., 48:1217.

Tardieu, A., Laporte, D., Licinio, P., Krop, B., and Delaye, M., 1986, Calf lens α-crystallin quarternary structure: a three-layer tetrahedral model, J. Mol. Biol., 192:711.

SYNCHROTRON BEAMLINES AT THE EUROPEAN MOLECULAR

BIOLOGY LABORATORY OUTSTATION IN HAMBURG

Keith S. Wilson

European Molecular Biology Laboratory
Hamburg Outstation, c/o DESY
Notkestrasse 85, 2000 Hamburg 52
West Germany

INTRODUCTION

The European Molecular Biology Laboratory (EMBL) outstation was set up
in 1974 to apply synchrotron radiation (SR) to problems in molecular biology. The facilities have been developed by a staff too large to be given
full credit here. The capabilities for biological structure research at the
outstation are given in Table 1. The purpose of this brief report is to
outline the present beamlines at the outstation. The beamlines are used for
in-house research by the groups in the outstation and by many outside visitors. Over 50 projects are studied by visiting groups in a typical year.

RING PARAMETERS

EMBL has five SR beamlines in Hamburg: two for protein crystallography,
two for small angle scattering and one for EXAFS. The lines use radiation
generated by the DORIS storage ring at the Deutsches Electronen Synchrotron
(DESY) site. The storage ring is run in two modes. In the first, or parasitic mode, electrons circulate in one direction in the ring and positrons
in the other. The prime function of the ring in this mode of operation is
that the paths of the two sets of particles cross at two critical points of
the orbit, the collisions being the object of study of high energy physics
groups. The synchrotron users, taking radiation generated by either of the
two sets of particles, are parasites in this mode. The running parameters
are dominated by the needs of the high energy physicists. The radiation is
nevertheless well suited to many of our experiments.

In the second mode of operation, the DORIS ring is dedicated to the
production of SR. This is therefore called "main-user" time. The beam
parameters are now dictated by the users of SR and various parameters such
as beam energy and lifetime are significantly different from parasitic time.
Only electrons circulate in the ring during main-user time. Some important
parameters of the ring under the two types of operating conditions are given
in Table 2.

Table 1. Capabilities of the Hamburg Outstation

---

1. Small Angle Scattering

    Solutions/Gel/Fibres

    Muscle

2. EXAFS

3. Protein Crystallography

4. Instrumentation and Development

BEAMLINES FOR PROTEIN CRYSTALLOGRAPHY

The two EMBL beamlines for protein crystallography are called X11 and X31 for historical reasons. Line X11 is situated in the EMBL building and takes X-rays from the circulating positrons. Thus it is only available during parasitic-user time. X11 is the high intensity line for protein crystallography.

The X11 line consists of a bent Fankuchen triangular Ge(111) crystal, segmented flat quartz mirrors, slits, and a moveable bench with a mount for either a rotation camera or a 2-D detector. All movable elements are remotely controlled from a computer. The distance from the source to the central mirror is 23.5 m and from the central mirror to focus 4 m, so that the demagnification is roughly 5:1. The monochromator is bent to achieve focussing in the horizontal plane. The available wavelength range is 0.95-2.3 Å. Changing the wavelength requires changing the take-off angle by rotating the whole bench. Vertical focussing is carried out by bending the

Table 2. Beam Parameters at DORIS

	Parasitic-User Time	Main-User Time
Users	High energy physics + SR	Dedicated to SR
Particles	Electrons and Positron	Electrons only
Beamlines	Hamburg Synchrotron Laboratory (HASYLAB) + EMBL	HASYLAB only
Particle energy	5.5 GeV	3.7 GeV
Fill length	1 Hour	3 Hours
Maximum current	40 mA	90 mA
Critical wavelength	λ_c=0.41 Å	λ_c=1.34 Å
Bunch structure	Single Bunch	Multi(4)Bunch
Availability	30 weeks p.a.	15 weeks p.a.

bench on which the mirror segments have been prealigned. The set-up is shown in Fig. 1.

The dimensions of the focussed beam are about 1.2 x 0.7 mm^2. A modified Arndt-Wonacott rotation camera is used for film data collection. It will be replaced by a 2-D detector at a later date. The crystallographic cradle on the bench can be optimally oriented into the beam by a fully automatic procedure taking 1-2 minutes. This optimization is achieved through the use of four motors which provide horizontal and vertical translation and rotation and inclination of the cradle about the first collimator slits. The intensity is monitored after the first collimator slits for the translation and after the second slits for the rotations.

The intensity of the beamline allows data collection at roughly 200 times the speed on a conventional source, which makes X11 one of the most intense beamlines currently available for protein crystallography.

The X31 beamline in Fig. 2, is in the Hamburg Synchrotron Laboratory (HASYLAB). The line takes radiation generated by the electrons and is available in both main- and parasitic mode. X31 consists of a channel cut Sl(111) crystal monochromator, double focussing segmented toroidal mirrors, slits, and the same type of automatically aligned crystallographic cradle with oscillation camera as X11. All elements are mounted on moveable benches and their movement is remotely controlled by computer. The distance from source to mirror centre is 16 m and is equal to the distance between mirrors and focus. Thus demagnification is 1:1. The line is characterised by easy wavelength tunability, but has relatively large focal dimensions. The rotation camera can be replaced by a detector and data collection has been carried out using an Enraf-Nonius FAST system by the Max Planck group of Hans Bartunik in Hamburg.

The overall intensity on the crystal is some 30 times weaker on X31 in comparison to X11. One advantage of the line is its high positional stability resulting from the large focus. In addition the wavelength can be very easily tuned, involving only a rotation of the monochromator crystal with the outgoing beam remaining parallel and merely undergoing a small vertical shift. This property, coupled with the narrow wavelength band-pass of the Si(111) channel cut crystal, makes the line ideal for multiple energy anomalous scattering studies.

At present, data collection for protein crystallography is restricted to photographic methods, although two-dimensional gas-filled detectors are used by the muscle and small angle scattering groups. A scanner for storage phosphor plates is currently being developed by the instrumentation group. A parallel electrode gas-filled detector with a high count rate capability is also under development.

SMALL ANGLE SCATTERING

There are two lines for small angle scattering: X33 and X13; the latter will become available late in 1988. X33 is in HASYLAB and receives radiation from the electron direction. X13 is in the EMBL bunker and uses positron generated radiation. Hence, as for protein crystallography, one line, X33, is available in both main-user and parasitic time; X13 only in parasitic time.

X33 is composed of a double focussing monochromator-mirror camera (Fig. 3). The line operates at fixed wavelength of 1.5 Å, but this can be changed by replacing the monochromator. A triangular monochromator deflects and focusses about 2 mrad of radiation in the horizontal plane. The lateral

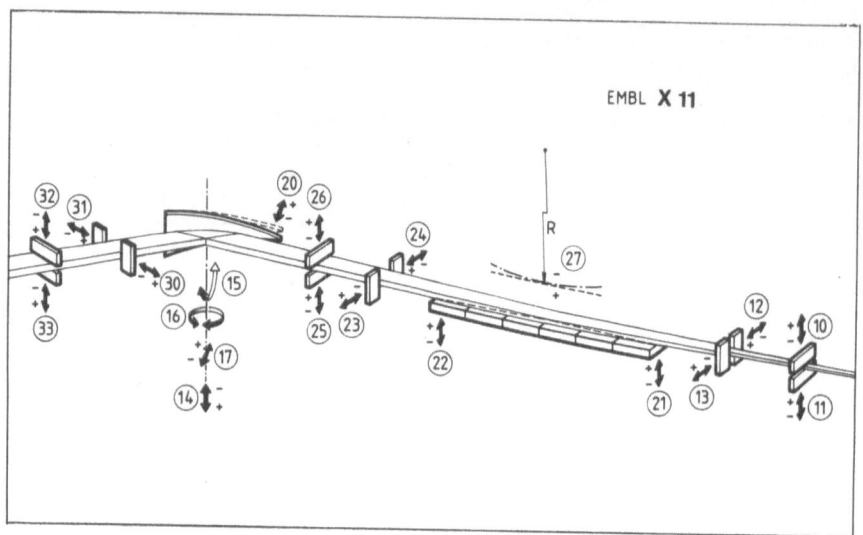

Fig. 1. The X-11 beamline at EMBL Hamburg.

deflection ensures a low overall background in the camera. The segmented mirror gives the vertical focussing and rejects harmonics. The several sets of slits help define the beam and reduce the background. The monochromator is a Ge(111) single crystal with a compressing Fankuchen cut. The demagnification is variable between 3 and 6, with a beam size at the focus of 5 x 2 mm^2.

Quadrant and area (256 x 256 pixels) wire chambers with delay-line readout, 1 mm resolution and 300 kHz total count rate are used on X33. In addition a linear detector with delay-line readout and 0.5 mm resolution or a linear parallel readout wire chamber with 128 channels, 1 mm resolution

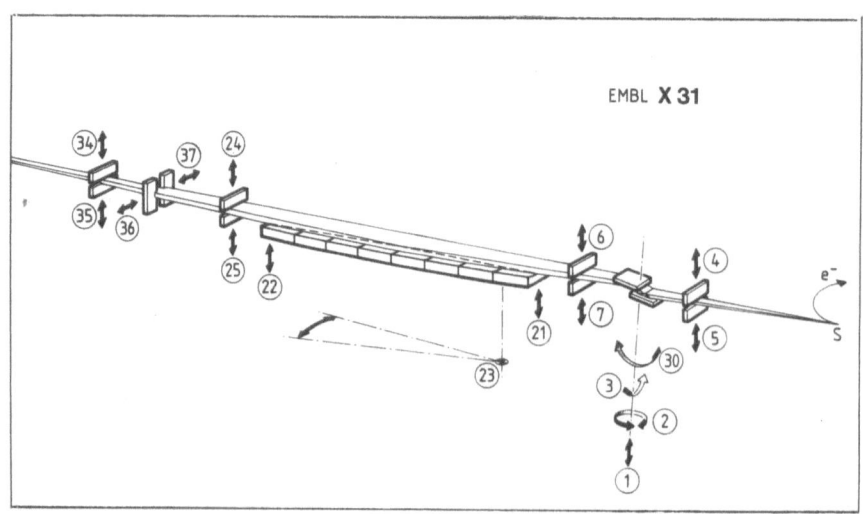

Fig. 2. The X-31 beamline at EMBL Hamburg.

and 20 MHz total count rate can be operated. Apparatus for time-resolved
studies include a stopped flow mixing cell and temperature jump equipment.
Samples studied include solution, gels and fibres by Michel Koch and his
colleagues, and muscle by Yuichir Maeda.

The second beamline for small angle scattering, X13, is similar to X11
in its optical elements. Horizontal focussing uses a bent Ge(111) monochromator, vertical focussing uses a segmented mirror. Plans for a continuous mirror are in hand. The major use of the line is intended for
single fibre muscle experiments and other studies needing the smallest focus
and highest intensity at the sample.

EXAFS

The EXAFs beamline (Fig. 4) in HASYLAB has been set up by Christoph
Hermes and Robert Pettifer for studies on biological samples and other very
dilute systems. The beamline is separated from the ring by two 0.4 mm thick
Be windows. A Si(220) double monochromator accepts 2 mrad of white radiation. Higher harmonics can be minimized by a piezoelectric fine-tuning
mechanism. Energy drift is reduced to a minimum by accurately maintaining
the first monochromator face at a constant temperature. The second optical
element is a segmented gold-coated toroidal mirror which slightly demagni-
fies the source, giving a focal spot of about 4.0 mm x 1.5 mm at the sample.

A unique feature of the EMBL instrument is the energy calibration. A
static Si crystal and detector after the second ionisation chamber allow the
simultaneous recording of Bragg reflections. This allows absolute energy
calibration, determination of the instrument function (energy resolution)
and beam diagnostics.

The line can cover the energy range 5 KeV to 20 KeV with focussed radiation and 5 KeV to 30 KeV unfocussed. Thus it covers many of the absorption
edges of metals found in biological systems. There are two closed, refil-
lable ionisation chambers. Two fluorescence detectors are available with

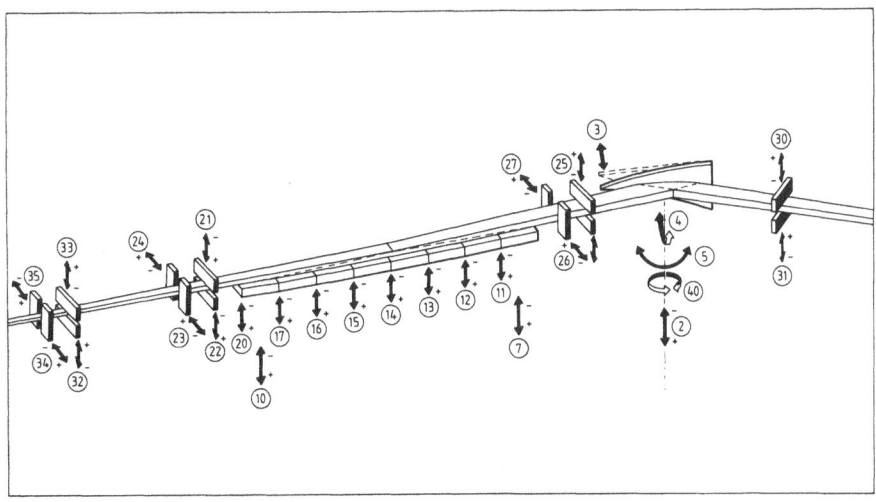

Fig. 3. The X-33 beamline at EMBL Hamburg.

high count rate capabilities. A two-stage Displex cryostat (17°K) allows simultaneous absorption and fluorescence measurements. Samples can be changed at cryogenic temperatures due to the top loading system and the samples can be scanned in He-exchange gas.

FUTURE PLANS FOR BEAMLINES

Further developments planned for the next years include two beamlines. The first is a line allowing Laue photographs to be recorded. These record the diffraction of the "white" beam from a stationary crystal sample without use of monochromator. This line would be constructed in the EMBL bunker.

The second development is the construction of a "by-pass" to one of the high energy physics stations on DORIS. This will provide seven straight sections in the ring where insertion devices, wigglers or undulators, can be installed. These lines will provide a gain of between one and two orders of magnitude in intensity at the sample compared to the present beamlines, given appropriate optical elements. The higher intensities lead to much shorter time scales for experiments being possible. It is planned that EMBL will build one of the beamlines on the by-pass. This line will be for small single crystal and time-resolved crystal work as well as for muscle experiments.

ALLOCATION OF BEAMTIME

The priorities committee of the EMBL is made of external scientists who meet to consider the proposals received after roughly every one year of beamtime. The EMBL staff scientists do not sit on the committee but do play

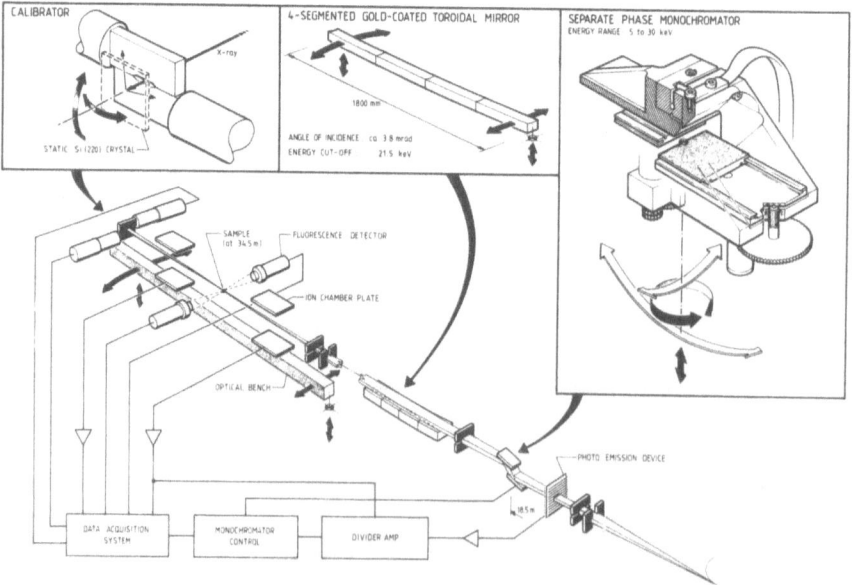

Fig. 4. The EXAFS beamline at EMBL Hamburg.

an advisory role regarding feasibility. The day-to-day allocation of beam-time is carried by the experimental groups in the outstation on the basis of the recommendations of the committee. Some beamtime is also reserved for new proposals received after the committee has met and judged by the staff to warrant urgent investigation. Beam proposal forms can be obtained from the author.

RESEARCH FACILITIES FOR BIOLOGY AT THE SYNCHROTRON RADIATION

SOURCE AT DARESBURY LABORATORY

J. Bordas

Daresbury Laboratory
Warrington WA4 4AD
Cheshire, England

INTRODUCTION

This report is an introduction to and brief description of the facilities available for research in biology and related subjects at the Synchrotron Radiation Source (SRS) at the Science and Engineering Research Council (SERC) at Daresbury. The SRS (a 2-GeV accelerator with a critical wavelength of ca. 0.39 nm) and the associated instrumentation and infrastructure has undergone continuous development over the last few years. Further development is still in progress. Among the recently completed projects with a major impact on the exploitation of synchrotron radiation in the general area of biology, biochemistry and biophysics, one must include the implementation of the High Brightness Lattice (HBL), the construction and equipping of the Biology Support Laboratory (BSL), which is jointly funded between the SERC and the Medical Research Council (MRC), and the successful implementation of a time-resolved x-ray diffraction station, incorporating a variety of one- and two-dimensional position sensitive x-ray detector systems. Besides these recent developments, there are now a number of established facilities for time-resolved spectroscopy, X-ray absorption spectroscopy and protein crystallography. The latter two subjects have stations making use of the radiation from a 5-Tesla superconducting wiggler as well as stations taking radiation from the dipole bending magnets.

ESTABLISHED SYNCHROTRON RADIATION INSTRUMENTATION

Time-Resolved Spectroscopy

The SRS is run in the single bunch mode for fluorescence lifetime and time-resolved spectroscopy measurements. In this mode of operation the SRS delivers a pulse of light which is 180 picoseconds long with a repetition rate of 3.125 MHz. There is an experimental station (12.1) devoted to these kinds of measurements. This station provides ca. 3×10^6 photons per pulse at the sample for a circulating current of 25 mA. These figures correspond to typical running conditions at present. Higher circulating currents can be achieved. The instrument operates at wavelengths longer than 120.0 nm and delivers monochromatic radiation with approximately a 0.1% bandpass. While the peak power is significantly lower than that attainable with lasers (10^{12} photons/pulse), the repetition rate and the ease of tunability makes these techniques highly competitive, e.g., one can tune to a specific fluorophore, thus allowing individual species to be resolved in a complex biochemical system. Also, the repetition rate is such that the data can be accumulated very rapidly (of the order of 10^4 counts/sec) while avoiding pile-up errors. Monro et al. (1985) and Munro and Martin (1988) give detailed reviews of instrumental aspects and applications (a large fraction of which concerns the study of membranes).

X-ray Spectroscopy

Four stations are available for X-ray spectroscopy. Three of these stations (3.4, 7.1 and 8.1) use radiation from bending magnets; the third one (9.2) is installed on one of the wiggler ports. For the study of hard edges in metalloproteins and enzymes, 9.2 is employed (e.g., Mo edge of Xanthine Oxidase), while 3.4, 7.1 and 8.1 are used for softer edges (usually Ni K-edge or softer). All the instruments are equipped with monochromators capable of harmonic rejection.

There is the option to use a double-focusing Pt-coated quartz mirror in station 7.1 which reduces the beam cross-section from 12x4 mm^2 down to 6x0.5 mm^2 at the expense of a somewhat deteriorated resolution of the instrument (normally $\Delta\lambda/\lambda = 3x10^{-4}$ at $\lambda = 0.15$ nm).

Station 8.1 (constructed as a result of the collaboration between the SERC and the Dutch NWO) has benefitted considerably from the development of a slitless, order-sorting, monochromator system (Van der Hoek et al., 1986; Dobson et al., 1986), which can simultaneously achieve high intensity, spectral purity and instrumental stability. All of these are crucial requirements in X-ray spectroscopy measurements in biochemical systems at low concentrations and with low metal content. Hasnain (1988) gives a review of recent applications to metal coordination biochemistry.

Protein Crystallography And High Resolution Fibre Diffraction

At the moment of writing this paper, protein crystallography and high angle fibre diffraction applications are carried out on two dedicated stations: 7.2 and 9.6, respectively (Helliwell et al., 1982; Nave et al., 1985; Helliwell et al., 1986). These stations are placed on a normal bending magnet (7.2) and on the wiggler (9.6). They both use double focusing, bent mirror-monochromator optics, and the latter has a vertically focused white beam option which is occasionally used. These stations utilize a variety of Si and Ge monochromators with different lattice planes and Fankuchen cuts. The stations can be used for multiple wavelength experiments, but they are more commonly set up at a fixed wavelength, with the wiggler station typically set up for data collection at shorter wavelengths ($\lambda<0.1$ nm). Station 9.6 is equipped with a FAST (Enraf Nonius) area detector diffractometer (Andrews et al., 1988). Besides, protein crystallography has a share of an unfocused white beam station (9.7) where the majority of the published Laue crystallography results have been obtained (Hajdu, 1987).

RECENT INSTRUMENTAL DEVELOPMENTS

The High Brightness Lattice

The implementation of the High Brightness Lattice (HBL) has significantly increased the possibilities for biological research at the SRS. The very nature of biological specimens (i.e., large unit cells, low scattering power, small dimensions, small quantities) imposes demands on the brilliance of the source that may not be an absolute requirement in other applications. The HBL has significantly increased the brilliance of the source with the consequent improvement in the possibilities for biological research (See below). The increase in brilliance (i.e., photons/sec/mrad^2/mm^2 in a certain wavelength bandwidth) relative to the old SRS is ca. 14.4 x 0.57 mm^2 to 2.6 x 0.24 mm. After a year of scheduled operations the standard operating current has increased to over 200 mA with 25-hour beam lifetime (Suller, unpublished).

X-ray Imaging Detectors And Data Acquisition Systems

Daresbury has a detector and data-acquisition development program. A number of devices have been completed (Lewis et al., 1988). One-dimensional linear detectors, two-dimensional area detectors and quadrant detectors, based on the proportional gas chamber approach and delay line read out, are routinely in operation. The associated data acquisition systems are fast (up to 400,000 events per second are routine) and the spatial resolution of all the detectors is about 300 micrometers. These detectors are meant for experiments in which it is vital that the corrections for detector response and background can be made over a wide dynamic range and with a high degree of reproducibility. This requires that the detector and associated electronics remain stable over the whole range of count rates encountered in an experiment and also throughout its duration. The systems have a reproducibility of about 1% over the whole detector area over 4 orders of magnitude in count rate.

Ultra-small Angle Scattering And Diffraction

This instrument is a scanning double crystal diffractometer equipped with a modified version of the Bonse-Hart multiple reflection small-angle scattering diffractometer (Nave et al., 1986). The main advantage of this instrument is that it permits the recording of diffraction features corresponding to spacings in excess of 2500.0 nm (i.e., it overlaps in resolution with the optical microscope). The main application to data has been the determination of the low resolution structure of frog sartorius muscle (Bordas et al., 1987).

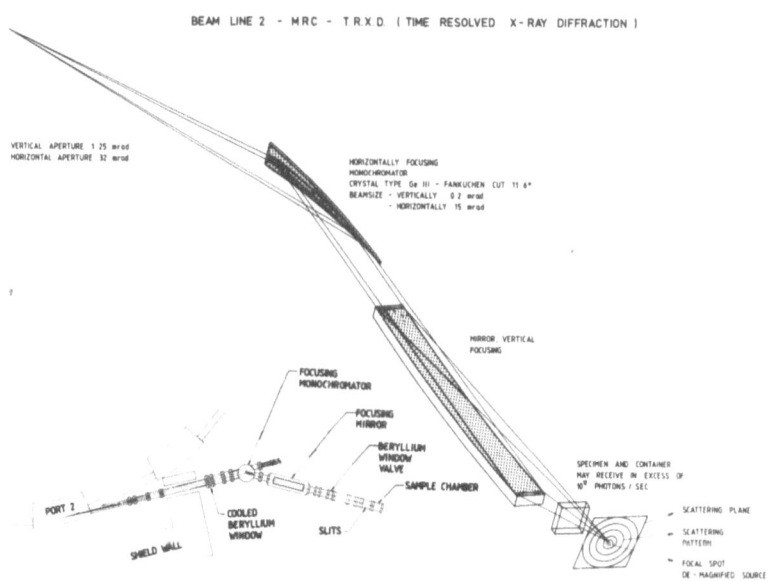

Fig. 1. Components of the time-resolved x-ray diffraction stations.

Time-Resolved X-ray Diffraction Station

This station is a double focusing monochromator-mirror system and was specifically designed around the specifications of the HBL, with particular emphasis on applications to time-resolved X-ray solution scattering and low angle fibre diffraction (e.g., studies of self-assembly in actin, tubulin, and clathrin and muscle research). A Ge (111) monochromator, with a compressing cut which reduces the horizontal beam cross section by a factor of 8, takes the white beam (as opposed to the more usual approach where the mirror is the first optical element), which is then vertically focused by a single, uncoated, quartz mirror (Fig. 1). The Ge monochromator is placed at about 9 meters from the tangent point and is able to collect up to 15 mrads of the horizontal aperture. The mirror (at about 10.5 meters from the tangent point) collects the whole vertical aperture of the monochromatic beam. The typical operating conditions are with a 1:1 focal distance in the horizontal plane, hence, the size of the focal spot in this plane is identical to the source size. The vertical dimensions are somewhat smaller than those of the source. We have found that this arrangement of the optics not only protects the mirror from radiation damage but also produces a much cleaner camera background without any sacrifice in the attainable intensity.

Other design features that improve the performance significantly are a set of conductances, placed inside the vacuum system within the radiation shield wall, that serve as acoustic delay lines in the event of a vacuum implosion and, more important from the point of view of use, these conductances also serve to collimate the white beam starting practically at the tangent point. This leads to an unusually clean background at the camera (for instance, a 2:1 signal to background ratio can easily be obtained in the scattering pattern from a solution of tubulin dimers at concentrations of ca. 1 mg/ml).

After the advent of the HBL the dimensions of the beam at the sample position are of the order of $8x0.7$ mm^2 (or smaller), while for the longest camera length (7 meters) one obtains dimensions of $2.6x0.3$ mm^2 (FWHM) at the

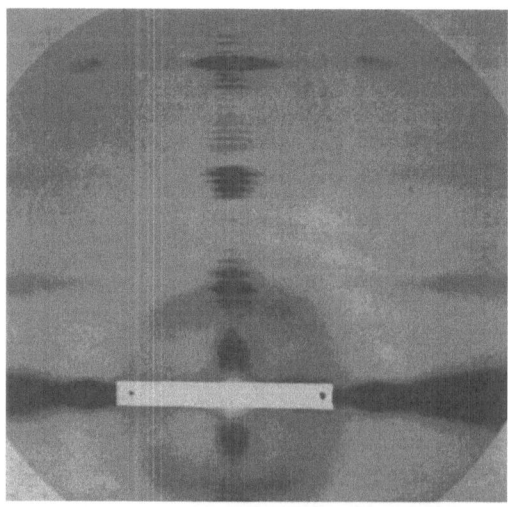

Fig. 2.   Digitally modified diffraction pattern of a muscle fibre taken on the time-resolving x-ray station.

focus (or smaller for shorter camera lengths). The design specifications of this device were to obtain in excess of 10^{12} photons/sec on the specimen with a circulating current of 200 mA, while attaining a background low enough to perform time-resolved measurements on solutions of protein at low concentrations (< 1 mg/ml for a "typical" molecule of ca. 40.000 D molecular weight) and resolving spacings corresponding to several 100 nm. The flux at the moment is 2.5 times below expectations, this is due to the poor quality of the Ge material available at present (it has proved extremely difficult to get a hold of perfect and large enough Ge crystals for the monochromator and as a result only ca. 4×10^{11} photons/sec are obtained in normal operation); otherwise all the other design specifications have been met.

Figs. 2 and 3 provide an illustration of the performance of this device. Fig. 2 shows a laser print of a diffraction diagram from frog semitendinosus muscle (Harris et al., 1988). The diagram was constructed from data recorded on a film, which was digitized, and every pixel in the pattern was multiplied by the value of the modulus of its reciprocal space position. This is necessary in order to view all the diffraction spots simultaneously in that it evens out the range of intensities (extending over several decades in the raw data). Because of this correction, the central scatter (all due to the muscle itself) appears much weaker than it is in the uncorrected data. Fig. 3 shows a logarithmic display of the meridional diffraction pattern against the reciprocal Z axis. Notable technical points are: a) The very low angle diffraction spots, indexing as the even orders of the sarcomere repeat (ca. 2550.0 nm in this example) are clearly resolved as from the 9th order (i.e., it is possible to resolve ca. 300.0 nm spacing). b) The order-to-order resolution corresponds to a unit cell of ca. 1275.0 nm or better.

From the point of view of muscle diffraction, it is worth pointing out the richness of the meridional diffraction diagram, clearly indicating that many other structural components besides the main contractile apparatus must be considered in its interpretation. Also notice the clear split in the 143-nm reflection, which is composed of at least two overlapping peaks (possibly three: notice the shoulder on the left side of the main peak), showing

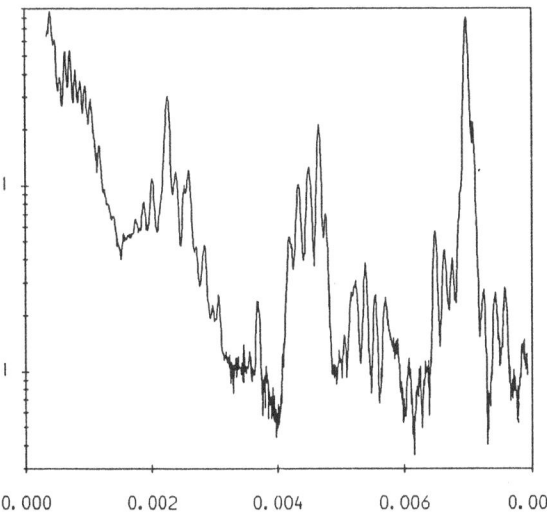

Fig. 3. Logarithmic plot of the meridional pattern in Fig. 2.

that intensity contributions arising for reasons other than the cross-bridge axial rise must be considered. Finally, notice the previously unreported layer line (with breadth comparable to the equatorial maxima) at about 58.0 nm spacing.

INSTRUMENTATION UNDER DEVELOPMENT

X-ray Microscope

X-ray microscopy utilizing zone plates are being developed in a collaboration between King's College London and Daresbury. The instrument has undergone a number of improvements and recently the first images showing the distribution of mass in unstained (but fixed) myofibrills were obtained. The contrast and resolution (ca. 150.0 nm) are good enough to reveal clearly the density in the I and A bands and, perhaps more significant, the Z lines are clearly resolved. It is expected that the instrument should soon be able to produce quantitative data on the native mass density distribution in muscle cells.

REFERENCES

Andrews, S. J., Papiz, M. Z., McMeeking, R. F., Blake, A. J., Lowe, B. M., Franklin, K. R., Helliwell, J. R., and Harding, M. M., 1988, Single micro-crystal structure determination with synchrotron radiation, Acta Cryst., B44:73.
Bordas, J., Mant, G. R., Diakun, G. P., and Nave, C., 1987, X-ray diffraction evidence for the existence of 102.0 and 230.0 nm transverse periodicities in striated muscle, J. Cell Biol., N3:1311.
Dobson, B. R., Hasnain, S. S., Hart, M., Van der Hoek, M. J., and Van Zuylen, P., 1986, An investigation of the performance of a novel double crystal X-ray monochromator for EXAFS and XANES measurements, J. de Phys., C8:121.
Hajdu, J., Machin, P. A., Campbell, J. W., Greenhough, T. R., Clifton, I. J., Zurek, S., Gover, S., Johnson, L. N., and Elder, M., 1987, Millisecond x-ray diffraction and the first electron density map from Laue photogaphs of a protein crystal, Nature, 329:178.
Harris, J. E., Diakun, G. P., Martin, M., Mant, G. R., and Bordas, J., 1988, The low resolution structure of striated muscle, in: "Proceedings of the Second International Conference on Synchrotron Radiation and Biophysics," England, (To be published).
Hasnain, S. S., 1988, Applications of EXAFS to biochemical systems, in: "Topics in Current Chemistry", Springer Verlag (In press).
Helliwell, J. R., Greenhough, T. J., Carr, P. D., Rule, S. A., Moore, P. R., Thompson, A. W., and Worgan, J. S., 1982, Central data collection facility for protein crystallography, small angle diffraction and scattering at the Daresbury Laboratory Synchrotron Radiation Source (SRS), England, J. Phys. E: Sci. Instrum., 15:1363.
Helliwell, J. R., Papiz, M. Z., Glover, I. D., Habash, J., Thompson, A. W., Moore, P. R., Harris, N., Croft, D., and Pantos, E., 1986, The wiggler protein crystallography workstation at the Daresbury SRS: progress and results, Nuc. Inst. and Methods in Phys. Res., A246:617.
Lewis, R., Sumner, I., Berry, A., Bordas, J., Cabriel, A., Mant, G., Parker, B., Roberts, K., and Worgan, J., 1988, Multiwire x-ray detector system at the Daresbury SRS, Nuc. Inst. and Methods, (in press).
Munro, I. H., and Martin, M. M., 1988, Synchrotron radiation for time-resolved fluorescence studies, in: "Fluorescence Spectroscopy, Vol. II, Biochemical Applications," J. R. Lakowicz, ed., Plenum, New York (In press).
Munro, I. H., Shaw, D., Jones, G. R., and Martin, M. M., 1985, Time-resolved fluorescence spectroscopy with synchrotron radition, Anal. Instrument., 14:465.

Nave, C., Diakun, G. P, Bordas, J., 1986, Ultra small angle x-ray diffraction from muscle, Nuc. Inst. and Methods, A246:609.

Nave, C., Helliwell, J. R., Moore, P. R., Thompson, A. W., Worgan, J. S., Greenall, R. J., Miller, A., Burley, S. K., Bradshaw, J., Pigram, W. J., Fuller, W., Siddons, D. P., Deutsch, M., and Tregear, R. T., (1985), Facilities for solution scattering and fiber diffraction at the Daresbury SRS, J. Appl. Cryst., 18:396.

Suller, V. P., Corlett, J. N., Dykes, D. M., Hughes, E. A., Poole, M. W., Quinn, P. D., MacKay, J. S., Thomson, S. L., and Walker, R. P., 1988, Performance of the Daresbury SRS with an increased brilliance optics, in: "Proceedings of the First European Particle Accelerator Conference," Rome, (To be published).

Van der Hoek, M. J., Werner, W., Van Zuylen, P., Dobson, B. R., Hasnain, S. S., Worgan, J. S., and Luijckx, G., 1986, A slitless double crystal monochromator for EXAFS and XANES measurements, Nuc. Inst. and Methods, A256:380.

FACILITIES AT THE NATIONAL SYNCHROTRON LIGHT SOURCE AT

BROOKHAVEN NATIONAL LABORATORY

Robert M. Sweet

Biology Department
Brookhaven National Laboratory
Upton, NY 11973

FACILITY DESCRIPTION

The National Synchrotron Light Source (NSLS) is the nation's largest facility dedicated solely to the production of synchrotron radiation. Located at Brookhaven National Laboratory, about 60 miles east of New York City, it is a national user facility supported by the U.S. Department of Energy. A plan view of the facility appears in Fig. 1.

The NSLS is available in general without charge to users. Its capabilities and the history of its development have been reported recently (Garrett, 1988). The facility has two electron storage rings. The vacuum ultraviolet (VUV) ring operates at an electron energy of 750 MeV and is designed for best radiation at energies from 10 ev to 1 keV. The X-ray ring operates at 2.5 GeV to optimize radiation from 1 keV to 20 keV. A total of 44 beam ports emanate from these rings. Each beam port is capable of supporting one to four experiments.

Operating parameters for both the VUV and the X-ray rings are shown in Table 1. Together they provide an intense, continuous source of radiation from 0.5 Å into the visible.

The VUV and X-ray rings presently accommodate over 800 scientists representing over 71 universities, industries, and government laboratories. Both basic and applied research are being done at the NSLS by groups from a variety of disciples which include physics, chemistry, materials science, metallurgy, biology, and medicine. Among the techniques used are EXAFS (extended X-ray absorption fine structure), scattering, diffraction, topography, fluorescence, gas phase spectroscopy, lithography, tomography, microscopy, and circular dichroism.

User Participation at the NSLS

There are several ways in which scientists can conduct research at the NSLS. Scientists who are not members of an existing beamline organization may submit proposals for experiments to the NSLS for beamtime. Termed general users, these scientists have available to them a portion of beamtime on all beamlines at the facility.

Table I. Selected NSLS Storage Ring Parameters

Ring	VUV	X-ray
Energy	0.75 GeV	2.5 GeV
Design Current	1 Å	0.5 Å
Circumference	51 meters	170.1 meters
Dipole beam ports	17	30
Insertion devices	2	5
Dipole critical energy	486 eV	5 keV
Dipole field	1.28 Tesla	1.22 Tesla
Max. number of bunches	9	31
Orbital period	170.2 nS	567.7 ns
Magnets	8 Dipole	16 Dipole
	24 Quadrupole	56 Quadrupole
	12 Sextupole	32 Sextupole
Power/Horizontal mRad	2.3 watts/Amp	80 watts/Amp
BM Source size σ_h, σ_v	0.5 mm, > 0.06 mm*	0.35 mm, 0.15 mm
Straight Section σ_h, σ_v	1.5 mm, > 0.025 mm*	0.35 mm, 0.02 mm

*Adjustable, typically runs at 0.2 mm in bend magnets and 0.07 mm in straight sections.

Many beamlines have been designed and constructed, and are administered and operated by Participating Research Teams (PRTs). PRTs are groups of scientists with comprehensive long-range programs. The PRTs are given priority for up to 75% of their beamline's operational time. The remaining time is made available to general users by proposal review.

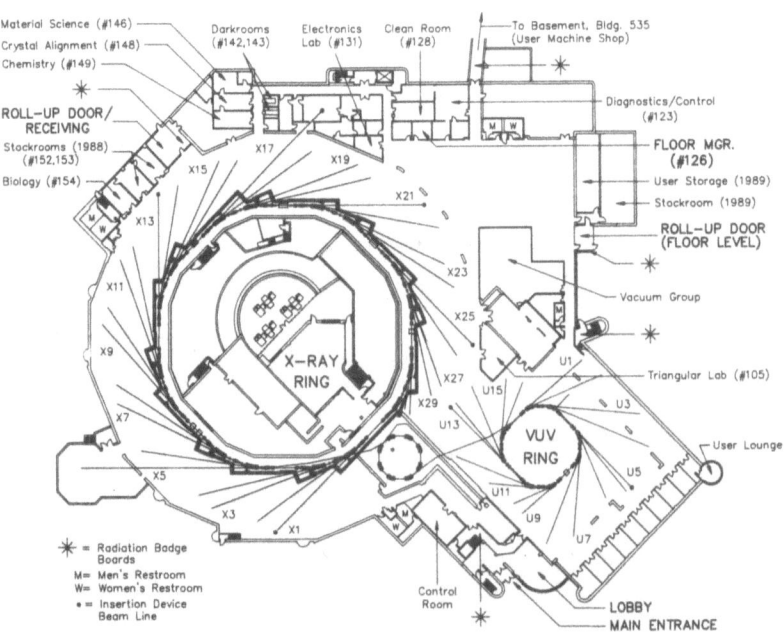

Fig. 1.   Schematic view of the NSLS facility. Only one beamline is indicated per port.

Insertion Device Teams (IDTs) are composed of groups of scientists who cooperate in the construction and use of wiggler and undulator beamlines. Some fraction of available beamtime is for use by the IDT members; additional time is available to general users, again, by proposal review.

General users are not required to enter into a scientific collaboration with PRT or IDT staff members as a condition for using a PRT beamline. Members of the NSLS or PRT staff are not expected to perform an experiment for a general user. General users are expected to have a thorough understanding of the experimental apparatus before they arrive to start work. They are also expected to provide adequate personnel for efficient use of the operation periods available to them.

Proprietary research may also be undertaken at the NSLS on a full cost-recovery basis. Scientists may have the option to retain title to inventions resulting from research at the NSLS. Users interested in doing proprietary work should contact the appropriate personnel.

Biological Studies at the NSLS

During recent years only a few beamlines have been used almost entirely for biological investigations. These include VUV circular dichroism, magnetic circular dichroism, and fluorescence lifetime studies on beamline U8, and soft x-ray microscopy on U15. On the x-ray ring, beamline X12-C has been dedicated to protein crystallographic investigations by rotation photography. Important studies by rotation photography have also been carried out at X13 and X21. Some biological small-angle scattering studies have been performed at X21. The bulk of the biological EXAFS performed at the NSLS occurs at beamline X11 (Sayers et al., 1983).

Three new facilities for biological investigations will be available in the near future. Beamline X12-B is being developed by the Biology Department at Brookhaven for small-angle x-ray scattering studies. At beamline X1 a multiple-magnet "undulator" has been installed. It will be optimized for soft x-ray microscopy and is being constructed by the same consortium who use U15. Also, at X8 a beamline is being constructed for protein crystallography. The PRT responsible is the Howard Hughes Medical Institute. The principal emphasis for the design of the facility is to provide data for multi-wavelength anomalous phasing.

Protein Crystallography

Most single-crystal diffraction studies on biological macromolecules are performed at beamline X12-C, developed by the Biology Department for that purpose. The monochromator is of the two-crystal sort, producing a beam that is stationary in the horizontal plane with a bandwidth near to $5 \cdot 10^{-4}$. A 2 mr-wide fan of radiation from this monochromator is focussed by an elastically bent cylindrical mirror. Data are collected by rotation photography with a highly automated camera system. A large and well-appointed darkroom lies a half-minute walk away from the experimental hutch.

During the first months of operation of this beamline, the emphasis has been on production of high-resolution data from poorly diffracting crystals. The monochromator system has energy resolution sufficient for multi-wavelength anomalous phasing. It will be used for that purpose after an electronic area detector has been installed.

X-Ray Absorption Studies

The Material Science EXAFS beamline at X11 has been constructed by a large and diverse consortium (Sayers et al., 1983). It produces a very intense x-ray beam in the range 4 - 20 keV. A four-crystal monochromator provides relative energy discrimination as good as $6 \cdot 10^{-5}$. Complete instrumentation is available for x-ray absorption measurements.

Biological research within the PRT includes studies of the iron core of ferritin, of porphyrin and chlorin-containing systems, and of non-heme oxygenases.

Synchrotron Ultraviolet Biophysics User Facility

Station U9B accepts 37 mrad of the radiation fan from port U9 on the VUV ring. A UHV window capable of supporting atmospheric pressure separates the experimental portion of the beamline from the ring. A rotatable mirror in the UHV section directs the photon beam towards one of two spectrometers. Thus two separate experiments time-share the beam over the course of a day without interference.

One spectrometer uses a vacuum monochromator. It is separated from the ring by a CaF_2 window which determines the short wavelength limit at 125 nm. For aqueous samples, however, the absorption of water results in a short wavelength limit of about 162 nm. The primary experiments performed are circular dichroism, magnetic circular dichroism, and fluorescence excitation and emission spectroscopy. Unpolarized absorption spectra are also recorded. The temperature of liquid samples can be varied over the range of -20 to 90°C. Small molecules (e.g. the purine and pyrimidine bases) can be isolated in solid argon or nitrogen matrices at temperatures < 30K.

The second spectrometer uses an unevacuated monochromator separated from the UHV by a quartz window. The short wavelength limit is about 200 nm. This instrument is dedicated to use of the time structure of the photon beam to measure fluorescence lifetimes and time-resolved fluorescence spectra.

REFERENCES

Garrett, R., 1988, The National Synchrotron Light Source, Synch. Radiat. News, 1:4.
Klaffky, R. W., ed., 1984, National Synchrotron Light Source, BNL 34710.
Sayers, D. E., Heald, S. M., Pick, M. A., Budnick, J. I., Stern, E. A., and Wong, J., 1983, X-Ray beamline at the NSLS for x-ray absorption studies in material science, Nucl. Instru. and Meth., 208:631.
White-DePace, S., and Gmur, N., eds., 1986, National Synchrotron Light Source annual report, BNL 52045.

SYNCHROTRON RADIATION TIME-RESOLVED SOLUTION X-RAY SCATTERING:

THE EXAMPLE OF CLATHRIN STRUCTURE AND ASSEMBLY

G. R. Jones[1], J. Bordas[2], D. Clarke[2], G. P. Diakun[2] and
G. R. Mant[2]

MRC/SERC Biology Support Laboratory[1]
SERC Daresbury Laboratory[2]
Warrington WA4 4AD
Cheshire, England

INTRODUCTION

High energy storage rings provide a source of X-rays with a brilliance
that exceeds that available from conventional sources by several orders of
magnitude. As a result it is now possible to observe dynamic processes in
biological systems by time-resolved X-ray scattering (Bordas, 1985). This
implies that the structural information, classically derived from X-ray
methods can now be combined with the kinetic information that is usually
obtained from measurements such as light scattering and fluorescence.

The implementation of a high intensity X-ray beamline equipped with the
necessary electronic recording devices for time-resolved X-ray scattering
experiments has been recently completed at the Daresbury SRS (Bordas, 1988).
This report illustrates the use of these methods with a few selected examples derived from an ongoing investigation of the structural dynamics of
clathrin assemblies (Kanaseki and Kadota, 1969; Pearse, 1975).

Clathrin is a large, soluble, polypeptide which is the main molecular
constituent of the polygonal network that forms the coat of coated pits and
vesicles (Pearse, 1975). These coats are formed from trimeric molecules of
clathrin, called triskelions, consisting of three heavy chains of ca. 180 kD
and three lights chains of ca. 30-40 kD. Besides the triskelions, the coats
contain a number of polypeptides (named assembly factors or clathrin associated polypetides: CAPs for short) with molecular weights of ca. 100-115 kD,
and 50 kD (Ungewickell and Branton, 1981; Pearse and Bretscher, 1981). These
coated vesicles are generally believed to play a central role in cellular
transport by receptor mediated endocytosis. This is the process whereby a
specific macromolecule (e.g., insulin) is captured by cell surface receptors
situated in coated pits (the coat proteins are placed on the cytoplasmic
side of the cell membrane). The coated pit then "pinches off" to form a
coated vesicle which carries its content into the cell. Once the coated
vesicle arrives at its target area, the coat is released from the vesicle
and its components are presumed to recycle to form new coated pits. The
vesicle is released so that fusion with another cellular membrane can occur
and the transported macromolecule is available for use by the cell.

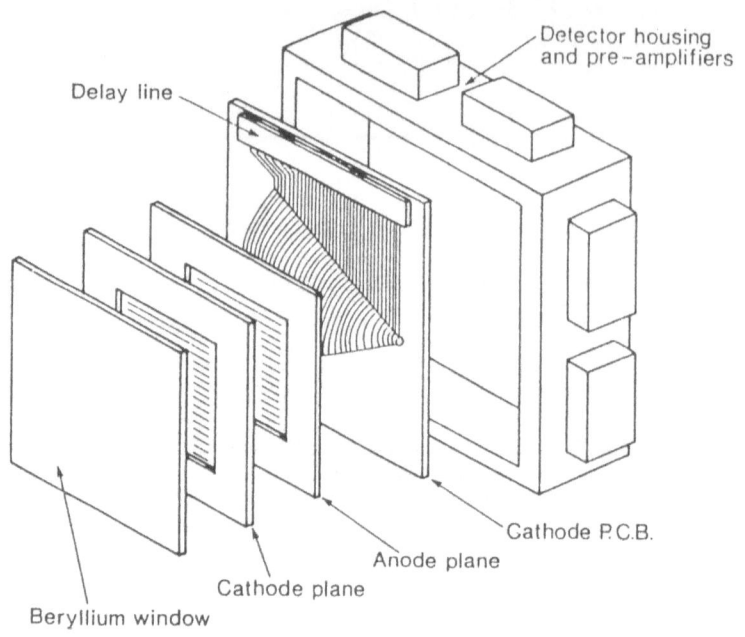

Fig. 1A. Exploded view of quadrant detector

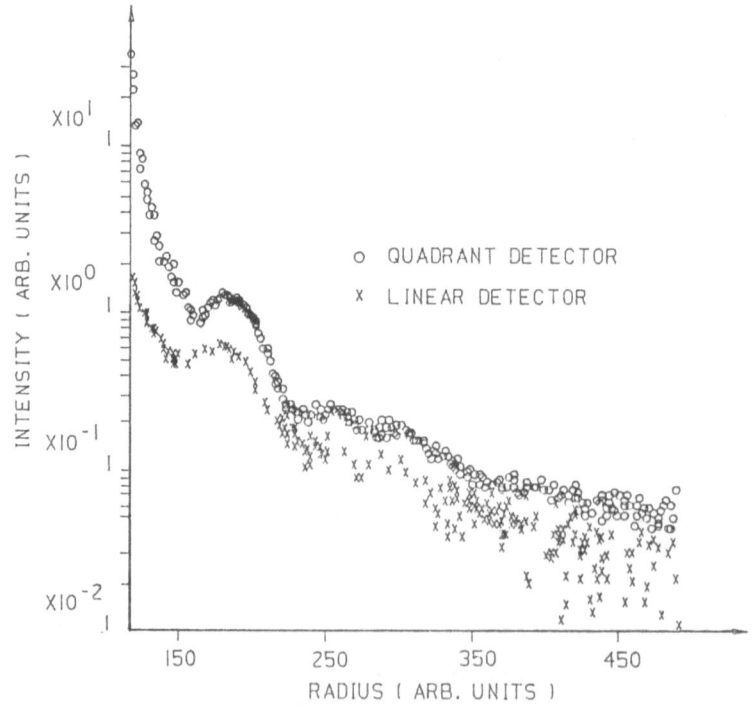

Fig. 1B. Comparison of output of quadrant and linear x-ray detectors

During receptor mediated endocytosis, clathrin is assembled and disassembled in a regulated fashion (Pearse and Bretscher, 1981; Hanover, et al., (1984). Under suitable biochemical conditions, clathrin or coat protein (i.e., clathrin plus the complement of CAPs) are able to undergo self-assembly into clathrates (Irace et al., 1982; Zaremba and Keen, 1983; Pearse and Robinson, 1984) which display a polygonal network and morphology which is essentially indistinguishable from that observed on the surface of coated vesicles.

We have started an investigation of the structure and assembly of clathrin and coat protein in the newly commissioned time-resolved X-ray scattering facility at the Daresbury SRS (Bordas, 1988). The results show that these methods are ideally suited for the study of this type of biological system. We describe below some instrumental aspects of the work, a method of interpretation of the X-ray solution scattering patterns and the effect of the CAPS on the stability of clathrates undergoing trypsin digestion.

Time-Resolved X-ray Solution Scattering From Solutions Of Clathrin And Coat Protein

The X-ray scattering patterns (static and time-resolved) from preparations of clathrin were recorded with a quadrant detector specially built for X-ray solution scattering experiments (Lewis et al., 1988). Fig. 1(A) shows an exploded view of this device. The main difference from any other position sensitive proportional gas chamber is that the cathode plane is built on a PCB with ring sectors which are then connected to a capacitively coupled delay line. The detector is placed with the apex of the quadrant coinciding with the position of the centre of the scattering pattern. The charge induced by a photon striking the chamber at any given radial distance will propagate along the corresponding ring and its position digitized, in the usual manner, by measuring its propagation time in the delay line. Consequently, by recording all the photons impinging on the quadrant at a given radial distance in a single pixel, this device ensures a front end polar integration in the scattering pattern, thus increasing the statistical accuracy at larger scattering angles where unavoidably the scattering patterns are weaker. In other words, the device has the ability to collect photons spread over an area (as in a two-dimensional position sensitive detector) while retaining the simplicity and speed of the electronics characteristic of a linear detector (Lewis et al., 1988).

Fig. 1(B) shows an example of the improved data quality achieved with this method. The logarithm of the X-ray intensities scattered by a solution of coat protein in assembly conditions are plotted against the radial distance in arbitrary units. The crosses correspond to the data obtained with a linear detector, while the circles show the scattering pattern obtained from the same specimen with the quadrant detector and for the same exposure time (See below for the interpretation of these patterns). The statistical accuracy is comparable at low angles; however, the quadrant detector provides a far superior statistical accuracy at high resolution.

Clathrin and coat protein were prepared from bovine brain by a modification of methods reported in the literature (Keen et al., 1979; Kirchausen et al., 1983). The purity of the preparation was assessed by SDS-PAGE analysis using ultrathin 7.5-12.5% gradient gels. Clathrin preparations show only the 180 kD, 33 kD and 36 kD bands of the heavy and light chains, while the gels from coat protein also show the bands at ca. 100 kD and 50 kD arising from the CAPs. The functionality of the protein was tested by electron microscopy. Fig. 2(A) shows an electron micrograph of a preparation of coats obtained by dialysing the protein against the assembly buffer (20 mM MES, 50 mM NaCl, pH 6.0), while Fig. 2(B) shows the same protein

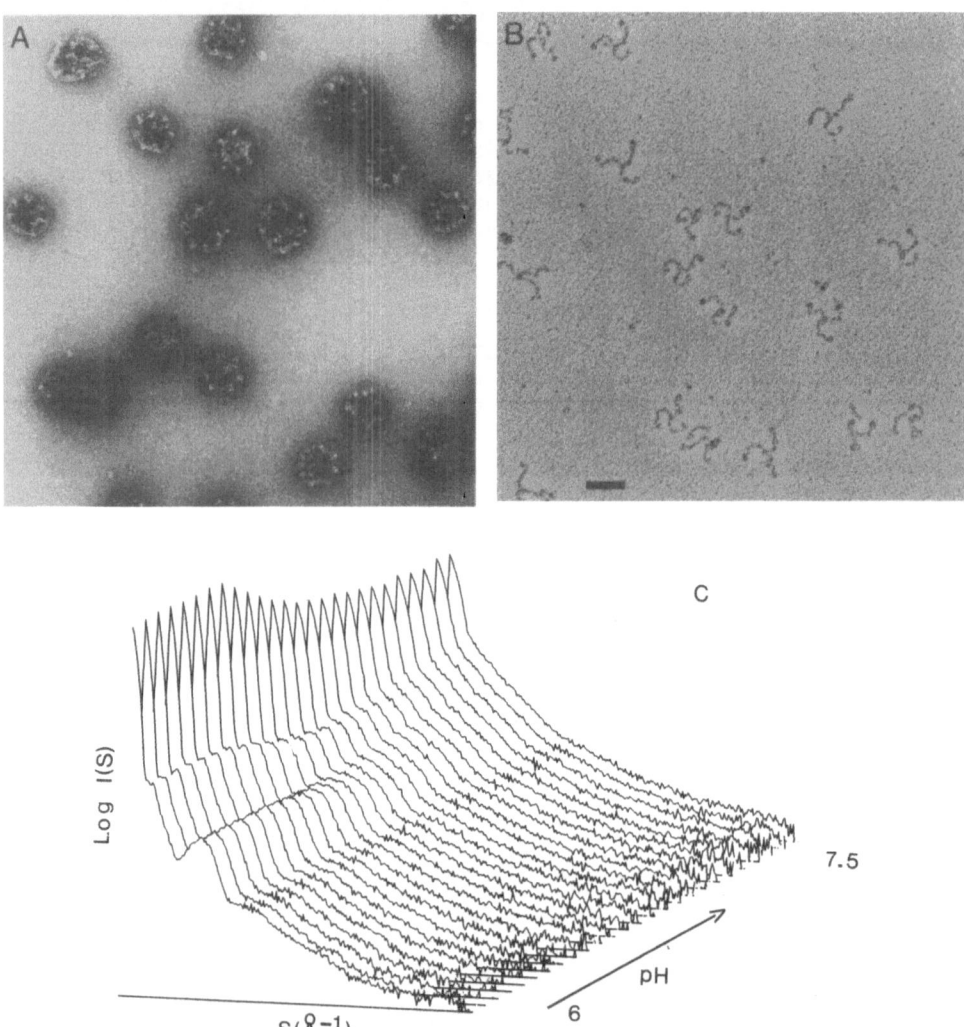

Fig. 2. A) Electron microscopic image of clathrin coats. B) Clathrin triskelions formed by the disassembly of clathrin coats. C) Time course of changes in scattered x-ray intensity (as a function of s, where s = 2sin θ/λ) of a solution of clathrin (5 mg/ml) as the pH is changed from 6 to 7.5. Every tenth frame from a series of 250 is shown. Each frame consists of a 40-second exposure. The range of the abscissa s is 0.0 to approximately 0.012 inverse Angstroms.

disassembled into triskelions after dialysis against disassembly buffer (20 mM PIPES, pH 7.5) followed by deletion.

Clathrin and to a lesser extent coat protein have proved to be extremely sensitive to radiation damage. Static X-ray scattering experiments on solutions of clathrin showed that, with the SRS running at 2 GeV, 200 mA and the X-ray camera optically aligned, the solutions scattering patterns revealed clear signs of structural damage induced by the irradiation after ca. 2 minutes exposure. This has proved a severe technical difficulty which was solved by resorting to continuous flow methods. This involved the construction of an X-ray cell through which the protein was continuously circulated by drawing the solution from a dialysis bag placed in a container with the appropriate buffer. The protein was circulated through the X-ray cell and dialysis bag with a peristaltic pump. This ensured that no portion of the protein solution was ever exposed by more than a few seconds.

The same device was used to carry out equilibrium studies of the protein conformation and state of polymerization against slowly shifting pH conditions. This was simply achieved by placing the desired dialysis buffer in the vessel containing the dialysis bag and simultaneously starting the recording of the X-ray scattering patterns against the changing pH. The pH was recorded by means of a microelectrode; the output of which was fed into a voltage to frequency converter and pulses counted with the same time framing protocol used for recording the time-resolved X-ray scattering patterns.

Fig. 2(C) shows a sequence of X-ray scattering patterns obtained from a solution of clathrin protein at ca. 5 mg/ml while the pH is slowly increasng. The exposure time was 40 seconds/frame and about 250 frames were recorded in a typical experiment. For ease of viewing, the figure shows every tenth frame. The data in these graphs and that shown below have been corrected by the same procedures as those described by Bordas et al., (1983). Experiments of this type, in which the equilibrium between the fully formed clathrates (as in Fig. 2(A)) is slowly shifting into a majority population of isolated triskelions (as in Fig. 2(B)), have revealed at least four different structural states in the overall disassembly pathways of coat and clathrin protein (Jones et al., unpublished data).

Assignment Of A Structural Origin To The Features In The X-ray Solution Scattering Patterns From Clathrates

Time-resolved X-ray solution scattering is a powerful method for detecting the presence of structural intermediates. Even without a structural interpretation of the X-ray scattering pattern, detection of intermediate structures can be achieved by correlation analysis methods (Bordas et al., 1983). However, the full potential of the method can only be realized if a structural interpretation of the X-ray scattering patterns is available. Besides the usual methods for the structural analysis of X-ray solution scattering patterns (Bordas and Mandelkow, 1986), one approach that we have found particularly useful in assigning a structural origin to the features in the X-ray solution scattering patterns is electron microscopy. Images of randomly oriented clathrates (Fig. 2A) are digitized and Fourier transformed. The modulus of the transform (shown in Fig. 3A) is then polar averaged in reciprocal space, yielding a function analogous to an x-ray scattering profile (shown in Fig. 3B). The result of these manipulations is shown by the bottom trace (i) in Fig. 3(B); the top trace (ii) shows the X-ray solution scattering pattern obtained from the same preparation of coats. In both cases the logarithm of the intensities is displayed (where S is defined as $2 \sin\theta / \lambda$, where θ is half the scattering angle and λ as a function of s is the wavelength, i.e., the data is plotted against reciprocal spacings). The X-ray pattern displays a central scatter decay, corresponding to the Guinier region, a relatively weak first subsidiary

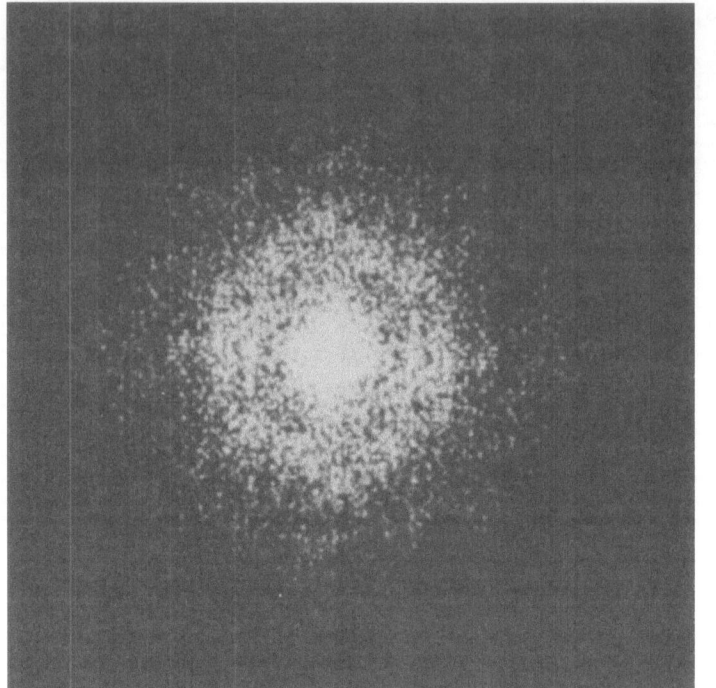

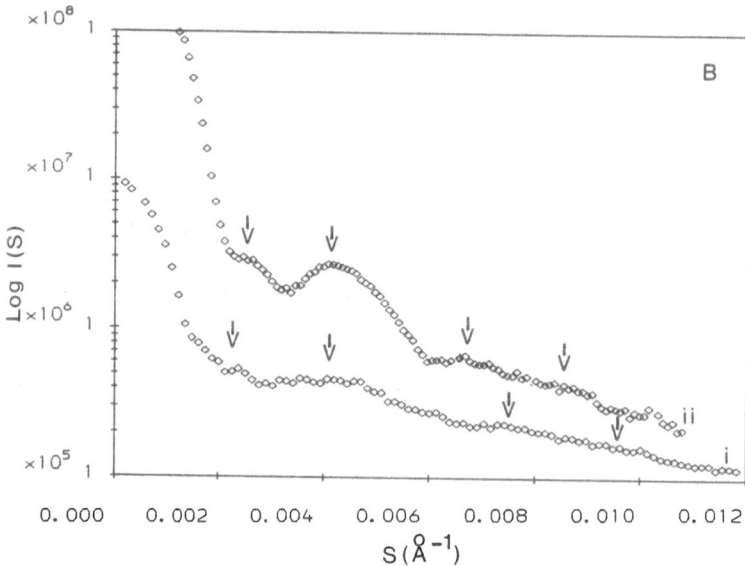

Fig. 3. A) Modulus of the Fourier transform of digitized electron microscopic images of clathrates deposited and dried on carbon grids.
B) Comparison of radial scattered x-ray intensity measured from
clathrin solutions (ii) with the reciprocal space polar average of
the modulus of the EM Fourier of Fig. 3A (i).

maximum at 0.0025 A^{-1}, a prominent band at ca. 0.0042 A^{-1}, followed by weaker ones at 0.007 and 0.009 A^{-1}. There is a one-to-one correspondence between these features and those in the equivalent pattern from the electron micrographs, although their absolute positions are somewhat different and their relative contrast is not commensurate. This is probably due to drying and staining artifacts associated with negative stain electron microscopy. Nevertheless from the correlation between the various bands it is possible to determine which structural features contribute most to the X-ray scattering bands. This can be done either directly from the spacing of the electron microscopy bands and referring to the initial image, or more systematically by Fourier filtering the rings of density corresponding to the radial distance of a given band, transforming back with the original set of phases and thus locating the features of the images which contribute to a given scattering maximum. For instance, filtering the central scatter and the first subsidiary maximum returns images in which only the outer profile of the clathrates is visible (showing, as expected, that these features are simply due to the low resolution transform of the clathrate which is approximated by a sphere), the filtered image obtained from the backtransform of the band at 0.0042 A^{-1} shows a punctate appearance in which the centroids of the density maxima coincide with the position of the vertices in the poligonal network at the surface, and so on. The conclusion from this type of data analysis is that the central scatter and the weaker subsidiary maxima are due to the transform of the hollow sphere approximation to the clathrate, while the prominent band at 0.0042 A^{-1} mainly arises from the surface lattice network. A full analysis of the scattering curves will be given elsewhere (Jones et al., unpublished).

Example Of Application To Digestion Studies

Once the origin of the X-ray scattering features is established, time-resolved X-ray solution scattering can be used as an assay to establish structural-functional relationships. For instance, Fig. 4(A) and (B) show the results of an experiment in which we addressed the question of the role that CAPs play in protecting the clathrates against digestive enzymes (trypsin in this example). The concentration of the protein was 10 mg/ml for coat protein and 8 mg/ml for clathrin protein. Trypsin at a final concentration of 1 and 0.8 mg/ml was added to the coat and clathrin protein solutions and the recording of the X-ray scattering patterns initiated. The pH of the solution was 6 (in order to ensure that a majority population of the starting structures were clathrates), while the high concentration of trypsin ensured enough activity for the digestion to occur.

The X-ray scattering patterns were recorded at frequent intervals while the digestion was proceeding. Fig. 4(A) and (B) only show a few time frames (30 seconds long) illustrating the development of the X-ray scattering patterns during the course of proteolysis. Fig. 4(A) shows the development of the X-ray patterns from a solution of coat protein (before digestion, i.e., time = 0 minutes and after 2, 5, 10, 40 and 80 minutes of adding the enzyme), while Fig. 4(B) shows the results of an equivalent measurement in a solution of clathrin protein (at t = 0, 2, 6, 11 and 44 minutes). The initial patterns show in both cases a clear surface lattice band, testifying for the integrity of the clathrates. In the case of clathrin (Fig. 4(B)) the scattering features begin to vanish immediately after the start of the digestion. After 2 minutes have elapsed the surface lattice band is already greatly reduced and is almost completely absent after 10 minutes. In marked contrast, the scattering patterns from solutions of coat protein show that the coats remain completely unaffected for a long period of time and only after more than an hour has elapsed it is possible to notice a slight reduction in the proportion of clathrates in the solutions. These results show that the CAPs, or at least some of them, must be blocking the main cleavage side of trypsin. It has been shown by electron microscopy (Vigers et al., 1986) that trypsin treatment appears to preferentially cut the terminal ends of the triskelion arms, which are arranged like fingers pointing towards the inside of the clathrates. Consequently, as CAPs increase the stability of the clathrates one can conclude from these results that not only the CAPs are located in the interior of the coats, in agreement with present views (Vigers, et al., 1986), but also that by keeping the terminal domains inaccessible to the action of the enzyme the clathrates are insensitive to the action of the enzyme.

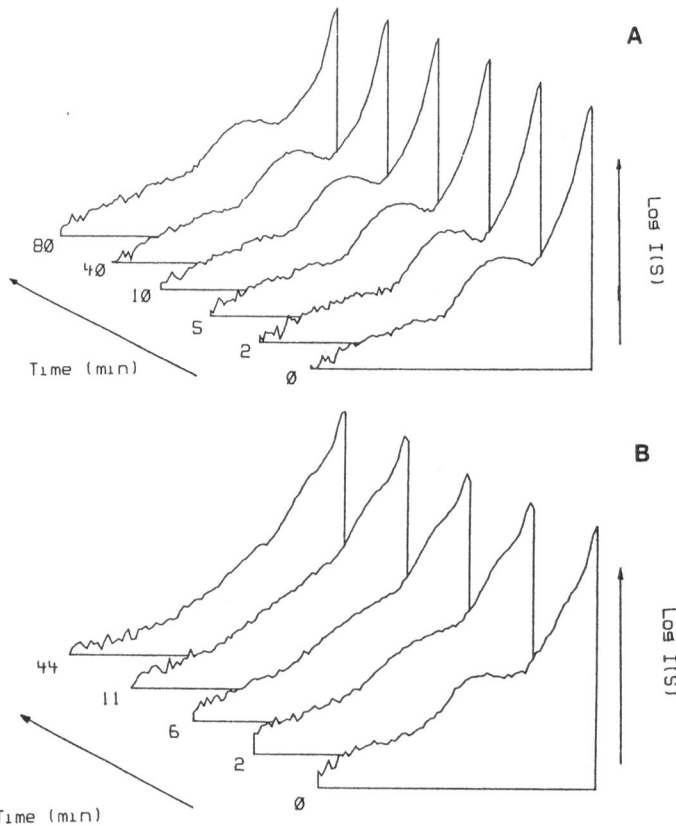

Fig. 4.   A) Time course of the scattered x-ray intensity of clathrates + CAPS undergoing trypsin digestion. B) Time course of the scattered intensity of clathrates without CAPS undergoing trypsin proteolysis.

REFERENCES

Bordas, J., 1985, Synchrotron radiation x-ray scattering and diffraction: applications to structural research in biology, Z. Phys. B- Condensed Matter, 61:389
Bordas, J., 1988, this volume.
Bordas, J., Mandelkow, E. M, and Mandelkow, E., 1983, Stages of tubulin assembly and disassembly studied by time-resolved synchrotron x-ray scattering, J. Mol. Biol., 164:7, 89.
Bordas, J., and Mandelkow, E., 1986, Time-resolved x-ray scattering, and diffraction: some aspects concerning the information content, in: "Structural Biological Applications of X-ray Absorption, Scattering

and Diffraction," D. T. Bartunik and B. Chance, eds., Academic Press, Orlando.

Hanover, J. A., Willingham, M. C., and Pastan, I., 1984, Kinetics of transit of transferrin and epidermal growth factor through clathrin-coated membranes, Cell, 39:283.

Irace, G., Lippoldt, R. E., Edelhoch, H., and Nandi, P. K., 1982, Properties of clathrin coat structures, Biochemistry, 21:5764.

Kanaseki, T., and Kadota K., 1969, Three-dimensional visualization of coated vesicle formation in fibro blasts, J. Cell Biol., 84:560.

Keen, J. H., Willingham, M. C., and Pastan, I., 1979, Clathrin-coated vesicles: isolation, dissociation and factor-dependent reassociation of clathrin baskets, Cell, 16:303.

Kirchausen, T., Harrison, S. C., Parham, P., and Brodsky, F. M., 1983, Location and distribution of the light chains in clathrin trimers, Proc. Nat. Acad. Sci. USA, 80:2481.

Lewis, R., Sumner, I., Berry, A., Bordas, J., Gabriel, A., Mant, G., Parker B., Roberts, K., and Worgan, J., 1988, Multiwire x-ray detector systems at the Daresbury SRS, Phys. Res., A273:773.

Pearse, B. M. F., 1975, Coated vesicles from pig brain: purification and biochemical characterization, J. Mol. Biol., 97:93.

Pearse, B. M. F., and Bretscher, M. B., 1981, Membrane recycling by coated vesicles, A. Rev. Biochem., 50:85.

Pearse, B. M. F., and Robinson, M. S., 1984, Purification and properties of 100-kd proteins from coated vesicles and their reconstitution with clathrin, The EMBO J., 3:1951.

Ungewickell, E., and Branton, D., 1981, Assembly units of clathrin coats, Nature, 289:420.

Vigers, G. P. A., Crowther R. A., and Pearse, B. M. F., 1986, Three-dimensional structure of clathrin cages in ice, The EMBO J., 5:529.

Vigers, G. P. A., Crowther, R. A., and Pearse, B. M. F., 1986, Location of the 100-kd -50kd accessory proteins in clathrin coats, The EMBO J., 5:2079.

Zaremba, S., and Keen, J. H., 1983, Assembly polypeptides from coated vesicles mediate reassembly of unique clathrin coats, J. Cell Biol., 97:1339.

LARGE-SCALE STRUCTURAL CHANGES IN THE SARCOPLASMIC RETICULUM

ATP-ASE ARE ESSENTIAL FOR CALCIUM ACTIVE TRANSPORT*

J. K. Blasie, D. Pascolini, F. Asturias, L. G. Herbette[1],
D. Pierce and A. Scarpa[2]

Department of Chemistry
University of Pennsylvania
Philadelphia, PA 19104.

[1]present address:
Departments of Radiology
Medicine and Biochemistry
University of Connecticut Health Center
Farmington, CT 06032

[2]present address:
Department of Physiology and Biophysics
Case Western Reserve University
Cleveland, OH 44106

INTRODUCTION

The active (energy-dependent) transport of calcium from the cytoplasm across the sarcoplasmic reticulum (SR) membrane into the sarcotubular system by the membrane Ca^{2+}ATPase is responsible for the relaxation phase of the contraction-relaxation cycle in striated and cardiac muscle (Ebashi et al., 1969). Since the sarcoplasmic reticulum membrane can be isolated in the form of closed unilamellar vesicles in which the Ca^{2+}ATPase comprises over 90% of the membrane protein (Meissner et al., 1973), the enzyme catalyzed chemical reactions involved in the calcium active transport process (DeMeis and Vianna, 1979), and the structure and organization of the membrane components (Blasie et al., 1985) have been relatively well characterized. As a result, this system provides an excellent prototype for the study of the critical structure-function relationships involved in the active transport of ions across biological membranes.

SUMMARY OF RECENT RESULTS

In the course of our structural and functional studies of the Ca^{2+}ATPase of the sarcoplasmic reticulum, we determined the separate profile structures of the phospholipid bilayer and the Ca^{2+}ATPase within

---

*Two Ca^{2+} ions per protein molecule are bound to all forms of the ATPase enzyme referred to in this paper. To simplifiy the notation, the Ca^{2+} ions have been omitted, i.e., the first phosphorylated intermediate, for example, is denoted as E_1~P instead of $(Ca^{2+})_2E_1$~P.

fully functional, isolated sarcoplasmic reticulum membranes (Herbette et al., 1985; Blasie et al., 1983). This work utilized meridional x-ray and neutron diffraction from oriented membrane multilayers coupled with the perdeuteration of the membrane phospholipids and H_2O/D_2O exchange and provided the profile structure of the ATPase to ~12 Å resolution, as cylindrically averaged about the membrane, normal by the smectic liquid-crystalline order in the membrane. We also analyzed the multiphasic kinetics of calcium active transport by these isolated SR membranes, employing flash-photolysis of caged ATP and double-beam spectrophotometry of metallochromic dyes localized in the extravesicular medium to initiate and monitor the calcium active transport by the ATPase ensemble (Pierce et al., 1983; Pierce et al., 1983). A fast phase of calcium uptake was associated with the formation of the first phosphorylated enzyme intermediate $E_1 \sim P$ [1] and the "occlusion" of the two high-affinity calcium binding sites. This was followed by a slow phase of calcium uptake associated with the translocation of calcium across the membrane profile as identified by its sensitivity to ionophores rendering the membrane permeable to calcium.

Using the above mentioned work as an essential basis, we then performed time-resolved x-ray diffraction studies to investigate whether large-scale changes in the (low-resolution) Ca^{2+}ATPase profile structure were associated with the ATP-induced calcium transport process (Blasie et al., 1985; Pascolini et al., 1988). These studies employed focused, monochromatic synchrotron x-radiation from both bending-magnet and wiggler-magnet beam-lines at the Stanford Synchrotron Radiation Laboratory. The calcium active transport process was synchronously (on the millisecond time scale) initiated among the ATPase ensemble in the oriented membrane multilayer by the flash-photolysis of caged ATP. An SIT image intensifier-based detector was used to record the meridional diffraction data in 0.2 sec to 5 sec time frames. Since the temperature-dependent lifetimes of the more interesting first and second phosphorylated intermediate forms of the enzyme, $E_1 \sim P$ and $E_2 \sim P$, vary from a few hundred milliseconds to a few seconds (Fernandez-Belda et al., 1984) over the useful temperature range for membrane multilayer stability (-2°C < T < 15°C), the inherently slow read-out time of the detector pixel array (many seconds) prevented "time-slicing" on this time scale (i.e., the recording of sequential time frames of 0.2 - 5.0 sec with only negligible delay between frames). Therefore, we focused our attention on the comparison of the "resting" E_1 versus the $E_1 \sim P$ forms of the enzyme. The thoroughly reproducible (multilayer to multilayer) and reversible (for each multilayer) changes in the profile structure of the SR membrane upon $E_1 \sim P$ formation in the oriented membrane multilayer at -2° to 0°C utilizing 2 to 5 sec time frames are shown in Fig. 1 (Pascolini et al., 1988). These changes are very similar to those observed at 6° to 8°C utilizing 0.2 to 0.5 sec timeframes. No such changes in the structure of the membrane profile are observable upon the flash photolysis of caged ADP under otherwise identical conditions in the multilayer, thereby providing a critical control against artifacts arising from the flash-photolysis reactions. These changes in the structure of the SR membrane profile observed upon $E_1 \sim P$ formation require a large-scale redistribution of ATPase mass from the ATPase "headpiece," which protrudes from the extravesicular surface of the membrane lipid bilayer into the lipid hydrocarbon core region of the membrane profile, namely a <u>net inward</u> movement of the ATPase mass into the nonpolar core of the membrane.

While the above mentioned time-resolved structural studies clearly identified the nature of the large-scale changes in the calcium ATPase profile structure associated with the ATP-induced formation of $E_1 \sim P$ from E_1, they could not alone address the significance of these changes to the actual calcium active transport process. The literature suggested that manipulation of several physical-chemical parameters available to this membrane multilayer system might be utilized to slow the ATP-induced formation and/or extend the lifetime (i.e., "transient trapping") of $E_1 \sim P$, the latter also

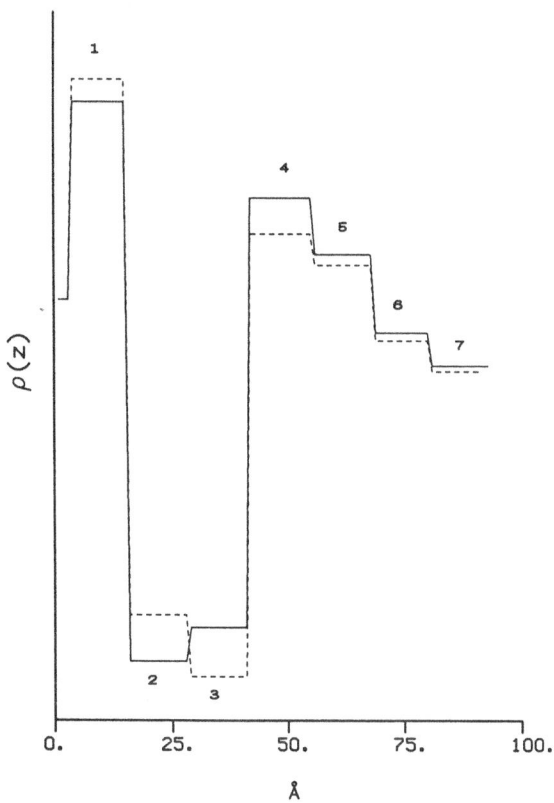

Fig. 1. Step-function model profiles at moderate resolution (17 Å) for a
single SR membrane profile (the multilayer unit cell profile con-
tains two apposed single membranes each within $0 \leq |z| \leq 106$ Å): the
solid line is the step-function model profile for the SR membrane
immediately before the UV flash-photolysis of caged ATP; the dashed
line is for the SR membrane immediately after the UV flash. The
profiles were derived from the analysis of the time-resolved x-ray
diffraction experiments performed at 0° to -2°C, with a time resolution of 2-5 sec. The numbers identify the following regions of
the membrane profile: 1 and 4, the lipid polar headgroup region of
the inner and outer monolayer, respectively; 2 and 3, the lipid
fatty acyl chain region of the inner and outer monolayer, respectively; 5-7 the extravesicular surface of the membrane (outside the
lipid bilayer) containing the ATPase "head-piece"; the region about
z=0 corresponds to the intravesicular water space. The changes in
the step-function model profile of the SR membrane upon the UV
flash photolysis of caged ATP under these conditions arise from
$E_1^+ \sim P$ formation from E_1^+ and include a decrease of density for
steps 3 and 4 and a corresponding gain of density for steps 1 and
2. Only a small decrease of density occurs for steps 5 through 7.
For the interpretation of these changes, see the text.

slowing the formation of $E_2 \sim P$ from $E_1 \sim P$, <u>without</u> totally inhibiting enzyme turnover. Further investigation of the ATP-induced calcium transport kinetics verified that low temperature at "high" $[Mg^{2+}]$ ($< 0°C$, ~25 mM); "low" $[Mg^{2+}]$ (~25 μM, t $< -10°C$); and the low water content of partially dehydrated, oriented multilayers (versus dispersions) all slowed the formation of $E_1 \sim P$ and extended its lifetime to varying degrees (Asturias and Blasie, 1988).

Subsequent investigation of the structure of the SR membrane profile for the "resting" E_1 form of the ATPase as a function of the same three thermodynamic variables: temperature, $[Mg^{2+}]$ and $[H_2O]$ in the oriented multi-layers, demonstrated a well-defined transition temperature for a large-scale change in the membrane profile. This change was associated with the onset of lipid lateral phase separation in the membrane plane (Asturias and Blasie, 1988; Pascolini and Blasie, 1988). The transition temperature for the structure of the membrane profile invariably occurred near the upper characteristic temperature t_h for the onset of lipid lateral phase separation (the formation of two-dimensionally crystalline domains of frozen lipid chains in the membrane plane). The upper characteristic temperature varies significantly as a function of the two other parameters ($[Mg^{2+}]$ and $[H_2O]$) employed. Comparision of these structural transition temperatures as a function of $[Mg^{2+}]$ and $[H_2O]$ with the corresponding ATP-induced calcium transport kinetics strongly indicates that it is this structural transition in the membrane profile which is responsible for the slowing of $E_1 \sim P$ formation and its "transient trapping". This structural transition in the SR membrane profile occurs, for example, at 2-3°C for "high" $[Mg^{2+}]$. Upon lowering the temperature through the transition, the structure of the membrane profile for the "resting" E_1 form of the ATPase changes to that corresponding to a different conformation for the ATPase, which we denote as E_1^+, as shown in Fig. 2. This requires a large-scale redistribution of ATPase mass <u>outward</u> from the lipid hydrocarbon core region of the membrane profile, into the "headpiece" region. Qualitatively (but not quantitatively), similar structural changes in the membrane profile occur upon lowering the temperature through the transition temperature as it varies with $[Mg^{2+}]$ and $[H_2O]$ in the multilayer.

DISCUSSION

On the basis of the published work from this laboratory as briefly described above, it has been firmly established that very similar large-scale structural changes occur within the SR membrane Ca^{2+}ATPase along with the ATP-induced formation of $E_1 \sim P$ from E_1 and with the slower formation of $E_1^+ \sim P$ from E_1^+. These changes that occur upon phosphorylation comprise a net <u>inward</u> movement of ATPase mass from the ATPase "headpiece" on the extravesicular surface into the lipid hydrocarbon core region of the membrane profile. Conversely, it has been similarly demonstrated that "large-scale" structural changes in the "resting" E_1 form of the ATPase, which result in the E_1^+ form, are responsible for slowing the kinetics of its ATP-induced phosphorylation to $E_1^+ \sim P$. These structural changes generally comprise a net <u>outward</u> movement of ATPase mass from the lipid hydrocarbon core to the ATPase "headpiece" on the extravesicular surface region of the membrane profile. Therefore, the structural changes in the "resting" E_1 form of the ATPase induced via manipulation of parameters of the membrane phase diagram (resulting in the E_1^+ form and in the slowing of ATP-induced $E_1^+ \sim P$ formation) are basically opposite to those structural changes that occur upon phosphorylation of either E_1 or E_1^+. As a result, the conformational changes in the "resting" E_1 form of the SR ATPase responsible for slower phosphorylation produce an ATPase structure (namely E_1^+), that is even more different from that of the corresponding phosphorylated intermediate $E_1^+ \sim P$, as compared with the differences between the E_1 and $E_1 \sim P$ forms. This

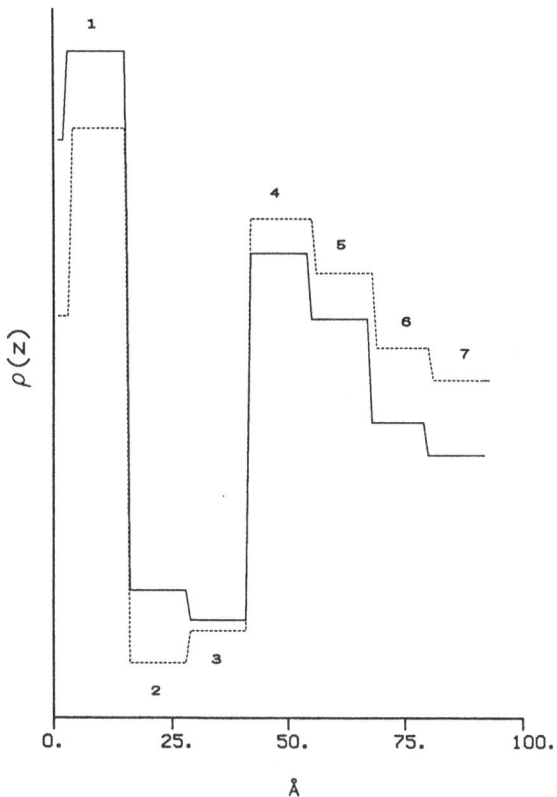

Fig. 2. Step-function model profiles at 16-17 Å resolution for a single SR
membrane profile for an oriented multilayer at high $[Mg^{2+}]$ (~25 mM)
as a function of temperature. The numbers (1 to 7) identify the
steps with the regions of a single SR membrane profile as in Fig.
1. The solid line is the step-function model profile for an SR
membrane multilayer partially dehydrated and in full equilibrium at
7.5°C, i.e., above the transition temperature (2-3°C). Partial de-
hydration of the multilayers at temperatures below the transition
temperature, e.g., -2° to 0°C, results in their full equilibration
at that temperature and the step-function model profile shown by
the dotted line. The changes in the SR membrane profile for the E_1
form of the ATPase upon lowering the temperature through the transition temperature, resulting in that for the E_1^+ form, include a
decrease of density for steps 1-3 and for the region about z=0,
i.e., the intravesicular water space, and an increase of density
for steps 4-6. For the interpretation of these changes, see the
text.

contrast holds true especially within the more conformationally restrictive lipid bilayer region of the membrane. Hence, the overall activation energy barrier in the ATPase polypeptide's configurational free energy surface is significantly greater for the formation of $E_1^+{\sim}P$ from E_1^+ as compared with that for the formation of $E_1{\sim}P$ from E_1. From the above considerations, it now appears that large-scale structural changes in the SR ATPase are indeed essential for its calcium active transport function.

REFERENCES

Asturias, F., and Blasie, J. K., Effect of Mg^{2+} concentration on the Ca^{2+} uptake kinetics and structure of the sarcoplasmic reticulum membrane, Biophys. J., (in press).

Blasie, J. K., Herbette, L., and Pachence, J. M., 1985, Biological membrane structure as "seen" by x-ray and neutron diffraction techniques. J. Membr. Biol., 86:1.

Blasie, J. K., Herbette, L., Pascolini, D., Skita, V., Pierce, D., and Scarpa, A., 1985, Time-resolved x-ray diffraction studies of the sarcoplasmic reticulum membrane during active transport, Biophys. J., 48:9.

Blasie, J. K., Pachence, J. M., and Herbette, L., 1983, Neutron diffraction and the decomposition of membrane scattering profiles into the scattering profiles of their molecular components, in: Neutrons in Biology, B. P. Schoenborn, ed., Plenum Press, New York.

DeMeis, L., and Vianna, A. L., 1979, Energy interconversion by the Ca^{2+}-dependent ATPase of the sarcoplasmic reticulum, Ann. Rev. Biochem., 48:275.

Ebashi, S., Endo, M., and Ohtsuki, I., 1969, Control of muscle contraction, Quart. Rev. Biophys., 2:351.

Fernandez-Belda, F., Kurzmack, M., and Inesi, G., 1984., A comparative study of calcium transients by isotopic tracer, metallochromic indicator, and intrinsic flourescence in sarcoplasmic reticulum ATPase., J. Biol. Chem., 259:9687.

Herbette, L., DeFoor, P., Fleischer, S., Pascolini, D., Scarpa, A., and Blasie, J. K., 1985., The separate profile structures of the functional calcium pump protein and phospholipid bilayer within isolated sarcoplasmic reticulum membranes determined by X-ray and neutron diffraction, Biochim. Biophys. Acta, 817:103.

Meissner, G., Conner, G. E., and Fleischer, S., 1973, Isolation of sarcoplasmic reticulum by zonal centrifugation and purification of Ca^{2+}-pump and Ca^{2+}-binding proteins, Biochim. Biophys. Acta., 298:246.

Pascolini, D., and Blasie, J. K., 1988, Moderate resolution profile structure of the sarcoplasmic reticulum membrane under low temperature conditioned for the transient trapping of $E_1{\sim}p$., Biophys. J., 54:669.

Pascolini, D., Herbette, L. G., Skita, V., Asturias, F., Scarpa, A., and Blasie, J. K., 1988, Changes in the sarcoplasmic reticulum membrane profile induced by enzyme phosphorylation to $E_1{\sim}p$. at 16Å resolution via time-resolved X-ray diffraction, Biophys. J., 54:679.

Pierce, D. H., Scarpa, A., Topp, M.R., and Blasie, J.K., 1983, Kinetics of calcium uptake by isolated sarcoplasmic reticulum vesicles using flash photolysis of caged adenosine 5'-triphosphate, Biochemistry, 22:5254.

Pierce, D. H., Scarpa, A., Trentham, D. R., Topp, M. R., and Blasie, J. K., 1983, Comparison of the kinetics of calcium transport in vesicular dispersions and oriented multilayers of isolated sarcoplasmic reticulum membranes, Biophys. J., 44:365.

ANOMALOUS SCATTERING IN MEMBRANE STUDIES

C. Boulin[1], G. Buldt[2], F. Dauvergne[1], A. Gabriel[1],
G. Goerigk[3], B. Munk[3], and H.B. Stuhrmann[3,4]

[1] EMBL Heidelberg and its Outstation at Grenoble
[2] Freie Universitat Berlin
[3] GKSS Forschungszentrum, D-2054 Geesthacht, F.R. Germany
[4] Johannes-Gutenberg-Universitat Mainz
(address correspondence to GKSS Forschungszentrum)

INTRODUCTION

Resonant (or anomalous) X-ray scattering is a nondestructive labelling technique of increasing usefulness in molecular structural analysis. Its use in macromolecular crystallography has been restricted to elements that show a reasonably strong contribution of resonant scattering at wavelengths below 2.5 Å. The experimental technique is governed by the symmetry of the diffraction pattern; i.e. by the presence or absence of anti-centrosymmetric diffraction features in the pattern. Any disorder (formation of twins or random orientation) may lead to centrosymmetrical scattering patterns.

Resonant X-ray diffraction from a single crystal with an asymmetric unit cell is largely determined by the dispersion of the imaginary part f" of the atomic form factor f. The parameter f" is related to the total cross section of photoelectric absorption δ by the optical theorem:

$$\delta = 2 \lambda \quad f''(0) \tag{1}$$

As X-ray absorption is quite a general phenomenon this is also the case for resonant X-ray scattering. For a given element, f" increases monotonically with the square of the wavelength but drops sharply near X-ray absorption edges (Fig. 1). On the short wavelength side of an absorption edge, the statistical accuracy of the diffraction data determines the relevance of the dispersion of f" to the analysis. Protein crystals with asymmetric unit cells will give only weak f" dispersion. In favorable cases the contribution of f" of sulfur to X-ray diffraction may be strong enough to become useful for phase determination of Bragg reflections. Crambin is a prominent example. Hendrickson and Teeter (1981) showed that the three disulfide bridges of crambin yield a measurable signal. They were able to measure Friedel pairs to 1.5 Å spacings on an X-ray diffractometer with CuK_α radiation (1.5418 Å), far away from the absorption edge of sulfur (5.02 Å). The Bijvoet-difference Patterson syntheses were cleanly interpretable and the sulfur structure was refined to 0.945 Å. The observed S-S distances were in good agreement with expectations. Clearly, as one approaches the absorption edge of sulfur, resonant scattering will have a greater effect on the phases (Hendrickson, 1986).

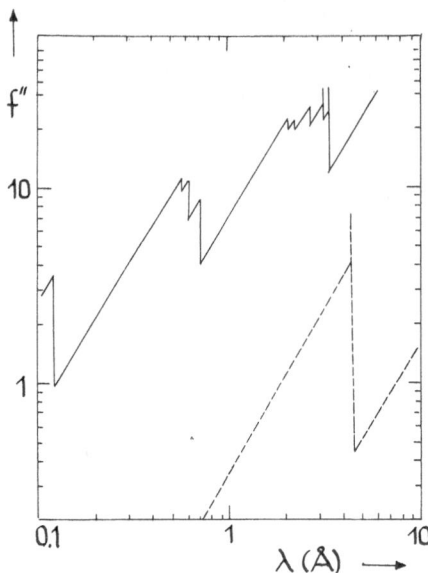

Fig. 1. The dispersion of the imaginary part f" of uranium and sulfur.

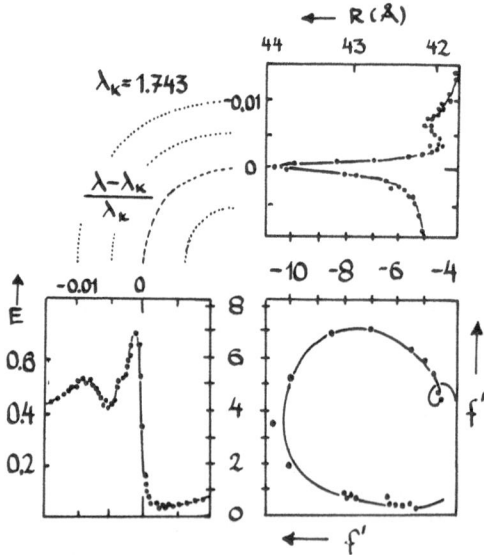

Fig. 2. The dispersion of f' and f" of iron in ferritin. f' is taken from
the dispersion of the apparent radius of gyration, R, of ferritin.
f" is proportional to the absorption of the ferritin solution at
the K-absorption edge of iron at λ = 1.743 Å. Elimination of the
wavelength in the upper and the left figure yields a plot of f'
against f", the Argand diagram (Stuhrmann, 1980).

The vast majority of elements show f" dispersion only as a second order effect. The main contribution of resonant scattering is due to the dispersion of the real resonant part f' of the atomic form factor. The value of f'is largest when the monotonic increase of f"-with wavelength changes to a drastic decrease (anomalous dispersion) within an absorption edge. Contrary to the case of f"-dependent X-ray scattering, the presence of f' in X-ray diffraction cannot be detected from measurements at a single wavelength. This means that anomalous diffraction relying on f' dispersion has to be carried out at wavelengths very near to an absorption edge. As an example we refer to the study of ferritin solutions by anomalous small-angle X-ray scattering if iron near the K-absorption edge at λ = 1.743 Å, the fundamental results of which are shown in Fig. 2.

X-rays with wavelength near the absorption edge of sulfur are strongly absorbed by any material. Their penetration depth in air is just a few cm. The thickness of a protein sample must be of the order of 0.01 mm. The questions related to the structural study of the sample must justify the technical effort. Membranes meet these requirements. We have chosen to study the purple membrane of <u>Halobacterium</u> <u>halobium</u>. Purple membrane can be prepared in pellets of nearly any thickness which diffract strongly up to the seventh order, even after several hours of irradiation by monochromatised soft X-ray synchrotron radiation.

THE BASIC SCATTERING FUNCTIONS OF CONTRAST VARIATION AND DIFFERENCE FOURIER ANALYSIS

Let the scattering amplitude F of an ensemble of atoms (e.g. of a molecule or of a whole unit cell) be

$$F = U + [f'(\lambda) + f''(\lambda)] V \qquad (2)$$

where vectors F,U and V are defined by the construction in Fig. 3. The scattering intensity then is

$$
\begin{aligned}
S = |F|^2 = \ &|U| \\
&+ 2 f'(\lambda) |U| |V| \cos(\phi_u - \phi_v) \\
&+ 2 f''(\lambda) |U| |V| \sin(\phi_u - \phi_v) \\
&+ [f'^2(\lambda) + f''^2(\lambda)] |V|
\end{aligned}
\qquad (3)
$$

The average S(av.) of F(Q) and F(-Q) no longer contains the sine term which is linked to the f" dispersion. Q is a vector in momentum space. In this form Eq. 2 resembles the equation of contrast variation given by Stuhrmann (1975):

$$S(Q) = |U(Q)| + 2 f'(\lambda) Re\{U(Q)V(Q)\} + [f'^2(\lambda) + f''^2(\lambda)] |V(Q)|$$

$$= S_u(Q) + f'(\lambda) S_{uv}(Q) + [f'^2(\lambda) + f''^2(\lambda)] S_v(Q) \qquad (4)$$

Very often the contribution of resonant scattering is small. The squares in f' and f" may be neglected. Dividing the scattered intensity by the absolute value |U| of the nonresonant scattering then yields

$$S(av.) / |U| = |U| + 2 f'(\lambda) |V| \cos(\phi_u - \phi_v) \qquad (5)$$

Fourier difference methods become feasible when the additional label of a derivative induces very small changes of the scattering intensity. Let S = |U| be the scattering intensity of the native molecule and SD = |F| that of the derivative. In the case of anomalous scattering, the derivative is created by the anomalous dispersion of one or few atoms of a

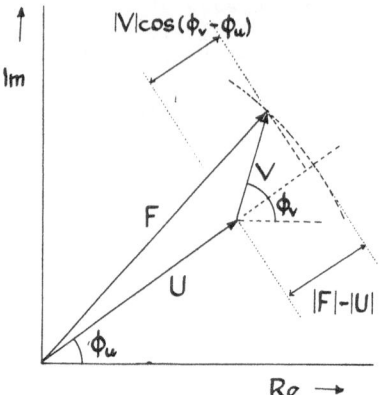

Fig. 3. The amplitudes of the 'native' molecule (U) of the derivative
(F) and of the label (V). The corresponding phase angles are
denoted by ϕ_u, ϕ_f and ϕ_v.

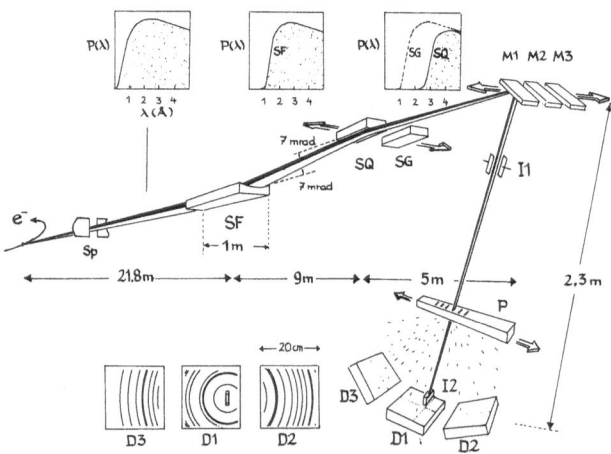

Fig. 4. Double mirror - single crystal camera for soft X-rays. Sp =
entrance slit. SF = gold coated double focusing mirror. SQ =
plane quartz mirror. One-half of it is coated with gold (= SG).
M1, M2, M3 are the crystal monochromators which can be used
alternatively. P = sample exchanger. I1, I2 = ionization
chambers. D1, D2, D3 are position sensitive counters. The upper
inserts show the actual synchrotron radiation spectrum. The lower
inserts show the diffraction pattern of bacteriorhodopsin as it
would appear on the three detectors using 5 Å photons. The
instrument is installed at beamline A1 of HASYLAB.

macromolecule. Fourier difference synthesis yields the electron density
$\rho(r)$.

$$\rho(r) = \Sigma \ [\ |F(Q)| \ - \ |U(Q)| \ ] \ \exp\{i\phi_u(Q)\} \ \exp\{-2\pi i \ Q \cdot r) \qquad (6)$$

Fig. 3 shows that $|F| - |U|$ is nearly $|V| \cos(\phi_u - \phi_v)$ in Eq. (4).
The latter term may be used in Fourier differences as well. Using the
definitions in Eq. 3 we obtain

$$\rho(r) \approx \Sigma \ S(Q) \ / \ [2\sqrt{S(Q)}] \ \exp\{i\phi_u(Q)\} \ \exp\{-2\pi i \ Q \cdot r\} \qquad (7)$$

THE INSTRUMENT

Diffraction experiments with soft X-rays face considerable difficulties
due to strong absorption by all materials. The design of our instrument at
HASYLAB is shown in Fig. 4. A brief description of its design follows.

Synchrotron radiation is deflected in the vertical plane by three
optical elements: two mirrors and a single crystal monochromator. The
choice of the vertical plane of deflection is obligatory as the incident
synchrotron radiation is linearly polarized in the plane of the orbit. The
polarization factor remains unity for the predominant component of polarization of synchrotron radiation for all angles of reflection by the crystal
monochromator. A single monochromator crystal is preferred to the double
crystal monochromator as the differential reflectivity decreases considerably with longer wavelengths. For the (111) reflection of a germanium
single crystal, reflectivity is 0.15 at wavelengths around 6 (Materlik,
personal communication). The price paid by this choice is increased complexity in the mechanism for the vertical rotation of the X-ray camera
around the axis of the crystal monochromator.

The double mirror system together with a single crystal monochromator
suppresses higher harmonics in the range of wavelengths from 1.2 to 7 Å.
Synchrotron radiation is reflected by a double focusing mirror of 80 mm
horizontal width and 1 m length at a mean glancing angle of 7 mrad. The
spectrum of the slightly converging beam then starts at wavelengths around
1.2 Å. Reflection by the gold coated surface of the second mirror (SG in
Fig. 4) has negligible effect upon the spectrum. The crystal monochromator
selects a very narrow wavelength band ($\Delta\lambda/\lambda < 5 \cdot 10^{-4}$ fhw), centered around λ
which is related to the glancing angle θ by

$$n \lambda \ = \ 2 \ d \ \sin \theta \qquad (8)$$

The fundamental wavelength λ (n = 1) is the only wavelength to be reflected by the (111) plane of a germanium single crystal as long as it does
not exceed 3 x 1.2 Å = 3.6 Å. The second order reflection is absent in
this case, whereas the third order harmonic is present along with wavelengths beyond this limit. At this point premonochromatization has to be
changed. Lateral displacement of the second mirror exposes its quartz surface to the radiation emerging from the focusing mirror. At a glancing
angle of 7 mrad the quartz surface shows significant reflectivity for X-rays
of wavelength greater than 3 Å. With this kind of premonochromatization the
use of the germanium single crystal monochromator can be extended to its
limit at $\lambda = 2$ d = 6.53 Å, (the case of back reflection). At present, we
encounter an increased reflectivity of the quartz surface at $\lambda > 2$ Å. The
reason for this is being investigated.

The glancing angle of the crystal is changed in steps of 0.8" while the
rotation of the camera proceeds in steps of 1.6". The camera can be rotated
into a deflection of the beam up to 160° (Fig. 5).

The photon flux of the primary beam is monitored by two ionization chambers. The first is between the monochromator and sample while the second counter is incorporated in the beam stop. The size of the beam stop is 12 mm in the vertical and 40 mm in the horizontal directions. The pressure inside the first chamber can be lowered to compensate for the increased absorption of soft X-rays.

The multiwire proportional counters from A. Gabriel (EMBL) are suitable for use with soft X-rays. The sensitive area of these detectors is 180 mm x 180 mm, with vertical and horizontal spatial resolution of 1 mm and 2 mm,

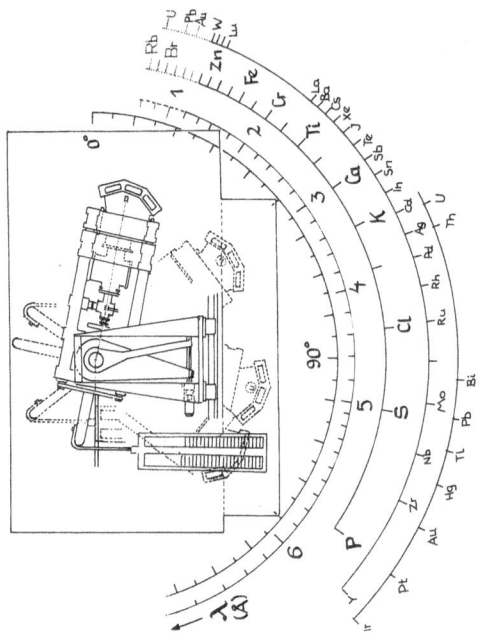

Fig. 5. The wavelength range of the diffractometer at beamline A1 of
         HASYLAB. Using the (111) plane of a germanium single crystal
         wavelengths up to 6.4 Å can be reached by rotating the crystal by
         θ = 80°. Thus the K-absorption edges down to phosphorus (Z=15),
         the L3-absorption edges down to Yttrium (Z=39) and the M5 absorp-
         tion edges down to iridium (Z=78) can be used. Replacing the
         germanium crystal with indium antimonide extends the wavelength
         range to 7.3 Å, thereby permitting use of the K_α absorption edge
         of silicon (Z=14). The dotted region on the short wavelength side
         is inaccessible.

respectively. Spatial resolution is determined by the spacing of the anode wires. The spatial resolution of the area detector deteriorates slightly with X-ray wavelengths greater than 5 Å. Up to 10^5 random events can be processed by each area detector.

With longer X-ray wavelengths the number of scattered photons recorded by the area counter depends on the absorption of various plastic windows in the beam path. These are between the ultra high vacuum of the storage ring and the evacuated X-ray camera (25 μm), at the first ionization chamber (2 x 6 μm), and at the end of the evacuated vessel (30 μm). The latter mylar window is supported by a steel grid. The mylar window of the area detector has a thickness of 20 μm. The total amount of plastic in the X-ray beam path amounts to nearly 87 μm not counting the thickness of the sample. The volume between the detector window and the window of the vacuum vessel is filled with helium. The transmission of the line for 5 Å X-ray photons is of the order of 0.01. It is nearly 1 for higher harmonics, if present. Nevertheless the count rate from purple membrane diffraction on the whole area of the detector was 18,000 cps and no higher harmonics were observed. Nevertheless, third order harmonics do appear at wavelengths near the K absorption edge of phosphorus (5.8 Å).

Anomalous Diffraction From Sulfur In Purple Membrane

A purple membrane pellet of low mosaicity may be obtained from slow sedimentation of flat crystallites on a thin mylar foil. The optimal thickness of about 40 μm is easily achieved. The sample has a low water content. Drying in vacuo hardly changes the diffraction pattern, shown in Fig. 6.

The X-ray absorption spectrum of purple membrane is shown in Fig. 7. Two peaks are observed. The absorption edge at 2470 eV is that of sulfur in methionine, whereas the second one at 2480 eV is due to glycolipid sulfates present in the lipid membrane. The regions of strong anomalous dispersion of sulfur in methionine and in the sulfate are well separated on the wavelength scale.

Anomalous Scattering Of Sulfur Bound To Methionine

The diffraction pattern of purple membrane was measured at two wavelengths (points marked by (1) and (2) Engelman and Zaccai, 1980; Hendrickson, 1986, in Fig. 7) up to the (4,1) reflection which corresponds to 11 Å resolution. These patterns are presented in Fig. 8. At these two wavelengths the difference in absorption and hence in f" is small whereas strong variation in the real resonant amplitudes is observed.

Fig. 9 presents a difference Fourier synthesis (Eq. 7) using the diffraction data. The broken contour lines show the projection of the spatial distribution of the sulfur atoms at 12 Å resolution in the membrane. The electron density map of the three bacteriorhodopsin molecules at 9 Å resolution is indicated in the same figure by slightly weaker full lines.

Most of the sulfur atoms, i.e., ligands of methionine are found in the central polar channels of the three bacteriorhodopsin molecules. Six out of the nine methionines per molecule are known to be associated with the seven alpha-helical sections of the protein chain. Our results are consistent with predictions of the positions of methionine obtained from hypothetical assignments of the helices of the electron density map to sections of the known amino acid sequence of bacteriorhodopsin (Engelman and Zaccai, 1980). The orientations of the helices of the helix-amino acid sequence correlation proposed by Engelman and Zaccai (1980) imply that four methionines are close to the polar channel. Further insight into the structure might be expected from higher resolution data.

Anomalous Scattering Of Sulfur In Sulfates

Anomalous X-ray diffraction due to sulfur in sulfates is deduced from diffraction data at wavelengths marked by (3) and (4) in Fig. 7. It differs considerably from that of methionine sulfur (see Fig. 8). However the number of reflections which could be obtained with reasonable accuracy is

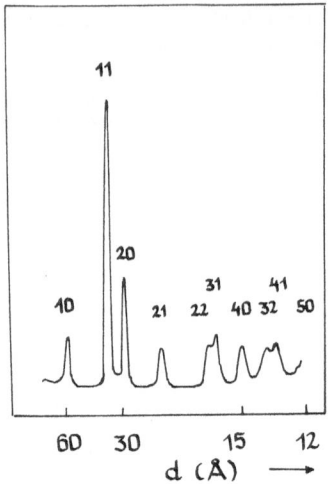

Fig. 6.   X-ray diffraction from purple membrane at a wavelength near the K-
absorption edge of sulfur at $\lambda = 5$ Å after correction for detector
response.

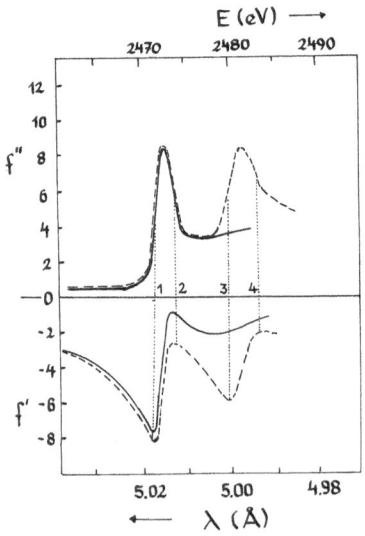

Fig. 7.   Anomalous dispersion of sulfur in purple membrane.   The imaginary
part of the scattering length f" (in units of electrons) is proportional to the X-ray absorption (Eq. 1).
----------   sulfur in methionine
- - - - - -   purple membrane

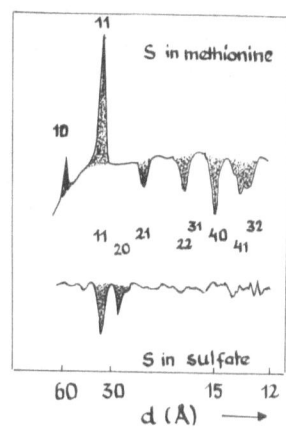

Fig. 8. Anomalous X-ray diffraction near the K absorption edges of sulfur
in methionine of the membrane protein bacteriorhodopsin. The data
were obtained as an intensity difference betwen wavelengths (1) and
(2) in Fig. 7.

much smaller. The data do not contradict to the assumption that all sulfates are in the lipid matrix.

CONCLUSION

The results from anomalous scattering of sulfur in purple membrane show
that light atoms may be used as labels in structural studies. The use of
the dispersion of phosphorus near the K-absorption edge at $\lambda = 5.8$ Å may
eventually prove feasible. These two elements are ideal probes to

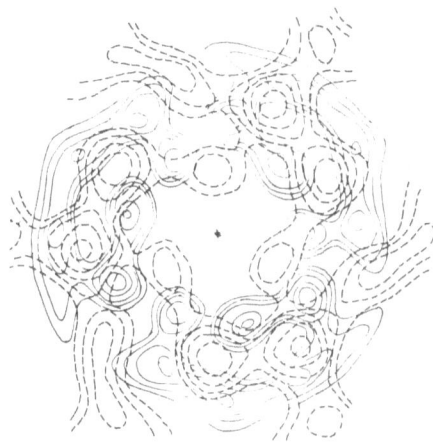

Fig. 9. Difference Fourier map seen in projection showing the distribution
of sulfur (-- - - -) at 11 Å resolution. The map of the bacteriorhodopsin molecules (----------) show the seven helices crossing the
membrane (8.7 Å resolution).

investigate the interaction of proteins with membranes and nucleic acids in that they are instrinsic labels. Contrary to phosphorus, sulfur in biomolecules may be present in different valence states giving rise to major shifts of the K absorption edge as a result of differences in sulfur valence. Even the formation of disulfide bridges (cysteine to cystine) leads to an additional peak at the absorption edge. The regions of anomalous dispersion of sulfides and sulfates are separated by 10 eV allowing their independent processing in structure determination. The structural discrimination between sulfur in cysteine and cystine may become possible as well.

ACKNOWLEDGMENT

We gratefully acknowledge the support (grant No. 05 353 FAI) by the Bundesministerium fur Forschung und Technologie, Bonn.

REFERENCES

Engelman, D. M., and Zaccai G., 1980, Bacteriorhodopsin is an inside-out protein, Proc. Nat. Acad. Sci. USA, 77:5894.

Hendrickson, W. A., 1986, Anomalous scattering in protein crystallography, in: "Structural Biological Applications of X-ray Absorption, Scattering and Diffraction," H. D. Bartunik and B. Chance, eds., Academic Press, Orlando.

Materlik, G., 1980, personal communication.

Stuhrmann, H. B., 1980, Anomalous dispersion of small-angle scattering of horse-spleen ferritin at the iron K absorption edge, Acta Cryst., A 36:996.

Unwin, P. T. N., and Henderson, R., 1975, Molecular structure determination by electron microscopy of unstained crystalline specimens, J. Mol. Biol., 94:425.

SYNCHROTRON X-RAY STUDIES OF BIOPOLYMERS:

SELF-ASSEMBLY AND OSCILLATIONS OF MICROTUBULES

E. Mandelkow, G. Lange and E.-M. Mandelkow

Max-Planck-Unit for Structural Molecular Biology
c/o DESY, Notkestraße 85
D-2000 Hamburg 52, F.R.G.

INTRODUCTION

Microtubules are fibrous structures that form part of the cytoskeletal network of eukaryotic cells. They are responsible for a variety of cellular functions, for example the static support of a cell (axostyle), movement against the external medium (cilia: directed streaming; flagella: swimming), intracellular transport (e.g., in nerve cells), or the separation of chromosomes during mitosis. Microtubules are hollow cylinders, about 25 nm wide, built of the subunit protein tubulin. There are two forms of tubulin (α and β) of molecular weight 50 kDa. Together they form a heterodimer (α-β) which is the effective assembly unit. A linear string of α-β-heterodimers is called a protofilament; thirteen of these associate laterally to form the microtubule cylinder (Fig. 1c, e). A variety of microtubule-associated proteins (MAPs) can be attached to the outside of the microtubule core (Fig. 1d). They tend to stabilize microtubules and enhance the efficiency of self-assembly. The mixture of tubulin and MAPs is called microtubule protein (Fig. 1a). Tubulin and MAPs can be isolated in large quantities from mammalian brain and induced to form microtubules in vitro. This process requires binding and hydrolysis of the nucleotide GTP, and it can be controlled conveniently by temperature (microtubules form at 37°C and fall apart at 4°C). Microtubule protein can assemble into a variety of polymorphic forms, for example, ring-like oligomers are prominent at low temperature (Fig. 1b). The rings also contain bound MAPs, and therefore ring-containing solutions assemble more efficiently than purified tubulin. In the past, several assembly models were proposed in which rings were considered as nucleating centers (Kirschner, 1978); the actual mechanism is different, as described below.

Microtubule assembly or oscillations can be studied by different methods, for example, UV light scattering, turbidity (Gaskin et al., 1974; Carlier et al., 1987), viscosimetry, ultracentrifugation, or synchrotron X-ray scattering. One advantage of the latter technique is that it yields kinetic and structural information simultaneously, in contrast to others which lack one or the other. These methods are sensitive to the average of all particles in solution. Since the particles are randomly oriented, the X-ray scattering can be analyzed by solution scattering theory. In cases where information on individual microtubules is required, one has to employ microscopic techniques, for example, electron microscopy, interference contrast or dark field microscopy (Horio and Hotani, 1986). In general, these approaches are complementary to one another.

93

X-ray studies on microtubules proceeded in several stages. The initial step was static X-ray fiber diffraction of oriented microtubules which yielded the main diffraction maxima and their structural interpretation (Mandelkow et al., 1977). After this step X-ray instruments, detectors, and data acquisition systems became available that were suitable for time-resolved X-ray diffraction of fibers (e.g., muscle) and solutions (Rosenbaum and Holmes, 1980; Koch and Bordas, 1983). All experiments described here were performed on the X-ray instruments of the European Molecular Biology Laboratory (EMBL) outstation at DESY, Hamburg. An initial study (Mandelkow et al., 1980) showed that a solution of assembling microtubules could indeed be monitored by X-ray scattering, and that the scattering patterns could be interpreted on the basis of the oriented fiber pattern and additional information derived from electron microscopy and image reconstruction. These experiments required sample cells that were compatible with X-ray scattering (~1 mm path length, 50 mm thick mica windows) and allowed rapid, reversible temperature jumps (up to 40°C). Temperature jumps were accomplished with a system of circulating water baths (Renner et al., 1983).

Tubulin can assemble not only into microtubules but also into ring-like oligomers. The rings played an important role in early assembly models, but there were conflicting views regarding their structure. The discrepancies could be resolved by X-ray scattering (Mandelkow et al., 1983a), showing that rings were equivalent to short coiled fragments of protofilaments. Subsequently, we employed very slow temperature scans to define a series of structural transitions between cold microtubule protein (containing rings, dimers, and other oligomers) and microtubules at 37°C (Bordas et al., 1983). This provided the ability to distinguish between prenucleation events, nucleation, elongation, and post-assembly events. The kinetics of prenucleation events and microtubule assembly were the subject of another

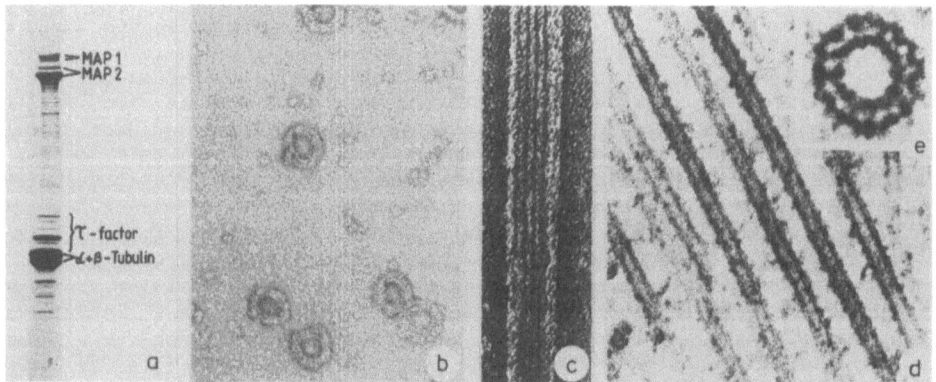

Fig. 1.   SDS gel and electron micrographs of microtubule protein. (a) SDS gel, showing the doublet of α- and β-tubulin (mol. wt. 50 kDal) and various coassembling microtubule-associated proteins, MAPs (e.g., tau protein, MAP1, MAP2, and others). (b) Microtubule protein at low temperature, showing rings (diameter 36 nm, containing tubulin and MAPs) and smaller oligomers. (c) Negatively stained microtubule. Note the longitudinal striations due to the protofilaments, spaced about 5 nm apart. (d) Microtubules embedded in plastic and thin sectioned (diameter about 25 nm). Note the projections of MAPs on the surface. (e) The inset shows a microtubule in cross section, with 13 protofilaments seen end-on.

study (Spann et al., 1987) whose main results will be described below. Finally, microtubule assembly has traditionally been described in terms of the theory of helical nucleation and condensation, in analogy with actin (Oosawa and Asakura, 1975; Johnson and Borisy, 1977); this predicts a pseudo-first order approach to a steady state of assembly. By contrast, microtubules can show distinct modes of nonsteady state assembly, including overshoots (Mandelkow et al., 1983b) and the synchronized oscillations (Mandelkow et al., 1988) to be described below. In this regard microtubules are unique among the self-assembling biopolymers.

MATERIALS AND METHODS

Protein Preparation

The protein was obtained from porcine brain (Mandelkow et al., 1985). X-ray experiments were performed either with microtubule protein (tubulin + MAPs) or with purified tubulin (lacking MAPs); this protein is obtained by phosphocellulose chromatography and therefore referred to as PC-tubulin. Concentrations typically range from 5 to 50 mg/ml.

Time-resolved X-ray Scattering Using Synchrotron Radiation

The X-ray experiments were performed at the EMBL Outstation at DESY, Hamburg, (Mandelkow et al., 1980; Bordas et al., 1983), using instruments X13 or X33 equipped with position sensitive detectors (linear or quadrant detectors) (Koch and Bordas, 1983). Assembly and disassembly were induced by temperature jumps between 0-5 and 37°C (half-time 4-10 sec; Renner et al., 1983). Reciprocal spacings were calibrated with respect to the collagen reflections from a sample of cornea or tendon. The scattering curves were normalized with respect to the incident intensity measured by an ionization chamber just upstream from the specimen chamber.

Data Analysis And Model Calculations

The observed intensities were corrected for background (measured from a buffer-filled cell), detector response, and variations in incident intensity due to the decay of the current in the storage ring. The scattering data were interpreted as described by Mandelkow et al. (1983a) and Bordas et al. (1983). Fig. 2 shows calculated scattering curves of several representative model structures, including microtubules, rings, protofilament fragments, and tubulin dimers. The curves are normalized to the same total protein concentration so that the forward scattering (S=0, not shown in the graph) represents the degree of polymerization. Microtubules show the highest central scatter and several subsidiary maxima separated by clear minima. Rings have a lower central scatter and side maxima that are slightly out of phase with respect to those of microtubules. In both cases the positions of the first side maxima are close to S=1.22/D, where D refers to the mean diameters of the respective hollow cylinders. Since microtubules have smaller diameters than rings, their subsidiary peaks occur at higher S-values. By contrast, both tubulin dimers and short protofilament fragments show smoothly decaying scattering curves. In practice one observes a mixture of several types of aggregate, especially at low temperature where rings, oligomers, and dimers are in equilibrium. The minimum preceding the first maximum of microtubules is particularly sensitive to the presence of oligomeric structures.

Because of the beam stop and parasitic scatter the lowest observable S-values are around 0.001 nm-1. In the following, the intensity integrated between about 0.01 and 0.015 nm-1 will be referred to as "central scatter" (this is only an approximation to the forward scattering). Integration between the scattered intensity 0.025 and 0.035 nm-1 yields the "ring scatter" since this overlaps with the first side maximum due to the form factor of rings. The region from 0.045 to 0.05 nm-1 will be called "microtubule scatter" (arrows C, R, and M in Fig. 2). Time-dependent changes in these regions allow one to assess the main features of assembly and disassembly. For example, a rise in the central scatter indicates overall assembly; a rise in the microtubule scatter usually means formation of microtubules and a rise in the ring scatter may mean ring formation and/or other structures (e.g., oligomers). The actual structures present may be analyzed from the scattering curves or difference plots at given time points.

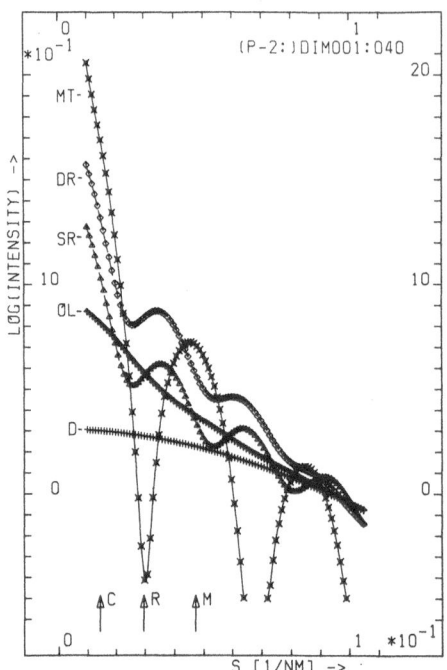

Fig. 2. Theoretical scattering curves of model structures. The curves were calculated on the basis of Debye's formula, using spheres of 4 nm diameter to represent the protein subunits. The scattering of mixtures of aggregates can be calculated from $I(S,t) = \Sigma x_k(t)p_k i_k(S)$ where the index k refers to the species of aggregate, $x_k(t)$ is the fraction of subunits assembled into the species k, p_k is the degree of polymerization, $i_k(S)$ is the form factor shown in the Figure, and $S=2\sin\theta/\text{lambda}$ is the Bragg scattering vector. From top to bottom on left-hand side: Microtubule (MT) with 13 protofilaments of mean diameter 26 nm; double concentric ring (DR) of 36 nm mean diameter (40 nm for outer, 32 nm for inner turn), with subunits spaced 4 nm apart (= coiled protofilaments); single ring (SR) of 36 nm mean diameter; oligomer (OL, protofilament fragment consisting of four dimers); and dimer (D). Note that the central scatter (left) increases with the degree of polymerization; the side maxima of rings and microtubules are roughly in antiphase; oligomers and dimers have smoothly decaying scattering curves, with the former contributing noticeably at small angles. The arrows marked C, R, and M indicate the regions sensitive to overall assembly: "central scatter," (C), presence of rings (R) or microtubules (M). They are used to monitor the time-dependence of the reactions (Spann et al., 1987).

RESULTS AND DISCUSSION

Prenucleation Events And Assembly

Fig. 3 illustrates one cycle of assembly and disassembly of microtubule protein in standard buffer conditions. The projection plot shows the time-dependence of the scattering traces. The initial state has a lower central scatter (left) and the side maxima is typical of rings of diameter 36 nm. The polymerized state has a higher central scatter and the side maxima characteristic of microtubules (mean diameter 25 nm). A detailed comparison with the calculated scattering of model structures (Bordas et al., 1983) shows that most observed traces can only be explained if one assumes mixtures of several types of aggregate. At high temperatures the patterns are dominated by the scattering from microtubules, with a minor contribution from species that are best explained in terms of oligomers and/or MAPs that contribute mainly at very small scattering angles and fill in the minima of the microtubule trace. The scattering at low temperature can be modelled by mixtures of single and double concentric rings plus oligomers. The scattering from double rings has a high first subsidiary peak but decays more

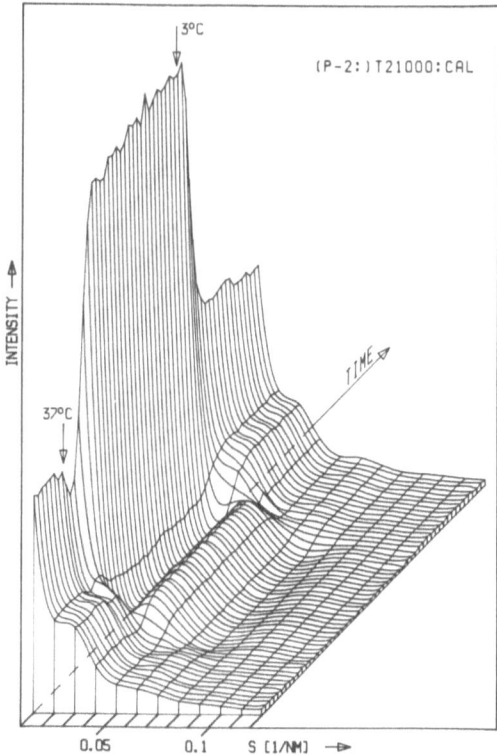

Fig. 3. Assembly of microtubule protein monitored by time-resolved X-ray scattering. The projection plot shows an experiment using the EMBL instrument X33, camera length 3 m, linear position-sensitive detector with 256 channels, 256 time frames of 3 sec per run (not all shown), temperature jumps from 3 to 37°C and back (arrows). Note the increase in central scatter during assembly and the change in side maxima. The side maximum of the cold solution is due to rings; that of the warm solution is due to microtubules (Spann et al., 1987).

quickly than the scattering from single rings (Fig. 2). As before, oligomeric species contribute mainly at small angles, with radii of gyration on
the order of 10 nm (determined from the slope of Guinier plots). The contribution from tubulin dimers is generally negligible until one reaches
higher scattering angles (second side maximum and beyond, see Fig. 2). However, the presence of species with unknown form factors permits only an approximate quantitation of the individual components, especially at low temperature and during the assembly phase. When the temperature is raised to
37°C there is a drop in intensity before the steep rise. This undershoot is
due to the partial disappearance of ring oligomers before microtubules are
formed and indicates that rings do not act as nucleating centers, as postulated in some earlier assembly models. However, in this experiment the
phases of oligomer dissolution and microtubule formation are not well separated so that it is difficult to determine the kinetic parameters of each
reaction separately.

Fig. 4 shows the time courses of the intensities in the regions designated in Fig. 2 as central scatter (C), ring scatter (R), and microtubule
scatter (M); the temperature (T) is shown in the bottom trace. In this experiment the solution conditions were adjusted to obtain a better separation
of prenucleation events and assembly (e.g., lower pH and salt concentration;
Spann et al., 1987). Following the temperature jump there is a pronounced
undershoot (best visible in curve C) which appears to level out before the
onset of the rise due to microtubule assembly. The magnitude of this

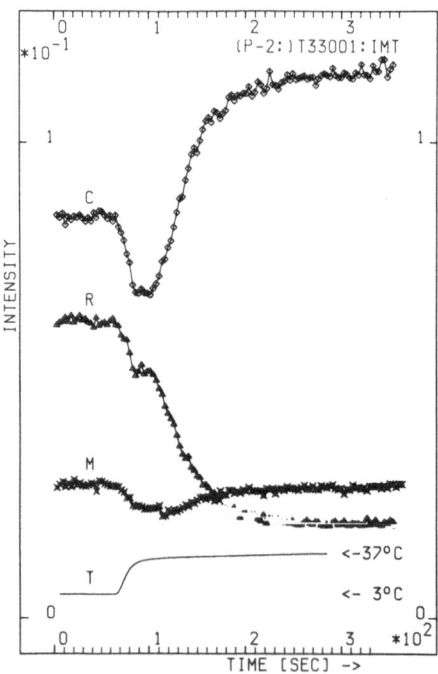

Fig. 4.   Another time-resolved experiment, showing the time course of the
intensities in the scattering regions C, R, and M (compare Fig. 2);
the bottom trace is the temperature T. Note the pronounced undershoot accompanying the temperature jump, due to the disappearance
of ring oligomers. The subsequent rise in region C is due to the
nucleation and growth of microtubules. The buffer conditions have
been chosen to maximize the separation of the prenucleation events
and assembly so that the rates of both can be determined independently (Spann et al., 1987).

undershoot is linearly correlated with the temperature rise. Thus, the observed rate does not reflect the intrinsic rate of oligomer dissolution, but rather a rapid temperature-dependent equilibrium. By contrast, microtubule assembly occurs well after the temperature has reached 37°C (Bordas et al., 1983). Other experiments show that the prenucleation phase occurs even in conditions where microtubule formation is inhibited, for example in the absence of GTP, or when the temperature is jumped from the cold to less than 20°C (not shown). The following conclusions can be drawn from these experiments (cf. Spann et al., 1987): (a) The formation of oligomers is exothermic; that of microtubules is endothermic. (b) The equilibrium between rings and smaller units (e.g., dimers) is rapid, while microtubule nucleation and elongation is slower. (c) The prenucleation events occur over a wide temperature range and do not depend on GTP binding or hydrolysis, in contrast to microtubule assembly.

These results mean that tubulin is capable of two different modes of aggregation: one leading to oligomers (e.g., at low temperature, without nucleotides) and one generating polymers (e.g., at high temperature and with GTP). The two types of reactions have distinct physico-chemical characteristics and are thus not directly related to one another. However, they draw on the same pool of subunits, i.e., tubulin dimers (Fig. 5).

Oscillations In Microtubule Assembly

The growth of microtubules, like that of other self-assembling biopolymers, is traditionally described by the endwise addition of subunits to the polymer at a roughly constant polymer number concentration (Oosawa and Asakura, 1975). Thus the elongation can be described by the reaction $MT_n + S \longleftrightarrow MT_{n+1}$). This scheme predicts an exponential approach to a steady state where microtubules are in equilibrium with a critical concentration of free subunits. In standard assembly conditions this is indeed observed (Figs. 3, 4). However, microtubules are capable of more complex reaction mechanisms that are not compatible with the above scheme.

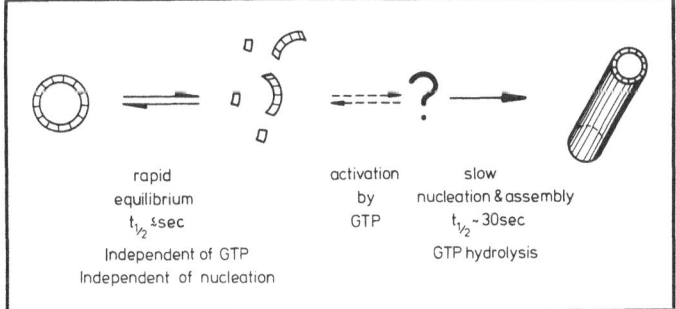

Fig. 5. Diagram summarizing the differences between prenucleation events and microtubule assembly. There is a rapid and reversible temperature-dependent equilibrium between rings, smaller oligomers, and dimers that is not directly related to microtubule assembly (left part of diagram). By contrast, microtubule nucleation and assembly are slow and take place above a threshold temperature (right part of diagram). Since the ring-forming and microtubule-forming modes of assembly draw on the same pool of building blocks (oligomers and dimers), one requires one or more intermediate steps during which the subunits are activated for assembly, e.g., by temperature and GTP. The structures of the activated dimers or oligomers are unknown but could correspond to the straight conformation of protofilaments, in contrast to the inactive coiled conformation (Spann et al., 1987).

Several of these mechanisms are illustrated in Fig. 6. Curve 1 was obtained with purified tubulin (largely devoid of the cold-stable oligomers because of the removal of MAPs). Its assembly shows the exponential approach to a steady state expected for simple endwise addition of subunits. Curve 2 shows an analogous experiment with microtubule protein (containing MAPs and therefore a larger fraction of cold-stable ring oligomers, similar to the experiments of Figs. 3 and 4). Here, too, one observes the exponential approach to steady state, aside from the undershoot due to ring dissolution preceeding assembly. Curve 3 shows a rapid assembly phase which in this case is followed by a slow decay. This experiment was performed with a low concentration of GTP so that a large fraction of the nucleotide was consumed during the assembly phase. Since the maintenance of steady state polymers requires free subunits with bound GTP, the observed slow decay can be explained by the lack of assembly-competent subunits so that the dissociation events are no longer balanced by association events. This is analogous to a single-turnover experiment in enzyme kinetics. The experiment illustrates that the tubulin.GTP complex must be regarded as the assembling unit. Curve 4 shows a further stage of complexity: assembly, decay, and recovery. In this case the Mg^{++} concentration was strongly reduced by addition of EDTA. The behavior can be explained by considering that nucleotide binding is mediated by Mg^{++}; in other words, the ternary complex of tubulin.Mg^{++}.GTP is the assembly-competent subunit. Mg^{++}.GTP is available in sufficient quantity for the initial assembly phase; after hydrolysis in the polymer, Mg^{++}.GDP is locked non-exchangeably in the microtubules. The free tubulin subunits have a reduced affinity for GTP in the virtual absence of Mg^{++}, leading to the slow decay of microtubules similar to that of Curve 3. After partial disassembly, Mg^{++} is released again from the subunits so that a new intermediate level of reassembly is attained.

The experiments described thus far set the scene for understanding the oscillations described below: Microtubules are intrinsically unstable after GTP hydrolysis; they can be stabilized by external factors (e.g., MAPs) or by continuously recharging the depolymerizing species Tubulin.GDP to the polymerizing species Tubulin.GTP. The recharging process Tubulin.GDP+GTP <<---->>Tubulin.GTP+GDP is rapid with tubulin subunits so that a steady state can be effectively maintained as long as GTP is supplied; however, the direct recharging of subunits incorporated into microtubules or oligomers is not possible since the nucleotide exchange rates are nearly zero in these structures.

Nonequilibrium features of microtubule assembly can be strongly enhanced by appropriate choices of assembly conditions. The interval between consecutive assembly phases can be shortened, and the number of assembly cycles can be increased. Curve 5 of Fig. 6 was obtained at particularly high protein concentrations in otherwise standard assembly buffer. Assembly becomes very rapid and runs into several damped oscillations before the system reaches a steady state at a lower degree of polymerization. A dramatic increase in oscillatory behavior is achieved by increasing the ionic strength ("oscillation buffer," Fig. 6, Curves 6-9). The periodicities are typically in the range of 1-3 min, and amplitudes can be up to 80% of the maximum. A projection plot obtained from an oscillating sample is shown in Fig. 7. The analysis of the scattering patterns (Fig. 8) shows that the reaction involves structures with distinct form factors so that different regions of the scattering pattern may swing either in phase or in antiphase (as in Fig. 6, Curves 7-9). The oscillations may continue for up to 20 min. They correlate with the consumption of GTP and the buildup of GDP. The damping depends upon many factors, including the GDP/GTP ratio. Increasing the ratio GDP/GTP leads to a lowered amplitude, with a relatively weak influence on the frequency.

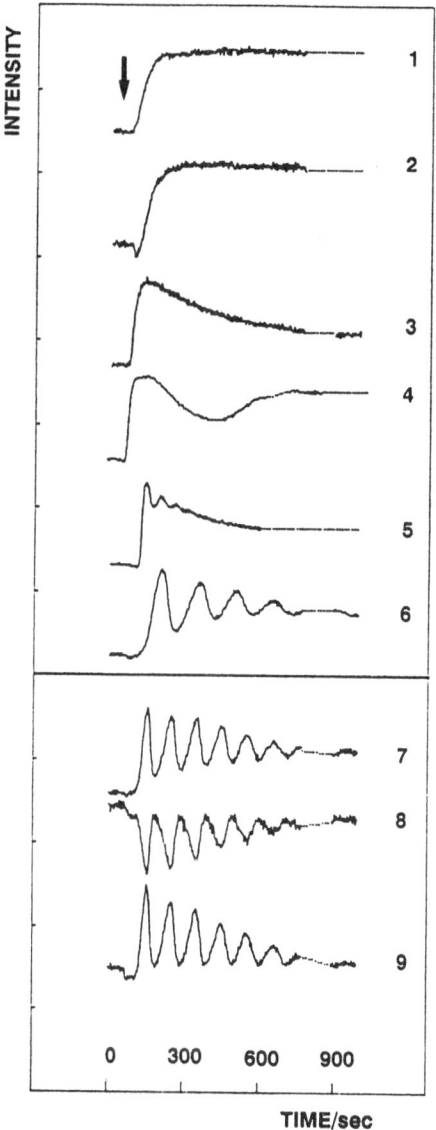

Fig. 6. (See page 102 for caption.)

Fig. 6. Different types of microtubule assembly reactions followed by the time course of X-ray scattering at selected scattering angles. Curves 1-7 show the intensity at very low angles sensitive to overall assembly (region C, see Fig. 2); Curve 8 is from a region sensitive mainly to rings and smaller oligomers (R in Fig. 2); Curve 9 is sensitive to the formation of microtubules (M in Fig. 2). The arrow marks the temperature jump from 4 to 37°C. The dotted regions represent breaks caused by data transfer to the computer.

Curve 1. PC-tubulin, concentration 35 mg/ml, in standard assembly buffer (Mandelkow et al., 1988). The initial rise in intensity following the T-jump is due to nucleation followed by elongation, showing a first-order approach to a stable steady state.

Curve 2. Microtubule protein, 27 mg/ml, standard buffer. The curve is similar to the previous one, except that there is an undershoot preceding assembly (due to the dissolution of rings into smaller oligomers and subunits, similar to Figs. 3, 4).

Curve 3. PC-tubulin at a low GTP/tubulin ratio (1 mM GTP, 36 mg/ml tubulin). The microtubule assembly phase is followed by a slow decay due to the depletion of GTP. The final state shows mainly oligomers.

Curve 4. Microtubule protein (17 mg/ml) in low Mg^{++} concentration (2 mM $MgSO_4$, 4 mM EDTA). The initial rise and decay is similar to Curve 3, in this case due to the transient depletion of $Mg^{++}GTP$; this is followed by a slow recovery of microtubule assembly to a new steady state.

Curve 5. PC-tubulin showing several oscillations at very high protein concentrations (46 mg/ml) in standard assembly buffer (compare Curve 3).

Curve 6. PC-tubulin showing pronounced oscillations (periodicity 2.5 min) in oscillation buffer.

Curves 7-9. Oscillating sample of microtubule protein monitored at three different scattering angles. Curve 7, region C (measuring overall assembly), Curve 8, region R (oligomers), Curve 9, region M (microtubules). The oscillations of Curves 7 and 9 are roughly in phase because overall assembly (Curve 7) correlates with microtubule assembly (Curve 9). Curve 8 is roughly in antiphase because this scattering region is dominated by oligomers.

What is the structural basis of the nonequilibrium behavior? We know from previous experiments that tubulin can exist in several states of aggregation that are interconvertible during assembly: tubulin dimers (the subunits of assembly), microtubules, and oligomers (short stretches of coiled protofilaments that tend to form closed rings when MAPs are present). Each of the structures has a characteristic X-ray pattern (Fig. 8). When comparing the patterns at various time points during the oscillations, one notes several nearly stationary isosbestic points (Fig. 9), indicating that the patterns are dominated by two main species. One of them is microtubules, as seen from their characteristic side maxima. Subtraction of the microtubule component shows the second component to be oligomers (while tubulin subunits, the third major component, scatter only weakly and are thus disguised when the other structures are present). The amplitude of the oscillations represents the fraction of microtubule mass participating in the reaction; microtubule assembly is maximal at the peaks of the central scatter and minimal at the troughs.

This conclusion was confirmed by complementary negative stain electron microscopy experiments (Fig. 10). Although this method is less quantitative, it shows the features expected from the X-ray patterns, i.e., an antiphasic increase and decrease of microtubules and oligomers, typically 20-100 nm in length.

What is the origin of the oligomers? In principle they could form from the disassembling subunits (containing bound GDP). However, since the nucleotide is exchangeable on free subunits they would rapidly become assembly-competent again by binding of GTP from the solution. The

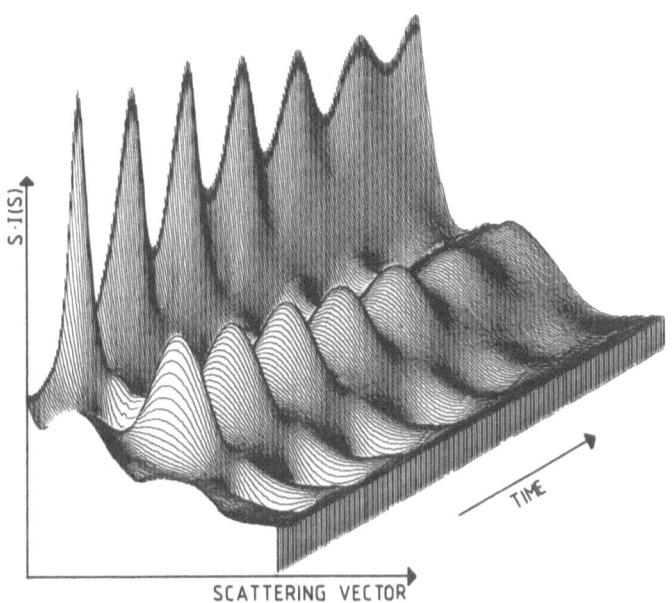

Fig. 7. Projection plot of the X-ray scattering from a solution of oscillating microtubules, showing the X-ray intensity (S.I(S), z-axis) as a function of scattering vector (S=2sinθ/lambda, x-axis) and time (y-axis, 3 sec scan interval). Microtubule protein, 32 mg/ml. The central scatter (left) indicates overall assembly; the subsidiary maximum arises from microtubules. The temperature jump is at time zero. The periodicity of the fluctuations is about 2 min. The final state (after disappearance of the oscillations, not shown) is dominated by the scattering from oligomers. The scattering curves here and in Figs. 8 and 9 have been smoothed by cubic splines (Mandelkow et al., 1988).

alternative is that microtubules disassemble not via release of dimers, but via oligomers. Direct evidence for this mechanism comes from earlier experiments where depolymerizing microtubule solutions in standard buffer were fixed by rapid freezing in amorphous ice and observed by cryo-electron microscopy (Mandelkow and Mandelkow, 1985). Combining these results with the earlier observation that oligomers tend to break down prior to assembly

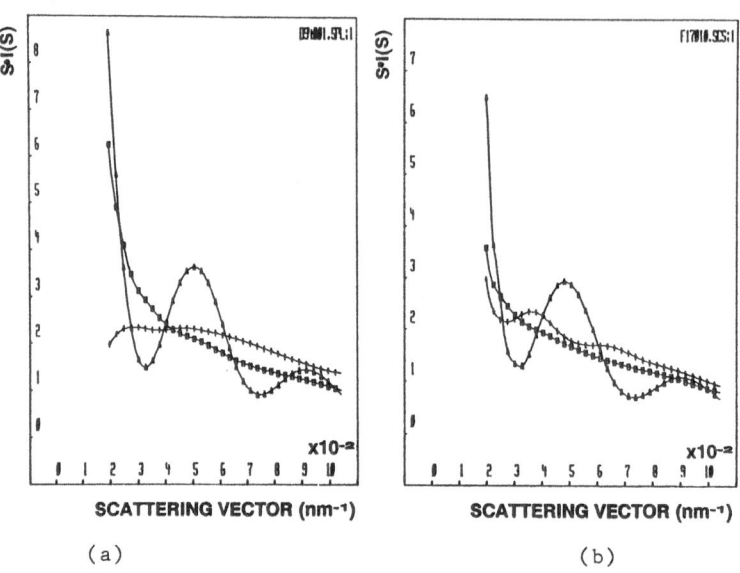

SCATTERING VECTOR (nm⁻¹)

(a)

SCATTERING VECTOR (nm⁻¹)

(b)

Fig. 8.   Scattering patterns of two oscillating samples of PC–tubulin and
microtubule protein. Vertical axis, I(S).S (corresponding to the
intensity integrated over circles of constant scattering angle 2θ);
horizontal axis, reciprocal spacing S in nm^{-1}. (a) PC–tubulin, 29
mg/ml. Crosses, initial solution before T–jump, containing mainly
tubulin dimers and a small fraction of oligomers. Triangles, microtubules at the first maximum of oscillations. Note the rise of
the central scatter (left) and of the microtubule maximum around
0.05 nm^{-1}. Rectangles, final state after the oscillations are
damped out, showing the smooth decay typical of non–ring oligomers.
(b) Microtubule protein, 30 mg/ml. Crosses, initial solution showing the scattering from oligomers and rings (the subsidiary maxima
around 0.033 and 0.065 nm^{-1} due to MAP–containing rings, not
present in PC–tubulin; compare (a). Triangles, microtubules at the
first maximum of oscillations. Rectangles, oligomers after damping
out of oscillations (mostly non–ring). Note that an increase in
the low angle scatter may be caused either by microtubules or by
oligomers; the distinction between them is based on the scattering
at higher angles.

(Fig. 5), we arrive at the following simple model for the cyclic part of the reaction:

```
Subunits--->>Microtubules--->>Oligomers
    |                              |
    ---------------<<-------------
```

When the conversion from oligomers to subunits is rapid, this reduces to the usual scheme of endwise assembly and disassembly, Subunits<<--->>Microtubules.

Any chemical oscillator requires an energy source which in the present case comes from GTP hydrolysis (data not shown; see Mandelkow et al., 1988): (a) Nonhydrolyzable GTP analogues support assembly but not oscillations; (b) the number of oscillations is limited by GTP, i.e., oscillations do not

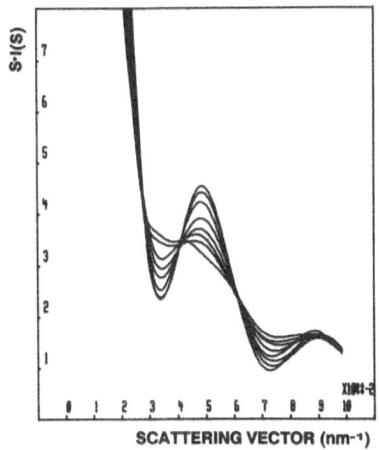

SCATTERING VECTOR (nm⁻¹)

Fig. 9.   Selected scattering traces observed during oscillations. The curve at the top of the central scatter (left) and at $S=0.05$ nm^{-1} is at the first assembly maximum; the bottom curve is at the subsequent valley; the others are at intermediate time points. Note the constant positions of the isosbestic points where the curves intersect, indicating that the patterns are dominated by two main contributions (from microtubules and oligomers) whose ratios vary cyclically during the oscillations.

outlast the GTP supply and (c) microtubules are formed with GTP and destabilized by GDP, whereas oligomers are stable with GDP. Thus one expects that the cyclic change of structures is accompanied by a cyclic change in the interaction between tubulin and GTP or GDP. This can be shown by analyzing oscillating solutions for protein-bound nucleotides by HPLC. Both the levels of GTP and GDP undergo oscillations with the same frequency as that of assembly, but in antiphase; when GTP shows a peak, GDP has a trough. Thus the energy of hydrolysis is not consumed continuously, but in bursts coupled to assembly. Control experiments showed that neither microtubules nor oligomers, but only tubulin subunits, were capable of exchanging nucleotide with the solution, and that the exchange on subunits is rapid.

These data are incorporated into the reaction mechanism of Fig. 11. From a structural point of view there are three species, tubulin subunits (Tu), microtubules (Mt), and oligomers (Ol), that are cyclically interconverted and can be distinguished by X-ray scattering. Functionally the number is five because tubulin and microtubules come in two forms, either with GTP or GDP bound. GTP-charged tubulin (Tu.GTP) is assembly-competent and in equilibrium with microtubule ends; tubulin with bound GDP (Tu.GDP) normally cannot assemble and is in equilibrium with oligomers. Free tubulin subunits quickly exchange their nucleotides according to the GTP/GDP ratio in solution; microtubules and oligomers do not exchange nucleotides. Microtubules are initially formed with bound GTP (Mt.GTP) which is then hydrolysed (Mt.GDP). Microtubules are destabilized by GTP hydrolysis; oligomers with bound GDP (Ol.GDP) break off and subsequently dissociate into Tu.GDP subunits which can then be recharged to Tu.GTP. Additional pathways listed in the scheme but less important for this discussion are: (a) the prenucleation events observed mainly with microtubule protein, rings--->>oligomers--->>Tubulin.GDP (Fig. 5); (b) nucleation during the first round of assembly, Tubulin.GTP--->>nuclei--->>microtubules. Subsequent cycles do not require new nucleation because microtubule ends are already present. This scheme

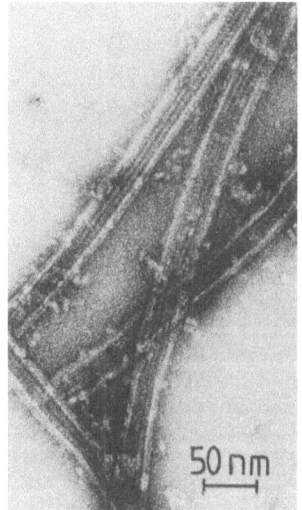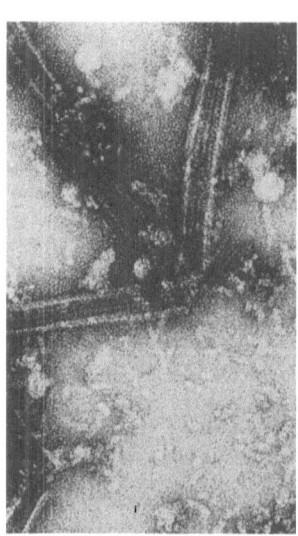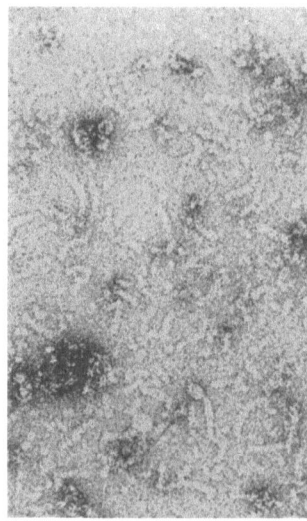

Fig. 10. Electron micrographs of PC-tubulin at different stages of oscillations. Left, maximal assembly. One observes microtubules of normal appearance and oligomeric material in the background. Center, phase of disassembly. There is a noticeable increase in oligomers, and microtubules tend to break. Right, field of oligomers at an assembly minimum. Typical lengths are between 20 and 100 nm, corresponding to 3-12 tubulin dimers. The longer ones clearly show the coiling characteristic of disassembled protofilament fragments that leads up to ring-like closure (see upper left). Complete rings are rare with PC-tubulin but frequent with microtubule protein due to the stabilization of oligomers by MAPs (Mandelkow et al., 1988).

can be applied to "classical" models of helical condensation as well as to oscillations.

The crucial point for microtubule assembly oscillations is that oligomers, the primary breakdown products of microtubules, have a measurable lifetime (i.e., the rate of step 4 is slow). This means that the reactivation of tubulin subunits is temporarily interrupted because oligomers act as a temporary storage form of nonpolymerizable tubulin. Since the pool of polymerizable Tubulin.GTP is nearly depleted during the assembly process (step 1) and cannot be replenished quickly, the destabilization of microtubules due to GTP hydrolysis (step 2) becomes noticeable. The depolymerization (step 3) proceeds until the oligomers have

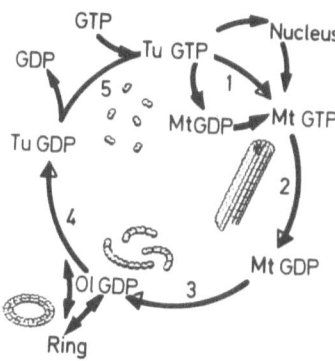

Fig. 11. Model of reaction cycle responsible for oscillations. Microtubules (Mt.GTP) are formed from active subunits (loaded with GTP,
Tu.GTP, step 1). GTP is hydrolyzed upon incorporation of the
subunits (Mt.GDP, step 2), leading to the destabilization of
microtubules and eventually to their disassembly into oligomers
loaded with GDP (Ol.GDP), which transiently lock the subunits in
an unpolymerizable state. When inactive subunits (Tu.GDP) are
released from oligomers (step 4), they can be recharged to Tu.GTP
(step 5), leading to the reassembly of microtubules. Side reactions are the dissolution of rings into oligomers and subunits
(observed with microtubule protein just after the T-jump and responsible for the undershoot, see Figs. 3, 4), and the nucleation
during the first round of assembly (Mandelkow et al., 1988).

released enough subunits to stop it and return to a new phase of assembly. GTP plays a dual role; its binding to subunits activates them for assembly, and its hydrolysis within the polymer induces their breakdown. In other words, GTP acts as an activator of the polymer; GDP activates the oligomer.

This view may be put into structural terms by distinguishing between two conformations of the subunits that affect the longitudinal bonds along protofilaments. The active complex (Tu.GTP) is capable of forming straight protofilaments which by virtue of their lateral bonding surfaces assemble into hollow cylinders. The inactive complex (Tu.GDP) can still form short protofilament fragments (oligomers), but they have an inside-out curvature rather than being straight. This prevents them from forming closed cylinders, and it destabilizes the cylinders after GTP hydrolysis. The destabilization is most effective at the microtubule ends which can be seen fraying apart during disassembly; internal regions are less sensitive because of the mutual interaction between protofilaments.

The model suggests the following conditions for observing assembly oscillations: (a) An appreciable difference in stability of microtubules before and after GTP hydrolysis; (b) stabilization of oligomers (i.e., slow release of subunits); (c) a sufficient supply of hydrolyzable GTP as the energy source; (d) initial synchronization of the particles in solution. These conditions can be met by a judicious choice of assembly buffer, protein concentration, and temperature jump, as discussed earlier.

Comparison With Chemical Oscillators And Implications For Microtubule Dynamics In Cells

Many examples of chemical oscillators are known, for example the Belousov-Zhabotinskii reaction or anaerobic glycolysis in yeast (Hess and Boiteux, 1971), and it is interesting to note some parallels and differences with the present one. Firstly, they depend on a rechargeable supply of energy; in this regard they differ from a mechanical pendulum which cyclically converts one form of energy into another (neglecting friction). The loss of energy in chemical systems is explained by irreversible thermodynamics; basically it stems from the fact that they are based on many particles that act in concert against entropy, rather than a single oscillator without entropy change. Secondly, biochemical oscillators are usually composed of a series of enzyme reactions, some of which are allosterically regulated by reaction intermediates or products, and this feedback leads to nonlinear behavior. In our case the enzyme analogue is the tubulin subunit whose ability to polymerize is regulated by binding of GTP or GDP, but tubulin does not function as an enzyme in the strict sense since hydrolysis occurs in the microtubule. Thus the buildup and breakdown of a structure corresponds to the rise and fall of enzyme activities in other systems.

The oscillatory behavior of microtubules in solution is related to their dynamic instability observed in living cells (Mitchison and Kirschner, 1984). During the cell cycle, microtubules must undergo a cyclic reorganization between different types of networks (e.g., the cytoskeleton of interphase and the mitotic spindle), which involves the periodic disassembly and reassembly. Moreover, even local regions of cells show microtubules switching between phases of growth and shrinkage. In these cases their behavior is not synchronized, and an observer recording the average degree of assembly would see an apparent steady state. In the experiments described here the whole microtubule population was synchronized by the rapid initiation of assembly. Although our solution conditions differ from the environment within a cell, the overall behavior is surprisingly similar (e.g., amplitudes and periodicities). In other words, it seems that the oscillations represent an amplified version of reactions that take place inside living cells. This opens the opportunity to study their basic properties in vitro.

ACKNOWLEDGMENTS

We are grateful to Dr. Michel Koch and the staff of the EMBL Outstation at DESY, Hamburg, for providing access to their X-ray facilities, to Annette Jagla for making the nucleotide exchange data available, to Petra Derr for expert technical assistance, and to Dr. Joan Bordas (Daresbury Laboratory, U.K.) for his contributions to the early stages of this work. This project was supported by the Bundesministerium fur Forschung und Technologie (grant 05-180MP-BO) and the Deutsche Forschungsgemeinschaft (grant MA 563/2).

REFERENCES

Bordas, J., Mandelkow, E.-M., and Mandelkow, E., 1983, Stages of tubulin assembly and disassembly studied by time-resolved synchrotron x-ray scattering, J. Mol. Biol., 164:89.

Carlier, M. F., Melki, R., Pantaloni, D., Hill, T. L., and Chen, Y., 1987, Synchronous oscillations in microtubule polymerization, Proc. Natl. Acad. Sci. U.S.A., 84:5257.

Gaskin, F., Cantor, C. R., and Shelanski, M. L., 1974, Turbidimetric studies of the in vitro assembly and disassembly of porcine neurotubules, J. Mol. Biol. 89:737.

Hess, B., and Boiteux, A., 1971, Oscillatory phenomena in biochemistry, Ann Rev. Biochem. 40:237.

Horio, T. and Hotani, H., 1986, Visualization of the dynamic instability of individual microtubules by dark-field microscopy, Nature, 321:605.

Johnson, K. A., and Borisy, G. G., 1977, Kinetic analysis of microtubule self-assembly in vitro, J. Mol. Biol. 117:1.

Kirschner, M. W., 1978, Microtubule assembly and nucleation, Int. Rev. Cytol., 54:1.

Koch, M. H. J., and Bordas, J., 1983, X-ray diffraction and scattering on disordered systems using synchrotron radiation, Nucl. Instrum. Meth. 208:461.

Mandelkow, E., Thomas, J., and Cohen, C., 1977a, Microtubule structure at low resolution by x-ray diffraction (tubulin/helical diffraction), Proc. Natl. Acad. Sci. U.S.A., 74:3370.

Mandelkow, E. -M., Harmsen, A., Mandelkow, E., and Bordas, J., 1980, X-ray kinetic studies of microtubule assembly using synchrotron radiation, Nature, 287:595.

Mandelkow, E., Mandelkow, E. M., and Bordas, J., 1983a, Structure of tubulin rings studied by x-ray scattering using synchrotron radiation, J. Mol. Biol., 167:179.

Mandelkow, E., Mandelkow, E. -M., and Bordas, J., 1983b, Synchrotron radiation as a tool for studying microtubule self-assembly, TIBS, 8:374.

Mandelkow, E. -M., and Mandelkow, E., 1985, Unstained microtubules studied by cyro-electron microscopy substructure, supertwist and disassembly, Mol. Biol., 181:123.

Mandelkow, E. -M., Herrmann, M., and Ruhl, U., 1985a, Tubulin domains probed by limited proteolysis and subunit-specific antibodies, J. Mol. Biol., 185:311.

Mandelkow, E. -M., Lange, G., Jagla, A., Spann, U., and Mandelkow, E., 1988, Dynamics of the microtubule oscillator: Role of nucleotides and tubulin-MAP interactions, EMBO J. 7:357.

Mitchison, T., and Kirschner, M., 1984, Dynamic instability of microtubule growth, Nature, 312:237.

Oosawa, F., and Asakura, S., 1976, "Thermodynamics of the polymerisation of protein," Academic Press, London.

Renner, W., Mandelkow, E.-M., Mandelkow, E., and Bordas, J., 1983, Self-assembly of microtubule protein studied by time-resolved x-ray scattering using temperature jump and stopped flow, Nucl. Instrum. Meth. 208:535.

Rosenbaum, G., and Holmes, K. C., 1980, Small angle diffraction of x-rays and the study of biolgical structures, in: "Synchrotron Radiation Research," Winick, H. and Doniach, S., eds., Plenum Press, New York.

Spann, U., Renner, W., Mandelkow, E.-M., Bordas, J., and Mandelkow, E., 1987, Tubulin oligomers and microtubule assembly studied by time-resolved x-ray scattering: separation of prenucleation and nucleation events, Biochem., 26:1123.

STRUCTURAL STUDIES OF RIBULOSE-1,5-BISPHOSPHATE

CARBOXYLASE/OXYGENASE FROM SPINACH

Stefan Knight[1], Inger Andersson[1], Carl-Ivar Branden[1],
and George Lorimer[2]

[1]Swedish University of Agricultural Sciences
Department of Molecular Biology
Uppsala Biomedical Center
P.O. Box 590, S-751 24
Uppsala, Sweden

[2]E.I. du Pont de Nemours and Company Inc.
Wilmington, DE 19898

INTRODUCTION

Ribulose-1,5-bisphosphate carboxylase/oxygenase, Rubisco, plays an important role in photosynthesis as well as in photorespiration. In photosynthesis, the carboxylase activity of the enzyme catalyzes the condensation of carbon dioxide and ribulose-1,5-bisphosphate to yield two moles of 3-D-phosphoglycerate (Miziorko and Lorimer, 1983; Andrews and Lorimer, 1987). In addition, Rubisco has an oxygenase activity whereby oxygen is added to ribulose 1,5-bisphosphate to yield 3-D-phosphoglycerate and 2-phosphoglycolate, the major substrate for photorespiration. As a result of the subsequent metabolism of phosphoglycolate one of the carbon atoms is oxidized to carbon dioxide and the released energy is lost as heat. Up to 50% of the photosynthetically reduced carbon may be oxidized through this pathway. The possibility to increase the carboxylase/oxygenase ratio has therefore attracted substantial interest. An understanding of the structural basis for the two activities would greatly facilitate the successful use of site directed mutagenesis techniques towards this end.

Rubisco from all higher plants as well as blue-green algae is built up from two types of subunits; eight large, L,(Mw=55 kD) and eight small, S, (Mw=15 kD), forming an L_8S_8 molecule of molecular weight around 550 kD. The enzyme is found in the chloroplast, where it can represent as much as 50% of the total protein. The large subunit is chloroplast encoded whereas the small subunit is coded in the nuclear genome. The small subunit is synthesized in the cytoplasm as a higher molecular weight precursor, which is transported into the chloroplast where assembly to the complete enzyme occurs (Musgrove and Ellis, 1986). The large subunit is responsible for the catalytic activity whereas the function of the small subunit is as yet unknown.

Common to all Rubisco molecules is an activation process, during which a lysine residue becomes carbamylated by an activator CO_2 molecule (Lorimer and Miziorko, 1980; Lorimer, 1981). The carbamate is further stabilized by a magnesium ion (Pierce and Reddy, 1986). This activation step is necessary for both the carboxylation and the oxygenation reactions. The activated ternary complex is able to bind the substrate ribulose 1,5-bisphosphate, such that it is subsequently attacked by either carbon dioxide or oxygen. CO_2 thus has a dual role in the reaction mechanism; as effector molecule and as substrate using two different binding sites for the two reactions.

CRYSTALS

Large single crystals of an activated quaternary complex of spinach Rubisco, Mg^{2+}, CO_2 and 2-C-carboxy-D-arabinitol 1,5-bisphosphate, CABP, have been obtained from ammonium sulphate solutions at pH 7.3 (Andersson and Branden, 1984). CABP, which is regarded as a transition state analogue, has a hydroxyl group at C-3 instead of the carbonyl group present in the six-carbon reaction intermediate (Pierce et al., 1980). The crystals belong to space group $C222_1$ with cell dimensions a=157.2 Å, b=157.2 Å, c=201.3 Å and diffract to at least 1.7 Å resolution. The asymmetric unit contains half the L_8S_8 molecule. The diffraction pattern to low resolution shows the presence of a pseudo fourfold axis parallel to c as well as pseudo F-centering. Rotation function calculations confirm the presence of the pseudo fourfold axis. The pseudo fourfold symmetry is also reflected in the similar lengths of the a and the b axes. From the position of the pseudo F-centered peak in a native Patterson map to high resolution it was obvious that the fourfold axis is offset from the molecular center at x=0, y=1/4, z=1/4 by about 1 Å in the a direction.

Some of our crystals exhibit merohedral twinning in which the a axis and the b axis of the twins become overlapping. We have defined a degree of tetragonality in the diffraction data by examining pairs of reflections which should have different intensities in an orthorhombic crystal, but equal intensities in a tetragonal crystal. This can be done by calculating an R value as:

$$R(tet) = sum \quad F(hkl) - F(khl) /0.5 (F(hkl) + F(khl))$$

This R value was 27% for all reflection pairs to 7 Å resolution for single crystals. Twinned crystals give much lower R(tet) values, particularly noticeable for reflections where h+ℓ=2n+1, an effect of the local symmetry. True tetragonal symmetry would give an R value around 4% corresponding to the accuracy of the measured data. A typical orthorhombic crystal (liver alcohol dehydrogenase) gave an R(tet) value of 63%. We have devised a screening procedure for twins in which the intensities of a number of selected reflection pairs that show large deviation from tetragonality in single crystals are measured. These measurements are routinely made for each crystal on a single counter four-circle diffractometer prior to data collection.

DATA COLLECTION

The native data set of the activated complex comprises 83000 independent reflections to a resolution of 2.4 Å. The data were collected on oscillation films at a wavelength of 0.86 Å using the wiggler beam line at the Daresbury synchrotron source. In addition to the native data set, a number of heavy atom derivatives of the complex have been investigated, including two mercury ($K_2Hg(CN)_4$ and ethyl mercuri thiosalicylate, EMTS) and one gold ($KAu(CN)_2$) derivative as well as a complex where the activator-magnesium atom has been substituted by cobalt. Due to the high brilliance of the beam of the wiggler line it was generally possible to collect entire data sets from one single crystal. The R-merge values, based on intensity for such data sets are around 5%.

LOCATION OF HEAVY ATOM SITES

The heavy atom sites of the derivatives were located using an automated Patterson search routine done in two steps. The first step was a single site search on a grid twice as fine as the Patterson grid. Here the pseudo symmetry was included so that a possible site represents a set of four positions related by the local symmetry. Scores for each position were calculated as a weighted sum of the peak densities at the expected vector positions. In the second step the best sites from this search were used as input to a cross vector search. The asymmetric unit was scanned, and for each point on the search grid, crossvectors to the input positions were calculated, this time using only space group symmetry. The peak densities found were summed up to give a score. Sets of positions related by the fourfold symmetry were then extracted from the resulting list and checked against the result from the single site search. Once the major sites had been located in this way they were refined and then used to compute MIR difference Fourier maps to locate minor sites. The sites thus obtained showed good agreement with the Patterson maps to low resolution, but to high resolution there was a slight offset of the peaks from the expected vector positions. We finally realised that this was due to a small translation component parallel to the fourfold rotation axis. By incorporating this translation component into our description of the pseudo symmetry a very good agreement between calculated and observed vector positions was obtained.

In this way 28 sites were found for each of the two mercury derivatives and 12 for the gold derivative. In addition, this search procedure was able to find the main gold sites from an anomalous Patterson map, as well as the four unique cobalt atoms from the difference Patterson map of the cobalt substituted enzyme. The mean peak height of the vectors generated by the four Co sites was 1.3 standard deviations. In the anomalous gold Patterson map the mean peak height of the vectors generated by the major four sites was 5.5 standard deviations. This was the top site in the single site search. The ability to interpret these maps reflect the high quality of the wiggler data, which presumably is due to the lack of absorption effects at the low wavelength used, as well as the possibility to collect entire data sets from one crystal. Thus, we strongly recommend the use of high brilliance radiation of short wavelength for high quality data collection.

The search program has now been expanded to a space group independent set of Patterson interpretation "tools," including Harker and cross vector search as described above.

INTERPRETATION OF THE ELECTRON DENSITY MAP

Initially an MIR map to 5.5 Å resolution using the mercury cyanide and gold cyanide derivatives was calculated. The mean figure of merit for the MIR phases was 0.56. This map was then solvent flattened using B.C. Wang's program package (Wang, 1985). After 5 cycles assuming 40% solvent, the refinement had converged to a mean figure of merit of 0.71 with an accumulated phase shift of 30.5 degrees. In this low resolution map we could identify helical features where the helices had the characteristic arrangement found in the eight-stranded alpha/beta barrel structure observed in many enzymes (Philips et al., 1978; Lindqvist and Branden, 1985; Pristle et al., 1987), including the dimeric L_2 Rubisco structure from Rhodospirillum rubrum (Schneider et al., 1986). The overall sequence homology between the Rh. rubrum enzyme and the large subunit of spinach Rubisco is around 31%. Thus, we expected the two chains to have similar folds. The orientation of the four barrel motifs in the asymmetric unit, together with the crystallographic twofold axis parallel to b, clearly showed that the Rh. rubrum dimer structure was essentially preserved in the spinach enzyme. Starting from the eight barrel helices the Rh. rubrum model was then fitted to the spinach density. A difference Fourier map of the cobalt substituted complex also was calculated using the phases from the solvent flattening. In this map there were four peaks at above 12 standard deviations of the map. No other peaks above five standard deviations were observed. The positions of these four high peaks agreed very well with what was expected from model building in the Rh. rubrum structure and with the solution of the difference Patterson map. The MIR phases were then extended to 2.8 Å comprising 49000 reflections, including anomalous phases from the gold derivative, as well as data from the EMTS derivative. The final figure of merit was 0.51. With the aid of the Rh. rubrum model, we could define good envelopes for the subunits which were then used in refining the initial MIR phases using G. Bricogne's real space averaging procedure (Bricogne, 1976). The envelopes were traced on plastic sheets and then transferred to the computer using a program written by M. Bergdoll. The local symmetry was refined first from the heavy atom positions and then using an electron density correlation routine written by G. Bricogne. From the resulting map, a model of the L-subunit with most of the side chains positioned in density could be built using A. Jones' interactive graphics program FRODO and starting from the Rh. rubrum structure. The map shows clear density for the transition state analogue, CABP, close to the metal binding site. There also is well defined density for the small subunit, which is located at the top and bottom of the molecule, in crevices between the large subunits. We can observe a core of beta-structure but in the present map there is some ambiguity in defining the connecting loop regions.

DESCRIPTION OF THE STRUCTURE

Structure of the Large Subunit

The spinach Rubisco large subunit is a two domain structure similar to the Rh. rubrum enzyme (Schneider et al., 1986) with a smaller N-terminal domain comprising residues 1 to 143 and a larger C-terminal domain which is built up from residues 144 to 475. Fig. 1 shows a schematic diagram of the topology of the large subunit. The central motif of the N-terminal domain is a five stranded mixed beta sheet with two alpha helices on one side and one helix on the other side of the sheet. The connection to the C-terminal domain is through a short helix followed by an extended piece of chain. The C-terminal domain has a parallel alpha/beta barrel structure, as found in triose phosphate isomerase (Philips et al., 1978), glycolate oxidase (Lindqvist and Branden, 1985) and a number of other enzymes. The domain starts with a short alpha helix, which is not a part of the alpha/beta barrel structure before entering the first strand of the barrel. This helix closes off the N-terminal side of the barrel. There are also other secondary structure elements not belonging to the barrel motif. After the sixth helix the chain forms two antiparallel strands before going back into strand number 7 of the barrel. These two strands are involved in domain-domain interactions. There is a small additional alpha helix in the loop between strand no 8 and helix no 8 of the alpha/beta barrel. At the C-terminal end of the polypeptide chain there are three consecutive alpha helices.

Quaternary Structure

The fundamental unit of the L_8 part of the molecule is a dimer, similar to the L_2 Rh. rubrum Rubisco molecule (see Fig. 1 in Schneider et al. 1986). A superposition of the C atoms of the two dimers gives an rms deviation of 1.6 Å for 434 residues in secondary structure elements. The dimer resembles an ellipsoid with approximate dimensions 50*72*105 Å and is held together by tight and extensive interactions between the subunits. Two main contact areas are present. One such area is between the C-terminal domains of the two subunits, which build up the core of the dimer. The second contact area is between the C-terminal domain of one subunit and the N-terminal domain of the second subunit. Amino acid residues from loops at the carboxy end of the beta strands of the barrel and a few loop regions in the N-terminal domain are involved in these contacts. Some of these loop regions show extensive amino acid sequence homology between the bacterial and higher plant enzymes. This is in agreement with our finding that not only the subunit structure but also the dimer interactions of the Rh. rubrum dimer is essentially preserved in the spinach molecule. In spite of the structural similarities with the Rh. rubrum enzyme, several details of the interactions are quite different. In particular an S-S bridge between cys 247 of the two subunits can be clearly seen in our map. In the Rh. rubrum enzyme this position is occupied by a proline.

Fig. 1. Schematic diagram of the structure of one subunit of Rubisco. Cylinders represent alpha helices and arrows represent beta strands. (Drawing by Ulla Uhlin)

Four dimers arranged around an approximate fourfold axis build up the L_8 octamer. The small subunits are positioned at the ends of the molecule in crevices between the large subunits. The direct interactions between the dimers are rather few and involve salt bridges. The small subunits thus seem to function as a 'glue' holding the L_2 dimers together. The overall shape of the molecule resembles a cylinder with a diameter of approximately 110 Å and height 100 Å. In the center of the molecule, between the four L_2 dimers is a channel approximately 20 Å wide. The quaternary structure of the spinach $L_8 S_8$ Rubisco molecule observed here agrees with the model for the tobacco enzyme described by Chapman et al. (1987) but is very different from the model for the enzyme from Alcaligenes eutrophus proposed by Holzenburg et al. (1987).

The Active Site

The active site is built up of residues from the C-terminal domain of one subunit and the N-terminal domain of the second subunit in the dimer. As in all other alpha/beta barrel structures the active site is located at the carboxy end of the eight parallel beta strands of the barrel. In particular, five of the loop regions connecting alpha/beta units of the barrel at the carboxy end of the strands provide residues for binding and catalysis. Furthermore, amino acid residues from two loop regions in the N-terminal domain of the second subunit participate in building up the active site. Three conserved lysine residues, 175, 201 and 334, have been suggested to be part of the active site. Lys 201 is involved in the activation of the enzyme. It has been suggested that lys 175 is the base which initiates catalysis by abstracting the C-3 proton from the substrate. Lys 334 has been identified as an active site residue by chemical modification.

All three lysine residues are found at the carboxy end of beta strands or in loops between these strands and the helices of the alpha/beta barrel. Lys 175 is located in the loop after strand no 1; lys 201 is the last residue in strand no 2 and lys 334 is found in loop no 6.

The function of lys 201 is best understood. This residue is involved in the activation process of the enzyme. During activation a carbamate is formed between the amino group of lys 201 and an activator carbon dioxide molecule. This carbamate then becomes stabilized by Mg. NMR studies have demonstrated that the metal ion is close to the carbamate carbon but not necessarily directly coordinated by the carbamate oxygen atoms (Pierce and Reddy, 1986). Direct coordination is, however, supported by site-directed mutagenesis experiments in which replacement of lys 201 with a glutamate residue rendered the enzyme incapable of tight binding of Mg^{2+} (Estelle et al., 1985). In our map there is clear density for the side chain of lys 201 as well as for the carbamate group pointing directly towards the metal ion site, thus supporting the idea of direct binding of the metal to one of the carboxyl oxygen atoms of the carbamate. It is also obvious from our model that the replacement of the carbamylated lysine side chain by the shorter glu side chain does not give a proper metal binding site. The oxygen atoms of the carboxyl group of glu 201 in the mutant are too far away from the metal for direct coordination.

Our map shows clear density for the side chain of lys 175. In our model the side chain is in a bent conformation with N close to one of the phosphate groups as well as the carboxyl group of CABP, but approximately 6 Å away from C-3 of CABP. It is possible, however, by model building, to bring N within bonding distance of C-3 by having the side chain in an extended conformation. Our present model is thus compatible with the proposed role for lys 175 in proton abstraction from C-3. However, this proton abstraction occurs prior to CO_2 addition, whereas the complex we have observed simulates the enzyme after CO_2 addition. In our model we can observe three other groups that could serve this role: lys 334, his 327 and the nitrogen atom of the carbamate group on lys 201. In Fig. 2 a picture of the active site with bound CABP is shown.

CABP binding

In our map there is well defined density for the two phosphate groups of the bound transition state analogue, CABP, with continuous density connecting the two phosphate groups. We have thus been able to place CABP in density quite accurately. The transition state analogue is bound in an extended conformation, the distance between the two phosphorus atoms being 9.6 Å. In a fully extended molecule this distance would be 12 Å. The distance between the phosphorus atom of phosphate group 1 to the metal ion is 6.6 Å while the other phosphorus atom is 7.8 Å away. This is in good agreement with the experiment by Pierce and Reddy (1986) in which the ratio of the two distances was shown to be 1.2.

The two phosphate binding sites are quite different. One site is formed by a combination of hydrogen bonds to the protein main chain and interactions with positively charged side chains. There are two hydrogen bonds to the NH-groups of gly 403 and gly 404. These residues are at the N-terminal of a short helix in loop 8 of the barrel which might also contribute a helical dipole moment to the phosphate binding site (Hol et al., 1978). In addition, the side chains of lys 175 and lys 334 are within 4 Å of one of the oxygen atoms of this phosphate. In close proximity are also two threonine residues (thr 65 and thr 67) in a long loop in the N-terminal domain of the other subunit. The second phosphate binding site is surrounded by a number of polar and charged residues from loops 5 and 6 of the barrel. We see direct interactions between this phosphate group and his 298, his 327, as well as arg 295. These three residues are conserved in all known Rubisco sequences. The carboxyl group on C-2 points towards the solution and is bound to the metal ion. The side chain of lys 334 is also in the vicinity of this group.

Metal Binding

Fig. 3 gives a view of the metal binding site in spinach Rubisco. The metal ion is coordinated to both the substrate CO_2 and the effector CO_2 groups. The effector CO_2, bound to lys 201 as a carbamate, and the carboxyl group on C-2 of the transition state analogue are bound on opposite sides of the metal atom, with the carbamate group between the metal and the bottom of the barrel. The substrate CO_2 as well as the rest of CABP close off the

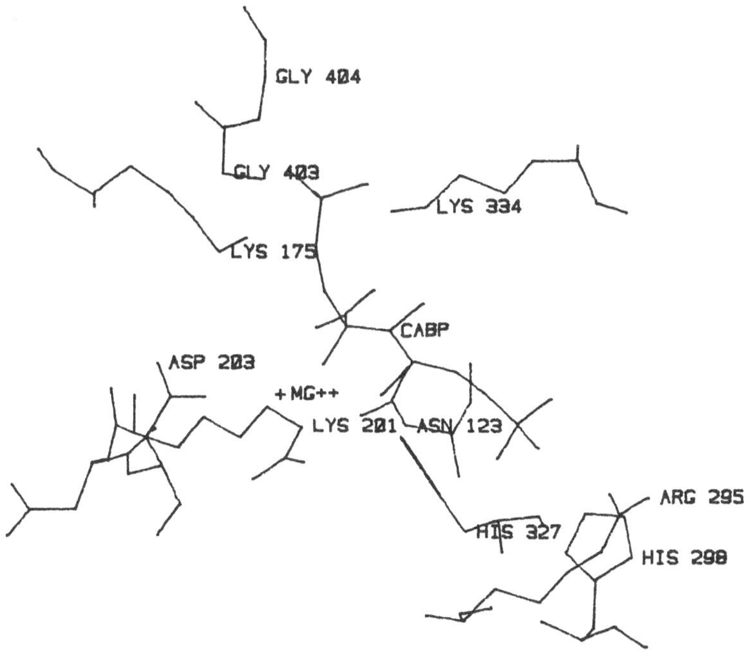

Fig. 2.   The active site of spinach Rubisco with bound CABP. The view is approximately down the barrel axis.

entrance to the active site. In addition there are two protein side chains
that are part of the coordination sphere of the metal atom: asn 123 in a
loop in the N-terminal domain and asp 203 in loop 2 of the barrel.

DISCUSSION

The active site of Rubisco lies in a cleft formed between the carboxy
end of the barrel from one subunit in a dimer and the N-terminal domain of
the second subunit. Most active site residues are located at the carboxy
end of strands or at the beginning of the loops following these strands and
connecting them to the next alpha/beta unit.

The transition state analogue, CABP, and by inference the substrate, is
bound to the enzyme in an extended conformation across the opening at the
carboxy end and towards one side of the alpha/beta barrel. The metal site
is buried by the substrate and thus shielded from the solution. The observed extreme stability of the quaternary complex enzyme-CO_2-Mg^{2+}-CABP can
thus be explained by the slower exchange rates for the metal ion and the
activator CO_2 in the quaternary complexes as compared to the ternary complexes, due to the inaccessability to solvent of these groups.

Loops 1,5,6 and 8 of the barrel are involved in phosphate binding with
loops 1,6 and 8 forming one phosphate binding site and loops 5 and 6 the
other. Loop number 8 contains a short additional alpha helix not belonging
to the barrel motif. The phosphate group on C-1 of the transition state
analogue is bound at the N-terminal of this helix through hydrogen bonds to
the main chain amide nitrogen atoms. In addition, the favorable interaction
of the alpha helix dipole with the negatively charged phosphate group may
also contribute to the binding site. It has long been known that proteins
frequently utilize the N-terminal of helices for binding of phosphate
moieties (Hol et al., 1978). Such phosphate binding sites share a common
'fingerprint' sequence, the main characteristic of which is the occurrence
of three conserved glycine residues in relative positions 1,3 and 6 of the
alpha helix (Wierenga et al., 1985). This part of the fingerprint is

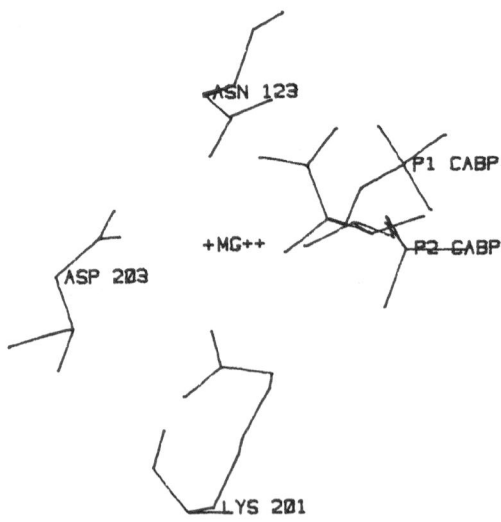

Fig. 3. A view of the metal binding site in Rubisco showing the carbamated
lys 201, asn 123, asp 203 as well as CABP.

present in the helix in loop 8, is conserved in all known Rubisco sequences and has been proposed to form a phosphate binding site in Rubisco (Chapman et al., 1986). Interestingly, phosphate binding sites at similar positions in the alpha/beta barrel are found in triose phosphate isomerase (Alder et al., 1987) and glycolate oxidase (Lindqvist and Branden, 1985).

Residues involved in metal binding are present in loop number 2 of the barrel domain. In addition asn 123 from a loop region in the N-terminal domain of the second subunit of the basic L_2 dimer is a metal ligand. Upon activation the positively charged side chain of lys 201 is converted into a negatively charged group by the addition of a carbon dioxide molecule to the N atom. The carbamate thus formed, together with asp 203 in loop 2 and asn 123, creates a binding site for the divalent metal ion. The well established reaction pathway for carboxylation of ribulose 1,5-bisphosphate involves five discrete steps: enolization, carboxylation, hydration, carbon-carbon cleavage and protonation (Miziorko and Lorimer, 1983). The reaction is initiated by the deprotonation of C-3 of ribulose 1,5-bisphosphate to give the 2,3-enediol. Carboxylation at C-2 of the enediol creates the 6-carbon reaction intermediate 2'-carboxy-3-keto-D-arabinitol 1,5-bisphosphate (CKABP). Hydration of CKABP and concomittant carbon-carbon cleavage between C-2 and C-3 yields one molecule of 3-phosphoglycerate (PGA) and a molecule of the corresponding carbanion which is then stereo-specifically protonated to give the second PGA molecule. During the carboxylation step an additional negative charge is introduced on the substrate which already bears four negative charges through the two phosphate groups. A positive group from the protein is thus clearly needed to stabilize the reaction intermediate. The metal ion, bound to the carbamate approximately halfway between the positive groups used for phosphate binding, provides the positive charge needed for stabilization of the 6-carbon reaction intermediate.

With as many as five negative charges present on the substrate at one time, a high concentration of positively charged groups from the protein is needed for binding and stabilizing the reaction intermediates. Within 12 Å of the metal binding site we find 8 positively charged and 4 negatively charged residues, as well as 17 polar side chains. All of the charged residues are conserved in all known Rubisco amino acid sequences. Out of the 17 polar side chains 7 are conserved. The charged residues are not evenly distributed among the loops; instead the positively charged side chains are located in loops 1, 5 and 6, while the negatively charged residues are contributed by loop number 2 of the barrel, as well as by the N-terminal domain of the second subunit in the dimer.

The shortest distance between two active site metal atoms in the molecule is approximately 36 Å. A direct interaction between the active sites as a basis for the negative cooperativity observed (Johal et al., 1985) is thus ruled out. There are no residues from the small subunit involved in or even close to the active site. It is thus obvious that if the small subunit plays any role in regulation of the activity of the enzyme this role most be mediated by indirect interactions, possibly involving parts of the amino end of the barrel which are close to the small subunit.

It is not known if the S-S bridge observed in our study is formed spontaneously or by a factor in the assembly process. However, it is clear that this S-S bridge stabilizes the dimer. Conservation of the correct subunit-subunit interactions within the dimers is an absolute necessity for activity of the enzyme.

The gene for the L_2 form of Rubisco from Rh. rubrum has been cloned and expressed as an active enzyme in E. coli (Somerville and Somerville, 1984; Nargang et al., 1984). Using recombinant enzyme, some site-directed mutagenesis experiments have been carried out without the knowledge of the three-dimensional structure of the enzyme, probing the active site (Hartman et al., 1987). The replacement of lys 191 (corresponding to lys 201 in the spinach enzyme) by a glutamine residue has been mentioned above. The conserved residue asp 188 (asp 198 in the spinach sequence) has been changed to glu 188 to probe the role for this side chain in metal binding (Gutteridge et al., 1984). From the crystal structure, it is now obvious that this residue is not involved in the binding of the metal ion. Asp 198 is located at the bottom of the alpha/beta barrel, more than 10 Å away from the metal binding site. In a recent experiment by Larimer et al. (1987) the involvement of two subunits in the active site of Rubisco has been proved. Two different mutant enzymes were prepared, one where glu 48 in the N-terminal domain had been replaced by gln and one with gly substituted for lys 166 in the C-terminal domain. Both of these enzymes are inactive. By simultaneously infecting E. coli with both plasmids containing the two different mutant genes, active dimeric enzyme molecules were produced. This can only be explained if the N-terminal domain of one subunit interacts with the C-terminal domain of the second subunit at the active site.

ACKNOWLEDGMENTS

This work was supported by grants from the Swedish Natural Science Research Council and from the Swedish Agricultural Research Council.

REFERENCES

Alber, T. C., Davenport, Jr., R. C., Giammona, D. A., Lolis, E., Petsko, G. A., and Ringe, D., 1987, Crystallography and site-directed mutagenesis of yeast triosephosphate isomerase: what can we learn about catalysis from a "simple" enzyme? Cold Spring Harbor Symp. Quant. Biol., 52:603.

Andersson, I., and Branden, C.-I., 1984, Large single crystals of spinach 1,5-bisphosphate carboxylase/oxygenase suitable for X-ray studies, J. Mol. Biol., 172:363.

Andrews, T. J., and Lorimer, G. H., 1987, Rubisco: structure, mechanisms, and prospects for improvement, in: "The Biochemistry of Plants," Hatch, M. D., ed, Academic Press, Orlando.

Bricogne, G., 1976, Methods and programs for direct-space exploitation of geometric redundancies, Acta Crystallogr., Sect. A, 32:832.

Chapman, M. S., Suh, S. W., Cascio, D., Smith, W. W., and Eisenberg, D., 1987, Sliding-layer conformational change limited by the quaternary structure of plant RuBisCO, Nature, 329:354.

Estelle, M., Hanks, J., McIntosh, L., and Somerville, C., 1985, Site-specific mutagenesis of ribulose-1,5-bisphosphate carboxylase/oxygenase, J. Biol. Chem., 260:9523.

Gutteridge, S., Sigal, I., Thomas, B., Arentzen, R., Cordova, A., and Lorimer, G., 1984, A site-specific mutation within the active site of ribulose-1,5-bisphosphate carboxylase from Rhodospirillum rubrum, EMBO J., 3:2737.

Hartman, F. C., Foote, R. S., Larimer, F. W., Lee, E. H., Machanoff, R., Milanez, S., Mitra, S., Mural, R. J., Niyogi, S. K., Smith, H. B., Soper, T. S., and Stringer, C. D., 1987, Function of active-site residues of ribulose bisphosphate carboxylase/oxygenase, in: "Plant Molecular Biology," von Wettstein, D. and Chua, N.-H., eds, Plenum Press, New York.

Hol, W. G. J., van Duijnen, P. T., and Berendsen, H. J. C., 1978, The α-helix dipole and the properties of proteins, Nature, 273:443.

Holzenburg, A., Mayer, F., Harauz, G., van Heel, M., Tokuoka, R., Ishida, T., Harata, K., Pal, G. P., and Saenger, W., 1987, Structure of D-ribulose-1,5-bisphosphate carboxylase/oxygenase from Alcaligenes eutrophyus H16, Nature, 325:730.

Johal, S., Partridge, B. E., and Chollet, R., 1985, Structural characterization and the determination of negative cooperativity in the tight binding of 2-carboxyarabinitol bisphosphate to higher plant ribulose bisphophate carboxylase, J. Biol. Chem., 260:9894.

Larimer, F. W., Lee, E. H., Mural, R. J., Soper, T. S., and Hartman, F. C., 1987, Intersubunit location of the active site of ribulose-bisphosphate carboxylase/oxygenase as determined by in vivo hybridization of site directed mutants, J. Biol. Chem., 262:15327.

Lindqvist, Y., and Branden, C.-I., 1985, Structure of glycolate oxidase from spinach, Proc. Natl. Acad. Sci. USA, 82:6855.

Lorimer, G., 1981, Ribulosebisphosphate carboxylase: amino acid sequence of a peptide bearing the activator carbon dioxide, Biochemistry, 20:1236.

Lorimer, G. and Miziorko, H. M., 1980, Carbamate formation on the -amino group of a lysyl residue as the basis for the activation of ribulose-bisphosphate carboxylase by CO_2 and Mg^{2+}, Biochemistry, 19:5321.

Miziorko, H. M., and Lorimer, G., 1983, Ribulose-1,5-bisphosphate carboxylase-oxygenase, Annu. Rev. Biochem., 52:507.

Musgrove, J. E., and Ellis, R. J., 1986, The Rubisco large subunit binding protein, Phil. Trans. Roy. Soc. Lond. B, 313:419.

Nargang, F., McIntosh L., and Somerville, C. R., 1984, Nucleotide sequence of the ribulosebisphosphate carboxylase gene from Rhodospirillum rubrum, Mol. Gen. Genet., 193:220.

Phillips, D. C., Sternberg, M. J. E., Thornton, J. M., and Wilson, I. A., 1978, An analysis of the structure of triose phosphate isomerase and its comparison with lactate dehydrogenase, J. Mol. Biol., 119:329.

Pierce, J., Tolbert, N. E., and Barker, R., 1980, Interaction of ribulose-bisphosphate carboxylase/oxygenase with transition state analogues, Biochemistry, 19:934.

Pierce, J., and Reddy, G. S., 1986, The sites for catalysis and activation of ribulosebisphosphate carboxylase share a common domain, Arch. Biochem. Biophys., 245:483.

Priestle, J. P., Grutter, M. G., White, J. L., Vincent, M. G., Kania, M., Wilson, E., Jardetzky, T. S., Kirschner, K., and Jansonius, J. N., 1987, Three-dimensional structure of the bifunctional enzyme N-(5'-phosphoribosyl)anthranilate isomerase-indole-3-glycerol-phosphate synthase from Escherichia coli, Proc. Natl. Acad. Sci. USA 84:5690.

Schneider, G., Lindqvist, Y., Branden, C.-I., and Lorimer, G., 1986, Three-dimensional structure of ribulose-1,5-bisphosphate carboxylase/oxygenase from Rhodospirillum rubrum at 2.9 Å resolution, EMBO J., 5:3409.

Somerville, C. R., and Somerville, S. C., 1984, Cloning and expression of the Rhodospirillum rubrum ribulosebisphosphate carboxylase gene in E. coli, Mol. Gen. Genet., 193:214.

Wang, B.-C., 1985, Resolution of phase ambiguity in macromolecular crystallography, Meth. Enzymol., 115:90.

Wierenga, R. K., De Maeyer, M. C. H., and Hol, W. G. J., 1985, Interaction of pyrophosphate moieties with α-helices in dinucleotide binding proteins, Biochemistry, 24:1346.

DATA COLLECTION FROM VERY THIN HLA CRYSTALS

USING SYNCHROTRON RADIATION

Pamela J. Bjorkman[1], William S. Bennett[2], and Don C. Wiley

Dept. of Biochemistry and Molecular Biology
Howard Hughes Medical Institute
Cambridge, MA 02138

[1]present address:
Dept. of Medical Microbiology
Stanford University
Stanford, CA 94305

[2]present address:
Max-Planck Institute fur Molekulare Genetik
Abteilung Wittmann
Berlin 33
FDR

INTRODUCTION

HLA molecules are polymorphic cell surface glycoproteins involved in the cellular immune response against viruses. Cytotoxic T-cells recognize viral peptides derived from intracellular processing that are complexed to HLA (Townsend et al., 1986; Maryanski et al., 1986). HLA class I molecules are heterodimers: the heavy chain has three domains (alpha1, alpha2, alpha3), and the light chain (beta2-microglobulin) consists of a single domain with homology to immunoglobulin constant regions. The heavy chain alpha1 and alpha2 domains are polymorphic between specificities, and the more constant alpha3 domain has homology to antibody constant domains and to the HLA light chain (reviewed in Hood et al., 1983). In order to understand how the polymorphic residues are distributed on the HLA structure, and how HLA interacts with peptides and T-cell receptors, we initiated a structure determination of HLA-A2, a human histocompatibility molecule. The structure of HLA-A2 and the implications for understanding how HLA molecules function have been previously described (Bjorkman et al., 1987a; Bjorkman,1987b). This report will concentrate only upon the data collection and processing of HLA films taken using synchrotron radiation as an X-ray source. Most of the HLA data were collected at a synchrotron facility in order to maximize the amount of diffraction data available from each crystal. Both HLA crystals and protein were in short supply, and HLA crystals are very thin (~20 microns).

CRYSTALLIZATION

The hydrophilic extracellular portion of the HLA molecule can be cleaved from the membrane by papain, releasing a soluble heterodimer retaining full serological reactivity (Nathenson and Shimada, 1968). HLA-A2 and HLA-A28 were purified from the cell membranes of homozygous human lymphoblastoid cell lines (Turner et al., 1975; Parham et al., 1977). Typical yields are 3-4 mgs. of purified protein from 200 liters of cells.

Crystals of HLA-A2 and HLA-A28 were grown from protein solutions (15 mg/ml in 25mM-2- (N-morpholino)ethane sulfonic acid or 100mM imidazole, pH 6.2 to 6.5) in two microliter drops with 15% (w/v) polyethylene glycol 6000 by vapor diffusion. (The crystallization and a proposed packing model for the two crystal forms of HLA were previously described (Bjorkman, 1984; Bjorkman et al., 1985). HLA-A2 crystallizes in two space groups: monoclinic $P2_1$, a=60.4Å, b=80.4Å, c=56.5Å, b=120.4° and orthorhombic $P2_12_12_1$, a=60.2Å, b=80.4Å, c=112.2Å, under similar conditions. Both crystal forms grow as thin plates (typical size 0.5mm x 0.4mm x 0.02mm) with the a and b axes in the plane of the plate. The orthorhombic crystals occasionally achieve a thickness of 0.1mm.

Crystals of both space groups are morphologically identical, and crystals of one form will grow when crushed seed crystals of the other form are added to the crystallization solution. These observations in addition to a similarity of the lattices (both in dimensions and transform), imply that the packing of molecules is similar in both space groups. As discussed elsewhere in more detail, (Bjorkman, 1984; Bjorkman et al., 1985) both crystal forms were proposed to result from planes of closely packed molecules similarly arranged along the a and b axes. In the direction of the c* axis (monoclinic), or c axis (orthorhombic), the planes were predicted to be more loosely stacked in the same direction with respect to c* (monoclinic), or in alternating directions with respect to c(orthorhombic). The packing model postulated that the crystal contacts in both space groups would be extensive along the a and b axes, but less extensive in the perpendicular direction, as also suggested by the plate – like morphology of the crystals (the direction perpendicular to the a and b axes is the thin dimension of the crystal). Some crystals show disordering due to stacking faults along the c or c* axis, and crystals that are mixtures of both space groups have been detected. The eventual electron density maps in the monoclinic and orthorhombic space groups showed extensive crystal contacts in the plane of the a and b axes, and fewer contacts in the perpendicular direction, as postulated by the packing model and suggesting that thinness of most of the crystals is an inherent property of the crystal packing (Bjorkman et al., unpublished observations).

DATA COLLECTION

Because of the limited amounts of HLA protein available, the difficulty of obtaining single well-ordered crystals, and the extreme thinness of the crystals, we chose to collect diffraction data using synchrotron radiation as an X-ray source. For some protein crystals, it is possible to collect more data from a single crystal before radiation decay becomes apparent using synchrotron radiation compared to using conventional sources (Bartunik et al., 1982). In the case of HLA crystals, between 5 and 10 times as much film data can be collected per crystal at a synchrotron facility compared to using a rotating anode as an X-ray source (Elliot GX-6; 40 KV, 20mA; 100 micron focusing cup, Franks focusing mirrors (Harrison, 1968)). Since crystals suitable for data collection are rare, this increase in data available from a single crystal using synchrotron radiation allowed us to collect complete data sets to high resolution from very few crystals. On initial photographs of HLA crystals, diffraction data can be observed to 2.2Å. The high resolution information rapidly decays, however, so that the data sets were processed at 2.6Å. Statistics on the monoclinic native and potential derivative data sets, and the orthorhombic native data set are presented

Table I.   Data Collection Summary (HLA-A2 Crystals)

a) P2$_1$

Data set name	heavy atom compound	where collected[*]	# crystals	wavelength (Å)	(to 2.6 Å res) # reflections total	unique	$\Sigma(I_{hkl} - I)/\Sigma(I)$ Rscale (%)
NCTI	--- (native)	DESY/CHESS	5	1.45/1.30/1.55	80891	13513	10.1
PTCL	K$_2$PtCl$_4$	DESY	2	1.49	42299	12808	9.5
PTAN	K$_2$PtCl$_4$	DESY	3	1.06	18460	9620	9.0
HGII	HgI$_2$	DESY	3	1.48	37101	12696	10.5
OSMI	K$_2$OsCl$_6$	DESY	2	1.48	30888	12110	9.1
PPMP	p-chloromercuri-phenol	SSRL/CHESS	5	1.55	34295	10247	11.4
CSHG	covalent modification of LYS by 2-chloromercuri-dinitrophenol	DESY	5	1.48	31168	10556	12.3
RBCL	RbCl	DESY	2	1.48	42701	13590	13.3
PTCN	K$_2$Pt(CN)$_4$	DESY	1	1.48	37868	13494	8.9
HGNP	2-chloromercuri-dinitrophenol	DESY	1	1.48	37045	13459	12.2
HGAN	Hg aniline	CHESS	4	1.55	21175	6449	8.8
HGSC	Hg(SCN)$_2$	CHESS	5	1.55	20256	10734	10.3

b) P2$_1$2$_1$2$_1$

Data set name	heavy atom compound	where collected[*]	# crystals	wavelength (Å)	(to 2.6 Å res) # reflections total	unique	Rscale (%)
NOTI	___ (native)	CHESS/XEN	4	1.55/1.54	30590	10805	8.7

Oscillation photographs (P2$_1$) of 5° or 2-3° (P2$_1$2$_1$2$_1$) were recorded, processed and scaled as described (Bjorkman, 1984). Three additional P2$_1$ derivative data sets not listed with comparable statistics were collected at CHESS. The orthorhombic Xentronics area detector data were recorded at 5' oscillation frames, then merged into batches of 20 frames and scaled to the native orthorhombic film data.

*DESY, Deutches Elektronen Synchrotron
 CHESS, Cornell High Energy Synchrotron Source
 SSRL, Stanford Synchrotron Radiation Laboratory
 XEN, Xentronics Area Detector

in Table I. Part of the orthorhombic native data, and a derivative data set were collected from the thickest orthorhombic crystals using a Xentronics Area detector (Durbin et al, 1986; Blum et al, 1987) mounted on an Elliot GX-13 rotating anode (40kV, 40mA, 100 micron focusing cup, Franks focusing mirrors (Harrison, 1968)).

Monoclinic crystals were mounted with the unique axis along the spindle, so that each photograph contains a reflection and its Friedel mate. In this orientation, 180° of rotation data are required to produce a complete data set which then has fourfold redundancy. For some data sets, a crystal was mounted with the a axis along the spindle and 20° of data were taken to fill the "blind region" that occurs when data are collected from crystals in a single orientation (Arndt and Wonacott, 1977). The usual orientation with the b axis along the spindle was chosen in order to collect Friedel related reflections (hkl and hk̄l) on the same film, in the hopes of using anomalous scattering to aid in phasing. Native and derivative crystals were mounted in the same orientation so as to minimize systematic differences between data sets which may be caused by differential absorption of the incident beam as it passes through different parts of the crystal.

In order to minimize the number of partially recorded reflections on each film, the largest oscillation range possible was chosen that did not produce a large number of overlapping reflections and that did not create an unusably high background. For monoclinic crystals, it was found that an oscillation range of 5° provided a good compromise between a reasonable background and a minimum of partially recorded reflections. Oscillation ranges of 2-3° were used for orthorhombic crystals. For a typical photograph, of 2200 reflections predicted to 2.4Å resolution, 300 reflections would be partially recorded and 200 would be overlapped. In practice, the resolution limits of most photographs were between 2.6Å and 2.8Å, so that the number of reflections actually recorded that are rejected during film processing because they are overlapped at high angle is less than 200. Partially recorded reflections were discarded.

The data photographs were processed using the program FILME written by Peter Schwager and Klaus Bartels. The spot integration was done over a fixed box size that had to be changed depending on the crystal size and the orientation of a given crystal with respect to the beam. The spot shapes on some HLA films differed between the top and the bottom of the film. These differences correlated with the position of the crystal with respect to the beam. When pictures were taken with the X-ray beam perpendicular to the flat face of the crystal, the reflections were approximately round and were the same shape on the top and bottom of the film. If the picture was taken like this (--> \) (where the arrow represents the beam and the slash represents the crystal), the reflections on the top of the film were round, and the reflections on the bottom of the film were elongated. If the picture was taken like this (--> /), the opposite situation occurred. We do not know the reason for the differences in spot shape on a film. It is possible that the spot shape on a particular part of the film is an image of the shape of the crystal in the direction that the beam traveled before reaching the film. (It is perhaps relevant to this argument that the beam (0.4mm^2 in diameter) is larger in some dimensions than the crystal.) On the films where the spot shapes differed, the whole film was evaluated once with a large box size to determine the missetting angles. The top of the film was then evaluated separately from the bottom of the film using the box sizes appropriate for each half. Reflections from individual films were scaled to each other using a linear scale factor and a resolution dependent temperature factor. Following this preliminary scaling, individual data sets were rescaled to the native using a bin-scaling or "local" scaling procedure in order to correct for absorption effects. Local scaling was found to improve the interpretability of both difference Patterson and difference Fourier calculations.

Heavy Atom Derivatives and Phasing

Potential heavy atom derivatives were evaluated by comparing screened precession photographs of crystals soaked in a heavy metal compound compared to a native photograph. Exposure times for a 6° screened precession (7.5Å resolution) photograph varied between 70 and 100 hours (Elliot GX-6 rotating anode; 40kV, 20 mA; Franks focusing mirrors (Harrison, 1968)). Pictures taken at higher angle were weak. Higher angle (9°; 5.2Å resolution) precession photographs were taken at the CHESS Synchrotron facility. Over 100 heavy atom compounds were tested in this way, leading to 14 candidates for derivatives from which full data sets were collected (Table I).

Heavy atom sites were located for four of the potential derivative data sets, using difference Patterson maps and checked with difference Fourier calculations. Phasing statistics are summarized in Table II. The multiple

Table II. Phasing Statistics

compound	site	x	y	z	occupancy (real/anomalous)	$B(\text{Å}^2)$ (fixed)
K_2PtCl_4	1	0.13	0.00*	0.26	0.99/3.40	20.
	2	0.45	0.12	0.82	0.59/1.80	20.
	3	0.46	0.36	0.79	0.37/0.89	20.
HgI_2	1	0.12	0.00	0.27	1.08/1.81	20.
	2	0.52	0.21	0.89	0.16/0.25	20.
K_2OsCl_6	1	0.80	0.24	0.99	0.70/1.23	20.
	2	0.22	0.25	0.80	0.19/0.01	20.
p-chloro-mercuri-phenol	1	0.97	0.23	0.16	0.88/1.80	20.

Resolution (Å)	10.6	8.2	6.7	5.7	4.9	4.3	3.9	3.5	total
# reflections	144	243	391	546	753	966	1228	1488	5759
m	0.66	0.66	0.68	0.67	0.63	0.62	0.58	0.53	0.60
K_2PtCl_4 Fc/Ei	1.31	1.79	2.06	3.04	1.89	1.71	1.73	1.56	1.78
HgI_2 Fc/Ei	1.29	1.58	1.29	1.18	1.04	0.89	0.77	0.73	0.92
K_2OsCl_6 Fc/Ei	0.46	0.63	0.64	0.69	0.59	0.54	0.47	0.42	0.52
p-chloromercuri phenol	0.97	0.85	1.18	1.04	1.14	1.05	0.88	0.87	0.97

* atom 1 y coordinate fixed at 0.0
m : figure of merit
Fc/Ei : heavy atom structure factor/residual lack of closure (phasing power)

Heavy atom sites and phasing statistics to 3.5Å are presented for 4 heavy atom derivatives of $P2_1$ crystals of HLA-A2. The most useful derivative (K_2PtCl_4) was the only data set with an overall phasing power (r.m.s. Fc/Et) greater than 1.0 to 3.5Å resolution. The other derivative data sets had phasing powers greater than 1.0 only at lower resolution (~6.0Å), and therefore were not useful in phasing. The 3.5Å MIR electron density map calculated from these 4 derivatives was not improved compared to an SIR plus anomalous map calculated from the K_2PtCl_4 derivative alone.

isomorphous replacement (MIR) electron density map was calculated from 4 derivatives and was not improved compared to a single isomorphous replacement (SIR) map calculated using the best derivative (K_2PtCl_4, Table II). The contributions from this derivative dominated the phases. The solution and refinement of the structure is described in more detail elsewhere (Bjorkman et al., 1987; Saper et al., in preparation) and will not be dealt with in this paper.

Anomalous Scattering Calculations

Anomalous difference Patterson maps indicated that the derivatives showed significant anomalous dispersion effects and anomalous data were therefore used in phasing and to establish the absolute hand of the molecule. Anomalous pairs were used only if recorded on the same films and were subjected to rejection criteria similar to those used by Hendrickson and Teeter (1981). Only ~20% of available anomalous data were used.

Two complete data sets were collected from the platinum derivative of HLA (Table I). These data were collected above (λ = 1.48Å; data set name: PTCL) and below (λ = 1.06Å; data set name: PTAN) the LIII absorption edge of platinum (λ = 1.07Å; Bearden, 1967). Because of the difficulties encountered in identifying useful isomorphous derivatives, we hoped to obtain additional phase information by data collection from derivative crystals at two different wavelengths.

For a protein that has been derivatized with a heavy metal, the total scattering vector of reflection is given by F_{pH} = Fp + f_H + f' + if", where Fp is the sum of scattering from the protein alone; f_H is the normal scattering from the heavy atom; and f' and f" are the dispersion and absorption components of the anomalous scattering factor of the heavy atom. The f' and f" terms of the anomalous scattering factor are wavelength dependent (Phillips et al., 1977 and Phillips and Hodgson, 1980). Near an absorption edge, f" is largest and therefore synchrotron radiation at the appropriate wavelength can be used to enhance the f" of a heavy atom derivative in order to increase the magnitude of the anomalous contribution to a standard phase determination. It is also possible to use the variation of f' with wavelength to obtain phase information by collecting data from the same heavy atom derivative at multiple wavelengths above and below the absorption edge of a heavy atom.

The PTAN K_2PtCl_4 data set (Table I) was collected slightly below the platinum absorption edge. We therefore hoped to see an enhancement of anomalous differences. However, it was apparent from the anomalous Patterson map and from the symmetry R factors on individual films (which should be higher compared to the native or PTCL data sets) that the anomalous signal had not been enhanced. It has been suggested that this lack of enhancement of the anomalous signal when data are collected at a wavelength very near the absorption edge indicates that the bandwidth of the monochromatized radiation is wider than the absorption edge (Phillips et al., 1977). Therefore the beam that hits the crystal is a mixture of "above edge" and "below edge" components that cancel each other to diminish the anomalous signal.

Although we see no enhancement of f" from data collected at a wavelength near the absorption edge, it is still possible to obtain phase information from two data sets collected at wavelengths at which f' differs. Therefore a 3.0Å Patterson map with coefficients ($F_{PTCL}-F_{PTAN}$)2 (where the structure factors are averages of Friedel pairs) was calculated. Of three expected Harker peaks, there are peaks on the two-wavelength Patterson in positions near those expected for sites 1 and 3 (Table II), and a very small peak in the location expected for site 2. The peak for the major site (site 1) is the second largest peak on the Harker section, but the peak for site 3 is one of the smallest of 20 peaks on the Harker section. Fig. 1 shows a comparison of Harker sections from this Patterson map and the 3.0Å platinum isomorphous difference Patterson map. The multiple wavelength Patterson also shows peaks near the expected positions for all six cross vector peaks, even though they are not the largest peaks of the map (Fig. 1).

Although there is apparently some phase information available from the intensity differences between the low wavelength and high wavelength platinum data sets, when combined with the existing SIR phases, this information was not sufficient to allow interpretation of a 3.0Å electron density map. The relatively poor quality of the HLA film data (typical R factors on intensities for 2.6Å data sets were ~10%) makes it difficult to accurately measure small intensity changes due to anomalous dispersion effects. However, after rejecting some Friedel pairs using the criteria of Hendrickson and Teeter (1981), anomalous differences from single wavelength derivative data sets were useful in phasing and eventually producing an interpretable electron density map. Techniques for precisely locating the absorption edge of a heavy atom derivatized to a protein crystal have improved since the time we collected the low wavelength platinum data set (December, 1982), so it is possible that further phase information could be extracted from derivatized HLA crystals using the multiple wavelength technique.

3.0 Å K_2PtCl_4 Isomorphous Difference Patterson

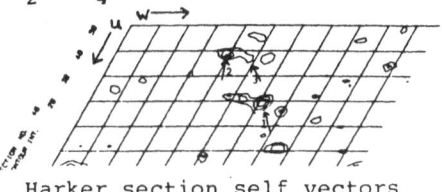

Harker section self vectors

Cross vectors

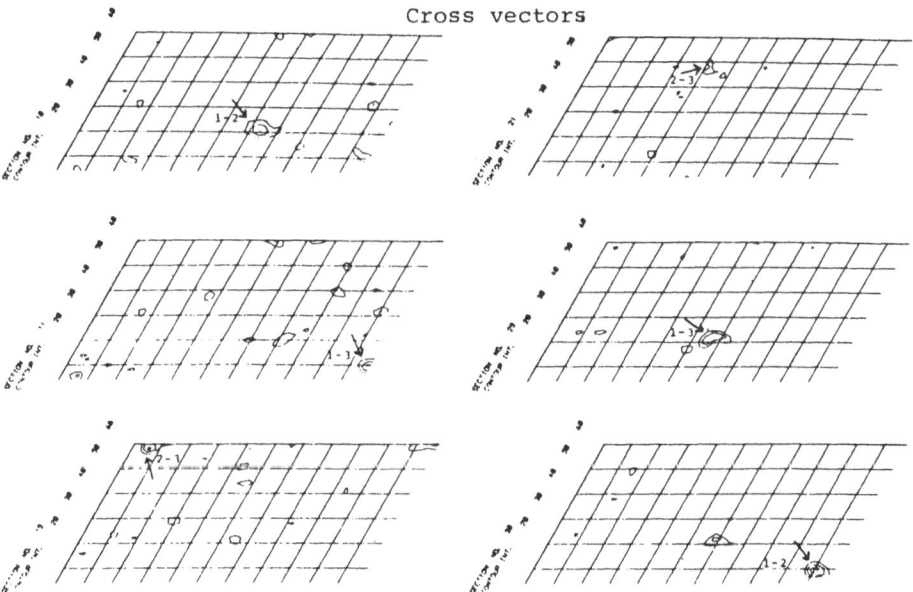

Fig. 1A.   Sections from the $P2_1$ 3.0Å isomorphous difference Patterson (part a) compared with 3.0Å platinum multiple wavelength Patterson (part b). The multiple wavelength Patterson function was calculated with the coefficients $(F_{PTAN}-F_{PTCL})^2$ as described in the text. (The PTAN platinum data set was collected at 1.06Å; the PTCL platinum data set was collected at 1.49Å). The maps were calculated with sections perpendicular to the y axis (z axis coordinate horizontal, x axis coordinate nearly vertical). Only the asymmetric unit was calculated (u=0-->1/2, v=0-->1/2, w=0-->1). The self vectors on the Harker section are labelled with arrows and numbers corresponding to the numbering of heavy atom sites in Table II. The cross vector peaks are labelled with arrows and the numbers of the two sites that generated the cross vector.

CONCLUSIONS

The use of synchrotron radiation allowed us to collect film data to
~2.5Å resolution from very thin (0.02mm) HLA crystals. We obtained 5-10
times the amount of data per crystal as compared to data collected using a
rotating anode source. Except for minor spot shape problems, the films were
processed similarly to films obtained from a conventional source. Derivative screening, using screened precession photography, which proved to be
nearly impossible using a rotating anode generator, was accomplished with
synchrotron radiation. Anomalous dispersion differences were significant
and aided in phasing.

ACKNOWLEDGMENTS

We thank Hans Bartunik and Klaus Bartels at DESY, Keith Moffat and
Wilfried Schildkamp at CHESS, and Paul Phizackerley and Ethan Merritt at
SSRL for help with synchrotron data collection. PJB held an American Cancer
Society postdoctoral fellowship during part of the work.

3.0 Å K_2PtCl_4 Multiple wavelength difference Patterson

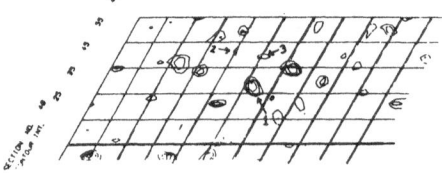

Harker section self vectors

Cross vectors

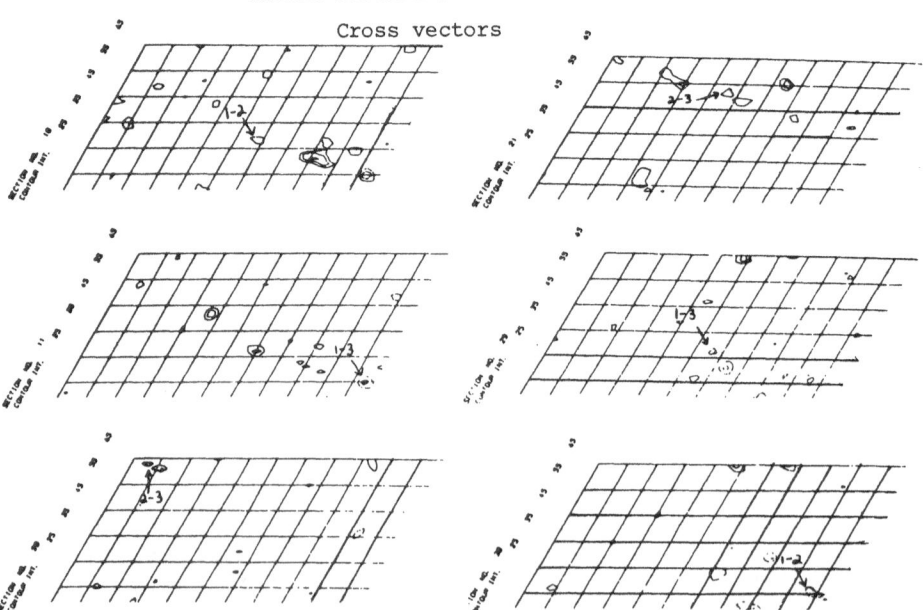

Fig. 1B. See the caption for Fig. 1A.

REFERENCES

Arndt, U. W., and Wonacott, A. J., 1977, "The Rotation Method in Crystallography," North Holland, Amsterdam.

Bartunik, H. D., Fourme, R., and Phillips, J. C., 1982, Macromolecular crystallography using synchrotron radiation, in: "Uses of Synchrotron Radiation in Biology," H. B. Stuhrmann, ed., Academic Press, London.

Bearden, J. A., 1967, X-ray wavelengths, Rev Mod Phys., 39:78.

Bjorkman, P. J., Crystallographic Studies of HLA, Ph.D. thesis, Harvard University 1984.

Bjorkman, P. J., Saper, M. A., Samraoui, B., Bennett, W. S., Strominger, J. L., and Wiley, D. C., 1987a, Structure of the human class histocompatibility antigen, HLA-A2, Nature, 329:506.

Bjorkman, P. J., Saper, M. A., Samraoui, B., Bennett, W. S., Strominger, J. L., and Wiley, D. C., 1987b, The foreign antigen binding site and T cell recognition regions of class I histocompatibility antigens, Nature, 329:512.

Bjorkman, P. J., Strominger, J. L., and Wiley, D. C., 1985, Crystallization and x-ray diffraction studies on the histocompatibility antigens HLA-A2 and A28 from human cell membranes, J. Mol. Biol., 186:205.

Blum, M., Metcalf, P., Harrison, S. C., and Wiley, D. C., 1987, A system for collection and on-line integration of x-ray diffraction data from a multiwire area detector, J. Appl. Cryst., 20:235.

Durbin, R. M., Burns, R., Moulai, J., Metcalf, P., Freymann, D., Blum, M., Anderson, J. E., Harrison, S. C., and Wiley, D. C., 1986, Protein, DNA and virus crystallography with a focused imaging proportional counter, Science, 232:1127.

Harrison, S. C., 1968, A point-focusing camera for single-crystal diffraction, J. Appl. Cryst., 1:84.

Hendrickson, W. A. and Teeter, M., 1981, Structure of the hydrophobic protein crambin determined directly from the anomalous scattering of sulphur. Nature, 290:107.

Hood, L., Steinmetz, M., and Malissen, B., 1983, Genes of the major histocompatibility complex of the mouse, Ann. Rev. Immunol. 1:59.

Maryanski, J. L., Pala, P., Corradin, G., Jordan, B. R., and Grottini, J-C, 1986, H-2-restricted cytolytic T-cells specific for HLA can recognize a synthetic HLA peptide, Nature, 324:578.

Nathenson, S. G., and Shimada, A., 1968, Papain solubilization of the mouse H-2 isoantigen: or improved method of wide applicability, Transplantation, 6:662.

Parham, P., Alpert, B. N., Orr, H. T., and Strominger, J. L., 1977, Carbohydrate moiety of HLA antigens, J. Biol. Chem., 252:7555.

Phillips, J. C., and Hodgson, K., 1980, The use of anomalous scattering effects to phase diffraction patterns from macromolecules, Acta Cryst., A36:856.

Phillips, J. C., Wlodawer, A., Goodfellow, J. M., Watenpaugh, K. A., Sieken, L. C., Jensen, L. H., and Hodson, K. O., 1977, Applications of synchrotron radiation to protein crystallography. II Anomalous Scattering, Absolute Intensity and Polarization, Acta Cryst., A33:445.

Townsend, A. R. M., Rothbard, J., Gotch, F. M., Bahadur, G., Wraith, D., and McMichael, A. J., 1986, The epitopes of influenza nucleoprotein recognized by cytotoxic T lymphocytes can be defined with short synthetic peptides, Cell, 44:959.

Turner, M. J., Cresswell, P., Parham, P., Strominger, J. L., Mann, D. L., and Sanderson, A. R., 1975, Purification of papain-solubilized histocompatibility antigens from a cultured human lymphoblastoid line RPMI 4265, J. Biol. Chem., 250:4512.

DOUBLE-STRANDED DNA BINDING PROTEIN HU

Keith Wilson

European Molecular Biology Laboratory
c/o DESY, Notkestrasse 85
2000 Hamburg 52, FRG

Isao Tanaka

Hokkaido University
Department of Polymer Science
Sapporo, Japan

Krzysztof Appelt

Agouron Pharmaceuticals Inc.
11025 North Torrey Pines Road
La Jolla, CA 92037

Stephen White

Biology Department
Brookhaven National Laboratory
Upton, NY 11973

INTRODUCTION

The genome of eukaryotes is highly organised, structurally and functionally. The supercoiling of the DNA, its packing around the histone core into nucleosomes, and the ordering of nucleosomes into chromatin results in highly condensed chromosomes in the cell nucleus (Felsenfeld and McGhee, 1986). Such condensation is essential for the orderly arrangement of the enormously long, linear molecules of DNA in the eukaryotic cell. The basis of this hierarchical superstructure is the histone proteins; small basic proteins that form the protein core of the nucleosome.

In prokaryotes, the ordering of the genetic material is less obvious. There is no clear nucleus, nor highly-compacted chromatin-like features. However, the bacterial genome is by no means without structure. Gentle disruption of the cell allows the removal and observation of a wealth of structural features, as was described by Griffith (1976). In these early micrographs, supercoiled and folded chromosome-like objects could be clearly seen.

There are several candidates for proteins that play histone-like roles in prokaryotes (Geider and Hoffmann-Berlin, 1981). The two most important are DNA binding proteins HU and Hl. HU is the subject of the present study. It has been previously called by a variety of names including Hu (Rouviere-Yaniv and Kjeldgaard, 1979), NS (Mende et al., 1978) and DNA-binding protein II (Geider and Hoffmann-Berlin, 1981): the terms generally reflect the interests and observations of the workers involved. The latter name was intended to indicate its nonsequence specificity for double-stranded DNA.

HU is a small, basic protein of molecular weight 9500 Daltons. It binds nonspecifically to double-stranded DNA, but will also bind less tightly to single-stranded DNA and to RNA (Drlica and Rouviere-Yaniv, 1979). There are some 30,000 dimers of HU in the eubacterial cell, sufficient to cover approximately one-sixth of the genome (Drlica and Rouviere-Yaniv, 1979).

HU is ubiquitous in prokaryotes and its amino acid sequence has been determined from a variety of organisms including <u>Escherichia</u> <u>coli</u>, <u>Pseudomonas</u> <u>aeruginosa</u>, <u>Rhizobium</u> <u>meliloti</u>, <u>Clostridium</u> <u>pasteurianum</u> and the archebacterium <u>Thermoplasm</u> <u>acidophilum</u>. There are two different protein chains for HU in <u>E</u>. <u>coli</u> (Rouviere-Yaniv and Kjeldgaard, 1979; Mende et al., 1978), but only one has been identified in the other bacteria. The significance of this finding is unclear. There is an extensive homology between the sequences and no deletions or insertions are required to give the optimum alignment. Only minor variation in chain length is observed (the occasional loss or gain of a single amino acid at the end of the chain). There are several highly conserved positions in the sequence and these will be discussed below in the context of the structural results.

HU functions as a dimer in solution, although this fact would have been difficult to deduce from hydrodynamic studies, since the molecule is elongated (see below). NMR studies of the protein suggested the importance of

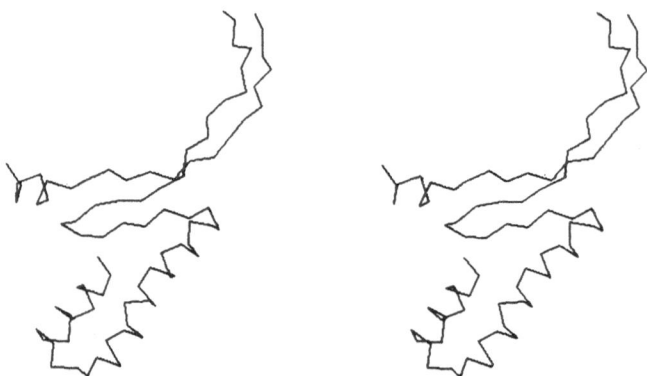

Fig. 1.   A stereo view of a monomer of HU. The molecule has two distinct halves; the N-terminal half (below) consists of two alpha helices, and the C-terminal half contains an anti-parallel 3-stranded sheet followed by a short alpha helix. All the secondary structural elements are connected by short turns except strands 2 and 3 which have an intervening 'arm' of some 25 residues. The 8 residues at the end of the arm are not visible in our structure but the others clearly adopt a beta ribbon conformation with a kink in the center. The arms contain several highly conserved positively charged residues which are thought to mediate the ionic interaction with the DNA backbone.

phenylalanine residues in the monomer-monomer interaction within the dimer and also of arginine residues in the binding to the DNA (Lammi et al., 1984).

The role of HU *in vivo* is not well understood. The protein can condense DNA *in vitro* into bead-like structures, similar in appearance to nucleosomes, which can be seen in the electron microscope (Rouviere-Yaniv et al., 1979). Also, HU is reported to have various effects on cellular processes such as DNA replication (Drlica and Rouviere-Yaniv, 1987). In contrast to the usual situation for proteins, considerably more is known about the structure of HU than its function. The protein has been crystallized from the moderate thermophile Bacillus stearothermophilus (Dijk et al., 1983), its structure analysed at a resolution of 3Å (Tanaka et al., 1984) and recently it was refined at 2Å (unpublished data). The B. stearothermophilus protein shows the expected extra thermostability over its mesophilic homologue from E. coli (Dijk et al., 1983), and indeed, the E. coli protein has resisted considerable efforts to crystallize it.

DATA COLLECTION AND STRUCTURE SOLUTION

The unit cell has symmetry P2, dimensions a=65.5Å, b=37.3Å, c=65.5Å and beta=114.5. The crystals grow as monoclinic needles up to 2.0mm in length, but only 0.15mm in the other two dimensions. The data to 3Å resolution were recorded on an Arndt-Wonacott oscillation camera mounted on a sealed tube X-ray generator in the Max-Planck-Institute for Molecular Genetics in West Berlin. Three Angstroms was effectively the limiting resolution using this experimental setup. Data were measured for the native protein and for one useful derivative, $K_3UO_2F_5$. Phases were calculated from this single derivative using anomalous dispersion data. The map clearly showed the molecular boundary.

The crystal lattice is of some interest and was of considerable help in eventually solving the structure of this protein. The unit cell contains three dimers and each sits on one of the four independent crystallographic twofold axes in space group P2. The empty fourth axis creates an extensive solvent channel in the crystal. As a consequence, monomers within each dimer are related by perfect twofold symmetry. However, the orientation of the three dimers in the cell is distinctly different, and they make unique intermolecular contacts within the crystal. This finding led to an immediate and unambiguous definition of the molecular dimer from the crystal structure. Moreover, the conformations of the three dimers are extremely similar lending further weight to the 'validity' of the crystal conformation. This similarity allowed the averaging of the electron density for the three molecules to improve the quality of the map. The resultant map allowed an unambiguous tracing of the chain and the positioning of most side chains.

Extension of the resolution beyond 3Å was not possible in Berlin. Instead, synchrotron radiation was used. Data were collected at station 7.2 in the SERC Daresbury Laboratory, England. In a single shift of 24 hours, a complete data set was recorded on film, rotating a single crystal 180° about the b axis with an oscillation range of 2°. The data extended to 2Å. The fine collimation of the beam and the length of the monoclinic needles was exploited by translating the crystal three times along its long axis during data collection to expose new portions of the needle to the beam, and reduce the effects of radiation damage. A second shift of 8 hours allowed the collection of 20° of data with a crystal mounted across the tube to fill in the blind region. The data were digitized with an Uptronics P-100 photoscan and integrated with the MOSCO film processing package. The final merging R at 2Å was about 7%. Such data could not be collected on a conventional source.

The structure has been refined using these data to a current R factor of less than 20% (unpublished results).

DESCRIPTION OF STRUCTURE

Figure 1 shows the structure of the HU monomer from B. stearothermophilus. The monomer of 90 residues is divided into two domains. The N-terminal domain is composed of two alpha-helices packed closely together, with a broad turn linking them. The C-terminal half of the protein is centered around an anti-parallel 3-stranded beta-pleated sheet, with a short alpha helix at the carboxy terminus. A two-stranded anti-parallel beta ribbon projects away from the main body of the sheet as a continuation of two of its strands to form an extended arm. In all three molecules in the unit cell there is no electron density apparent for residues 58-71 at the end of these extended arms. These residues are presumed to be disordered in the crystal, and are probably flexible in the native protein in solution.

The two domains of the monomer are well separated, with few contacts between them which is somewhat surprising for a globular protein. Furthermore, four phenylalanine residues form a hydrophobic patch on the surface of the monomer. These two observations are clearly explained by the organisation of the dimer shown in Fig. 2. The two helices from each monomer pack closely together, burying the eight aromatic side chains at the core of the dimer. The two three-stranded beta-sheets lie side by side above this alpha-helical base, but do not form an H-bonded six-stranded sheet. The dimer thus takes on a characteristic globular appearance, with the exception of the two extended arms which form a deep concave cleft at the top of the molecule. The alpha-helical base has a pronounced V-shape.

Many of the highly conserved residues are rationalized by this structure. The alanines at positions 11 and 24 are involved in close contacts between the alpha helices. The majority of the conserved hydrophobic residues, especially the phenylalanines, give the dimer its "oily" core. The glycines at positions 39, 46 and 48 are involved in sharp turns. Other residues which are conserved are probably important for the function of the protein.

STRUCTURE-FUNCTION OF HU

A model for the interaction of HU with DNA was proposed based on a computer graphics study of the docking of the two macromolecules (Tanaka et al., 1984). The deep cleft at the top of the molecule has an appropriate diameter (25Å) that exactly accomodates a double-stranded DNA helix, with the protein predominantly interacting with DNA via the sugar-phosphate backbone. The model aligned the twofold axis of the dimer with the pseudo twofold axis of the DNA perpendicular to the double helix axis. The DNA-protein interaction was optimized by rotating and translating the protein with respect to the DNA. Models were obtained with the extended beta-ribbons of the arms lying in either the wide or the narrow groove of the DNA, with the extended side chains of conserved arginine residues in the outgoing strand positioned to interact with the phosphates of the DNA backbone. The beta ribbon was originally placed over the wide groove of the DNA, but the possibility of binding to the narrow groove was not excluded, particularly when the flexibility of B-DNA and of the arms is taken into account. Later biochemical results for the homologous protein Integration Host Factor (IHF), as described in the next section clearly indicated that the proteins of this family do, in fact, bind to the narrow groove. This result is compatible with our model and we have fitted the HU to this groove in recent modelling experiments.

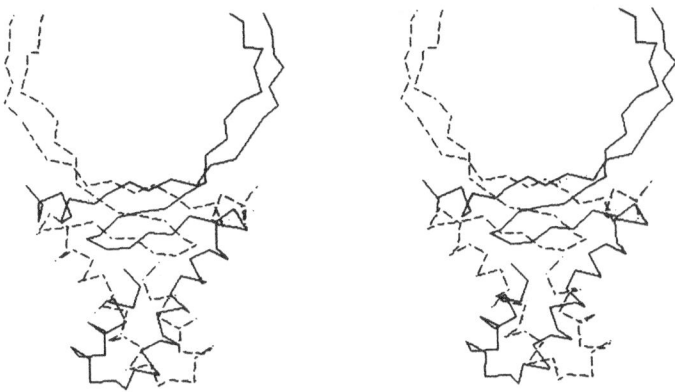

Fig. 2. A stereo view of a dimer of HU. The individual monomers are re-
        presented by continuous and dashed lines for clarity. The N-
        terminal helices create a wedge-shaped base and the two sheets form
        a 'lid' on top with the C-terminal helices on either side. The
        inner strands of the sheet together with the two arms make a heli-
        cal motif that complements the DNA helix.

     Further modelling involved the conformation of the disordered arm re-
gion that was not defined from the crystal structure. This region has been
built as a continuation of the two-stranded beta ribbon, with a kink at the
point where the electron density becomes invisible. Residues 60-65 at the
center of the disordered arm are highly conserved and these probably create
the turn at the end of the beta-ribbon. Proline residues are frequently
found at turns in protein structures; there is one located at position 63.
In the final model, the two arms of the dimer totally encircle one turn of
the DNA double helix (Fig. 3).

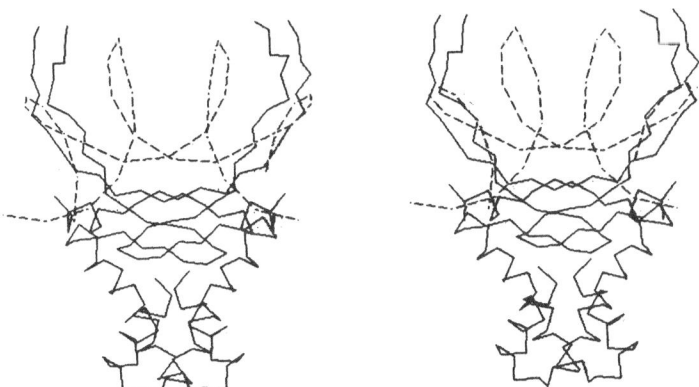

Fig. 3. A stereo view of a model for the interaction of HU with double-
        stranded DNA. Note that consistent with recent biochemical results
        the DNA is bent and the arms are positioned over the minor groove.
        This indeed results in a more favorable interaction with positively
        charged residues in the body of the molecule. It is proposed that
        in a nucleosome-like particle, the wedge-shaped HU monomers form a
        protein core and the DNA is wound around the outside.

Direct confirmation of this model requires a structural study of a protein-oligonucleotide complex. Oligonucleotides cannot be bound in the present crystals as the proposed DNA site is occupied by the bottom of the protein dimer lying above it on the twofold axis. Co-crystallization experiments with oligonucleotides are being carried out. However, there is much indirect evidence for the validity of the model:

(1) The relevant amino acids are all highly conserved.
(2) The obvious flexibility of the arms is in keeping with their function of wrapping around the DNA to form precise contacts with the sugar-phosphate backbone.
(3) The arms contain four conserved arginine residues, which have been implicated in DNA binding from NMR studies.
(4) Nonsequence specific DNA-binding by HU is in agreement with primarily electrostatic interactions via the sugar-phosphate backbone.
(5) The binding is inhibited by high salt concentrations.

Finally, the model suggests a means for cooperative dimer-dimer interaction during DNA supercoiling. If several dimers are bound along the DNA and kinks introduced between them to generate a supercoil, adjacent dimers are ideally placed to interact through the carboxy-terminal helix of one dimer binding to a pocket on the next.

HOMOLOGOUS PROTEINS

HU differs from such double-stranded DNA-binding proteins as CAP, cro and lambda repressor in that it binds nonsequence-specifically to DNA through a beta ribbon, while the latter recognise specific sequences of bases through the precise interactions of alpha helical side chains protruding into the wide groove of the double helix (Ollis and White, 1987). At first sight this would be a suitable distinction between proteins that bind specifically and nonspecifically to DNA. However, sequences of two proteins have been reported that clearly share the same structure as HU but which do not bind DNA nonspecifically.

The B. subtilis phage SP01 codes for a protein called transcription factor 1 (TF1) that binds 10-20 times more strongly to DNA that contains 5-hydroxymethyluracil (hmU) instead of thymine (Greene et al., 1984; Greene and Geiduscheck, 1985). SP01 contains hmU DNA and, although the precise function of TF1 is unclear, it must have some role in sequestering phage DNA from host DNA. The similarity between the sequences of TF1 and HU, both as regards structural and functional amino acids (Drlica and Rouviere-Yaniv, 1987), indicates that where (there are) differences, they are likely to explain the different DNA-binding properties of the proteins.

The striking differences occur in two of the important arginine residues and the carboxyterminus. Arginines 53 and 61 are replaced by a valine and a phenylalanine, respectively. These changes would certainly reduce the binding affinity of TF1 to DNA, and it has been suggested (Greene et al., 1986) that the aromatic side chain could intercalate between the bases of the less thermostable hmU DNA. TF1 has an extra nine residues at the C-terminus when compared to HU and these have been shown to be involved in the interaction with DNA (Sayre and Geiduschek, 1988).

The second protein is Integration Host Factor (IHF) from E. coli. IHF is a heterodimer of alpha and beta subunits and was discovered as a factor necessary for the site-specific recombination of phage lambda (Nash and Robertson, 1981). The normal role of IHF is to help in the formation of folded DNA structures by binding to specific sites and introducing bends or kinks at that site (Stenzel et al., 1987; Prentki et al., 1987; Robertson and Nash, 1988). This function is reminiscent of that of HU, which has to perform a similar role in creating the nucleosome-like complex. Like TF1, the sequences of the two IHF monomers are sufficiently similar to HU to be sure of near-identical structures. However, the typical alanine and proline residues in the kink of the beta ribbon arm are replaced by amino acids that could determine the specificity by hydrogen bond interactions. This interpretation is supported by the concensus sequence of the DNA binding site and our model which shows the important bases to be close to the kinks.

The determination of the sequences of IHF and TF1, homologous to those of HU, suggests the possibility that these may be members of an extensive family of prokaryotic and bacteriophage proteins, binding through beta-ribbon structures to DNA. The proteins are all dimeric with symmetric or, for the heterodimer, pseudosymmetric structures. IHF and TF1 show a degree of specificity in their binding, showing that DNA-protein interaction through such beta ribbons is not restricted to nonsequence-specific contacts believed to occur with HU. The functional significance of the pseudosymmetric heterodimers in E. coli and IHF perhaps lies in the fine-tuning of the interaction of these proteins with DNA or cooperatively with other protein molecules.

ACKNOWLEDGMENTS

This work was begun in the Max-Planck-Institute for Molecular Genetics in West Berlin and we are grateful for the support and continuing interest of Professor H.-G. Wittmann. We also thank the staff of the SERC Daresbury Laboratory for providing us with such excellent data collection facilities at Daresbury. EMBO provided travel funds for three of us to travel to Daresbury from West Berlin. S. W. White was supported by the Office of Health and Environmental Research of the United States Department of Energy.

REFERENCES

Dijk, J., White, S. W., Wilson, K. S., and Appelt, K., 1983, On the DNA-binding protein II from Bacillus stearothermophilus. Purification, studies in solution and crystallization, J. Biol. Chem., 258:4003.

Drlica, K., and Rouviere-Yaniv, J., 1987, Histone-like proteins of bacteria, Microbiol. Rev., 51:301.

Felsenfeld, G., and McGhee, J. D., 1986, The structure of the 30 nanometer chromatin fiber, Cell, 44:375.

Geider, K., and Hoffmann-Berlin, H., 1981, Proteins controlling the helical structure of DNA, Ann. Rev. Biochem., 50:233.

Greene, J. R., Brennan, S. M., Andrew, D. J., Thompson, C. C., Richards, S. H., Heinrikson, R. L., and Geiduschek, E. P., 1985, Sequence of the bacteriophage SP01 gene coding for transcription factor 1, a viral homologue of the bacterial type II DNA-binding proteins, Proc. Natl. Acad. Sci. USA, 81:7031.

Greene, J. R., and Geiduscheck, E. P., 1985, Site-specific DNA binding by the bacteriophage SP01-encoded type II DNA-binding protein, EMBO J., 4:1345.

Greene, J. R., Morrissey, L. M., Foster, L. M., and Geiduschek, E. P., 1986, DNA binding by the bacteriophage SP01-encoded type II DNA-binding protein, transcription factor I, J. Biol. Chem., 261:12820.

Griffith, J. D., 1976, Visualization of prokaryotic DNA in a regularly condensed chromatin-like fiber, Proc. Natl. Acad. Sci. USA, 73:563.

Lammi, M., Paci, M., and Gualerzi, C. B., 1984, Proteins from the prokaryotic nucleoid. The interaction between protein NS and DNA involves the oligomeric form of the protein and at least one Arg residue, FEBS Lett., 170:99.

Mende, L., Timm, B., and Subramanian, A. R., 1978, Primary structures of two homologous ribosome-associated DNA-binding proteins of Escherichia coli, FEBS Lett., 96:395.

Nash, H. A., and Robertson, C.A., 1981, Purification and properties of the Escherichia coli protein factor required for lambda integrative recombination, J. Biol. Chem., 256:9246.

Ollis, D. L., and White, S. W. ,1987, Structural basis of protein-nucleic acid interactions, Chem. Rev., 87:981.

Prentki, P., Chandler, M., and Galas, D. J., 1987, Escherichia coli integration host factor bends the DNA at the ends of IS1 and in an insertion hotspot with multiple IHF binding sites, EMBO J., 6:2479.

Robertson, C. A., and Nash, H. A., 1988, Bending of the bacteriophage 1 attachment site by Escherichia coli integration host factor, J. Biol. Chem., 263:3554.

Rouviere-Yaniv, J., Germond, J., and Yaniv, M., 1979, Escherichia coli DNA-binding protein HU forms nucleosome-like structure with circular double-stranded DNA, Cell, 17:265.

Rouviere-Yaniv, J., and Kjeldgaard, N., 1979, Native Escherichia coli HU protein is a heterotypic dimer, FEBS Lett., 106:297.

Sayre, M. H., and Geiduschek, E. P., 1988, TF1, the bacteriophage SP01-encoded type II DNA-binding protein, is essential for viral multiplication, J. Virology, 62:3455.

Stenzel, T. T., Patel, P., and Bastia, D., 1987, The integration host factor of Escherichia coli binds to bent DNA at the origin of replication of the plasmid pSC101, Cell, 49:709.

Tanaka, I., Appelt, K., Dijk, J., White, S. W., and Wilson, K. S., 1984, 3Å resolution structure of a protein with histone-like properties in prokaryotes, Nature, 310:376.

QUATERNARY AND TERTIARY STRUCTURES OF ISOMETRIC RNA VIRUSES

J. E. Johnson, Z. Chen, Y. Li, T. Schmidt, C. Stauffacher,
J. P. Wery, M. V. Hosur[1] and P. C. Sehnke

Department of Biological Sciences, Purdue University
West Lafayette, Indiana 47907

[1]Present Address: Neutron Physics Division, Bha Bha Atomic
Research Center, Trombay, Bombay 400085, India

INTRODUCTION

Small spherical RNA viruses infecting members of all five biological
kingdoms have been subjects of biophysical studies for decades (Kaper, 1975;
Argos and Johnson, 1984). Isolated from their hosts, these obligate parasites are homogeneous chemical entities that are now studied at atomic resolution using x-ray crystallography. In the crystal the virus exists in a
resting or dormant state, however, particles released from dissolved crystals are fully infectious. Many viruses form crystalline inclusion bodies
within their hosts (Martelli and Russo, 1977), suggesting that crystalline
aggregates are a natural and stable state for storing virus particles. In
the dormant state the viral capsid protects the nucleic acid from degradation and is essentially a storage protein. During other stages of the virus
life cycle, the capsid protein participates in a variety of functions; some
are listed in Table I. Although relatively few viruses have been investigated at atomic resolution (Table II), a clear pattern has emerged relating
the quaternary structures of different virus capsids (Fig. 1) and the tertiary structures from different virus subunits (Fig. 2). Beyond the striking similarities there are differences in these virus structures that reflect unique strategies evolved for accomplishing required functions. In
this paper the current understanding of the relationship between the structures of simple RNA viruses and their function will be discussed using, as
examples, three structures recently determined in our laboratory (Hosur et
al., 1987; Stauffacher et al., 1987; Chen et al., 1988). An introductory
section on the structure determination of one of these viruses (beanpod
mottle virus) will describe some of the modern methods of virus x-ray crystallography.

Virus Crystallography

Methods for crystallizing purified viruses are identical to those used
to crystallize proteins (McPherson, 1982). Conditions for crystallizing a
variety of spherical viruses were recently reported by Sehnke et al.
(1988a). All the virus particles studied at atomic resolution have particle
diameters of less than 350 Å, although crystals of viruses with diameters
greater than 400 Å have been produced and, in one case, diffraction patterns

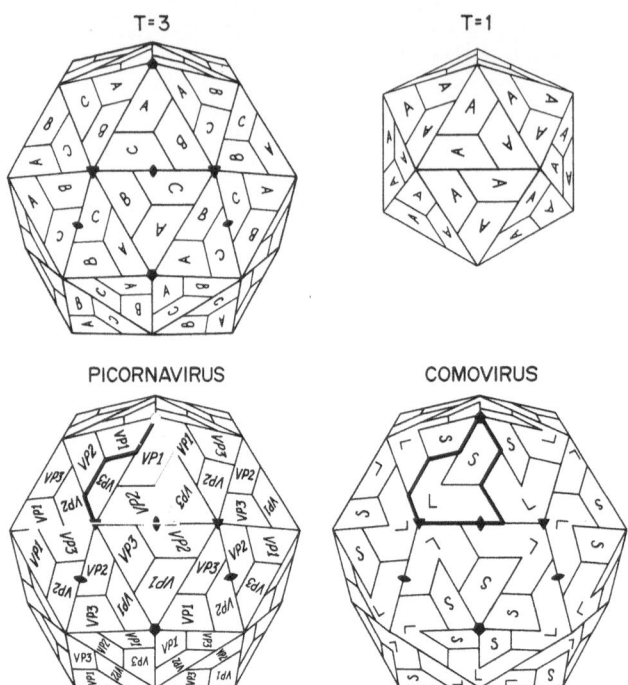

Fig. 1. Different capsids observed in high resolution virus structures.
All the shells have exact icosahedral symmetry. The T=1 shell
contains 60 subunits related by icosahedral symmetry. Each subunit
is represented by a trapezoid which is the approximate shape of the
β-barrel (shown in Fig. 2) when viewed from the top. All subunits
in the T=1 capsid are identical and labeled A for comparison with
the T=3 capsid. The icosahedral asymmetric unit is one subunit and
the threefold symmetry in the central triangle is exact. Satellite
tobacco necrosis virus and reassembly products of SBMV and AMV coat
proteins are the only T=1 structures determined. The asymmetric
unit of the T=3 capsid is the central triangle containing subunits
A, B and C. These subunits have the same amino acid sequence but
are located in slightly different environments. While this central
triangle is similar to the triangle in the T=1 structure, the
threefold axis relating A, B and C is not exact. Note that it re-
lates quasi-sixfold axes (left and right vertices of the triangle)
to a fivefold axis (top vertex). Like the T=1 structure, the T=3
structures are formed by identical subunits with the β-barrel
fold. The T=3 particle will accommodate more than four times the
RNA accommodated by a T=1 shell, assuming subunits of the same size
form both shells. The picornavirus shell, technically a T=1 particle, is closely related to the T=3 shell, being formed by 180
β-barrel domains. The three subunits in the central triangle
(labeled VP1, VP2 and VP3) are, however, distinct proteins. The
similarity in tertiary structure between subunits forming T=3
shells and the 3 subunit types forming picornavirus shells suggests
a triplication of a gene that originally formed T=3 shells, followed by independent evolution of each subunit type. The comovirus
capsid is very similar to the picornavirus capsid, with 180
β-barrels forming the shell; however, there are only two protein
types. The large protein (labeled L in the figure) is composed of
two β-barrel domains (the equivalent of VP2 at the amino terminus
and VP3 at the carboxy terminus) covalently linked together. The
small subunit (S) is a single β-barrel domain.

Table I. Functions of the Capsid Protein of Simple RNA Viruses

_____

Assemble to form a protective shell for the RNA.

Specifically package viral RNA.

May actively participate in virus infection processes, e.g.

    Binding to receptors and mediating cell entry (animal).

    Virion transport within the host (plant).

May mutate to avoid immune recognition in mammalian host.

May function as a primer for viral RNA replication.

_____

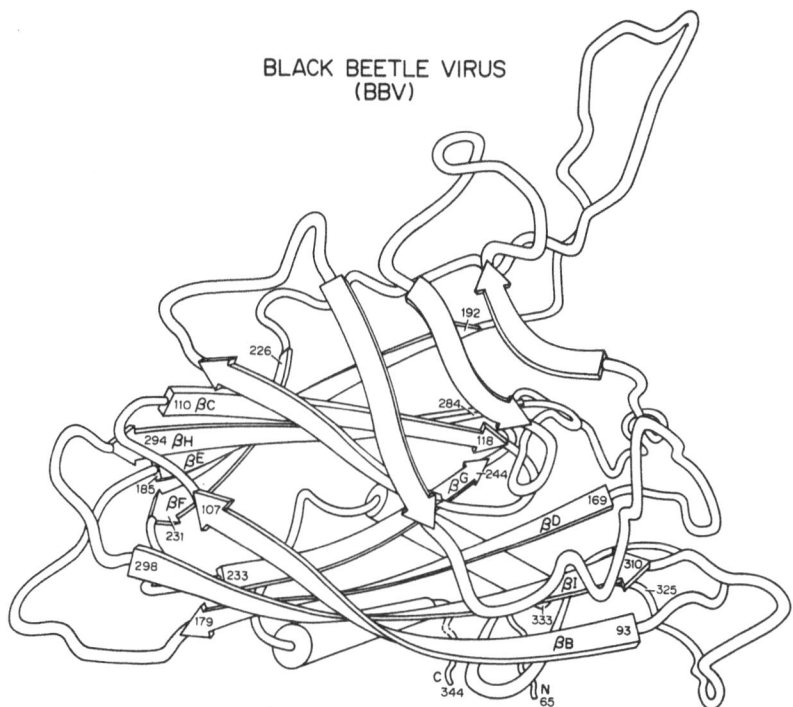

Fig. 2. An example of a viral capsid protein. Like all other icosahedral
viral capsid proteins visualized, the BBV subunit contains an
eight-stranded β-barrel structure. Strands of the barrel are
labeled βB to βI and are in the lower part of the protein. The
subunit is wedge-shaped with the narrow end to the left (short connections between strands) and the broad end to the right where the
strands splay out and large insertions, unique to the BBV subunit,
are accommodated. These insertions contribute the 4 unlabeled
strands on the upper part of the molecule and the extended chain at
the upper right. The visible carboxy terminus lies within the
shell and is preceded by two helices. Residues 1-65 and 345-407
are crystallographically disordered. BBV contains 180 subunits of
this type with the narrow end of the wedge packed against fivefold
or quasi-sixfold axes and the broad end forming trimeric contacts
(at the quasi-threefold axes) with neighboring subunits as illustrated in the T=3 capsid of Fig. 1.

Table II. Virus Structures Determined Using X-ray Crystallography

Plant viruses	Capsid Type (Fig. 3)	Resolution	Reference
TBSV	T = 3	2.9 Å	Olson et al., 1983
Swollen TBSV	T = 3	8.0 Å	Robinson and Harrison, 1982
TCV	T = 3	3.2 Å	Hogle et al., 1986
SBMV	T = 3	2.9 Å	Abad-Zapatero et al., 1980; Silva and Rossmann, 1987
CPMV	como	3.0 Å	Stauffacher et al., 1987
AMV reassembled protein	T = 1	4.5 Å	Fukuyama et al., 1983
SBMV reassembled protein	T = 1	7.5 Å	Erickson et al., 1985
STNV	T = 1	2.5 Å	Jones and Liljas, 1984

Animal viruses			
Rhinovirus 14	picorna	3.0 Å	Rossmann et al., 1985
Poliovirus	picorna	2.9 Å	Hogle et al., 1985
Mengo virus	picorna	3.0 Å	Luo et al., 1987
Polyomavirus	T = 7	22.5 Å	Rayment et al., 1982
Black beetle virus	T = 3	3.0 Å	Hosur et al., 1987

Virus components		
Influenzavirus Haemagglutinin	3.0 Å	Wilson et al., 1981
Influenzavirus Neuraminidase	2.9 Å	Varghese et al., 1983
Adenovirus hexon	2.9 Å	Roberts et al., 1986

TBSV - tomato bushy stunt virus
TCV - turnip crinkle virus
SBMV - southern bean mosaic virus
CPMV - cowpea mosaic virus
AMV - alfalfa mosaic virus
STNV - satellite tobacco necrosis virus

from a crystalline virus of this size extending to 2.8 Å have been recorded (Sehnke et al., 1988a).

The size and symmetry of beanpod mottle virus (BPMV) are typical of many RNA viruses, and its structure was recently determined at 2.8 Å (Chen et al., 1988). The solution of this virus structure will be presented in some detail to illustrate the impact that synchrotron radiation, super computers and specialized crystallographic methods have had on the atomic resolution studies of viruses.

BPMV preparations consist of three major components that are identical in size but differ in mass (Bancroft, 1962). The components are readily separated in a cesium chloride gradient (Fig. 3). The top component is an empty capsid, the middle component encapsulates one portion of the split gene (RNA2, ~3500 nucleotides) and the bottom component contains the other portion (RNA1, ~6000 nucleotides). Both nucleoprotein components are required for infection of susceptible plants. The proportion of components found in a crystal and in the solution from which the crystal was grown are identical. Crystals grown from a mixture of the components display the same high order found in crystals of an isolated component. Crystals grown from isolated components or a mixture of components are isomorphous. All the

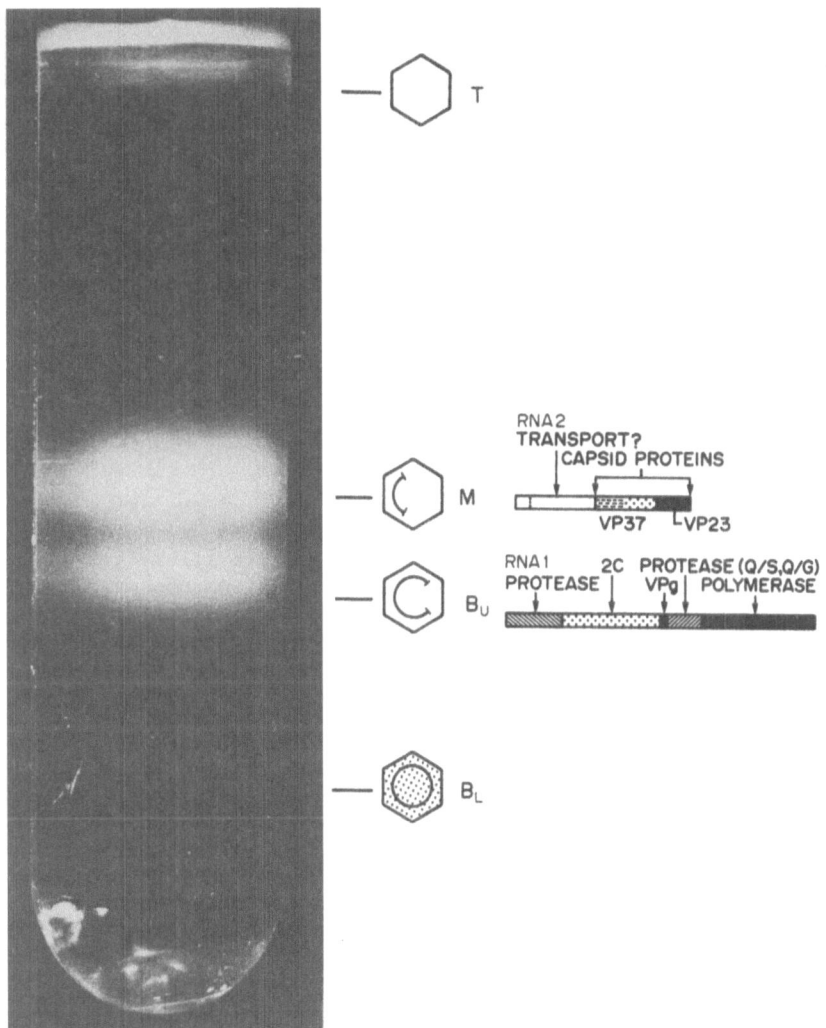

Fig. 3. A cesium chloride gradient with components of beanpod mottle virus banded at different densities. The components are identical isometric particles except for their RNA content. T is the top component (ρ=1.30 g/cc) and it contains no RNA. M is the middle component (ρ=1.40 g/cc) and it contains RNA2 (~3500 nucleotides) which codes for the capsid proteins VP37 and VP23, L and S respectively in Fig. 1. B_U is the bottom upper component (ρ=1.42 g/cc) and it contains RNA1 (~6000 nucleotides) which codes for enzymes associated with viral replication and assembly. B_L is bottom lower component (ρ=1.47 g/cc) and it is identical to B_L in RNA content, but cesium has permeated the capsid and displaced the natural counterions, polyamines, that neutralize the RNA.

Table III. Percentages of Theoretically Possible Reflections Measured
$(I/\sigma(I)\geq 4.0)$

Resolution	Reflections Measured (percent)
30	
	100
15	
	97
10	
	95
7.5	
	93
5.0	
	89
3.5	
	79
3.0	
	42
2.8	

studies described in this report were performed with purified middle component particles.

Crystals of BPMV middle component were grown using the vapor diffusion
method. Solution A, composed of middle component at 15 mg/ml in 0.1 M potassium phosphate buffer at pH 7.0, was mixed 1:1 with solution B, composed
of 2% (w/v) polyethylene glycol 8000 in 0.02 M potassium phosphate buffer,
and the mixture (30 μl total volume in a sitting drop) was allowed to
equilibrate through the vapor phase with 20 ml of solution B in a sealed
container. Elongated crystals appeared in 5-7 days and commonly grew to a
size of 0.4 mm x 0.4 mm x 1.5 mm. The crystals were characterized using 2°
precession photographs. The lattice was orthorhombic with unit cell dimensions, a=311.2, b=284.2, c=350.5 Å. At 22 Å resolution (the limit of the
precession photographs) only even reflections were visible along the axial
lines of the reciprocal lattice suggesting space group $P2_12_12_1$. The volume
of the unit cell was sufficient to accomodate only two virus particles given
the particle diameter of roughly 300 Å and molecular weight of 5.16×10^6
daltons (V_m=3.0 Å3/daltons for 2 particles/cell). After collection and processing of high resolution data, the space group was determined to be
$P22_12_1$. Crystals diffracted beyond 2.8 Å resolution when the Cornell High
Energy Synchrotron Source (CHESS) was used and a full data set was collected
using the oscillation method (Fig. 4) during a 48 hour period. The films
were processed using the program developed by Rossmann (1979) and 182 A/B
film pairs were included in the final data set. There were a total of
1,465,270 accepted reflections; 766,356 whole reflections and 698,914
partial reflections. The entire data set was postrefined (Rossmann et al.,
1979) with film setting parameters (ϕx, ϕy, ϕz) and anisotropic mosaicity
refined for each film. The lattice constants were refined, but constrained
to be the same for all the photographs. The final R_{merge}, using only whole
reflections, was 9.6% for a data set containing 578,044 unique reflections.
Table III shows the percentage of data measured as a function of resolution.

Crystallographic methods for solving virus structures include isomorphous and molecular replacement (Argos and Rossmann, 1980). Both methods
use the noncrystallographic symmetry present in icosahedral capsids. The
first step in solving a virus structure is to determine the orientation of
the icosahedral symmetry axes relative to the crystallographic axes. The

space group of the BPMV crystals requires that one particle twofold symmetry axis be coincident with the lattice twofold axis which is the a axis. In this case, the orientation problem is one-dimensional. Rotation functions (Rossmann and Blow, 1962) were computed for K=72°, 144°, 120° and 180°. The rotation function was computed at 1° intervals about the a axis. Using the convention for spherical polar coordinates defined by Rossmann and Blow (1962) (Fig. 5a), this corresponds to fixing ϕ to -90° and varying ψ from 0 to 90°. Fig. 6 shows the one-dimensional rotation functions computed about each crystal axis. Fig. 7 shows the particle packing with the icosahedra rotated by 8.4° about the x axis from the standard orientation shown in Fig. 5b.

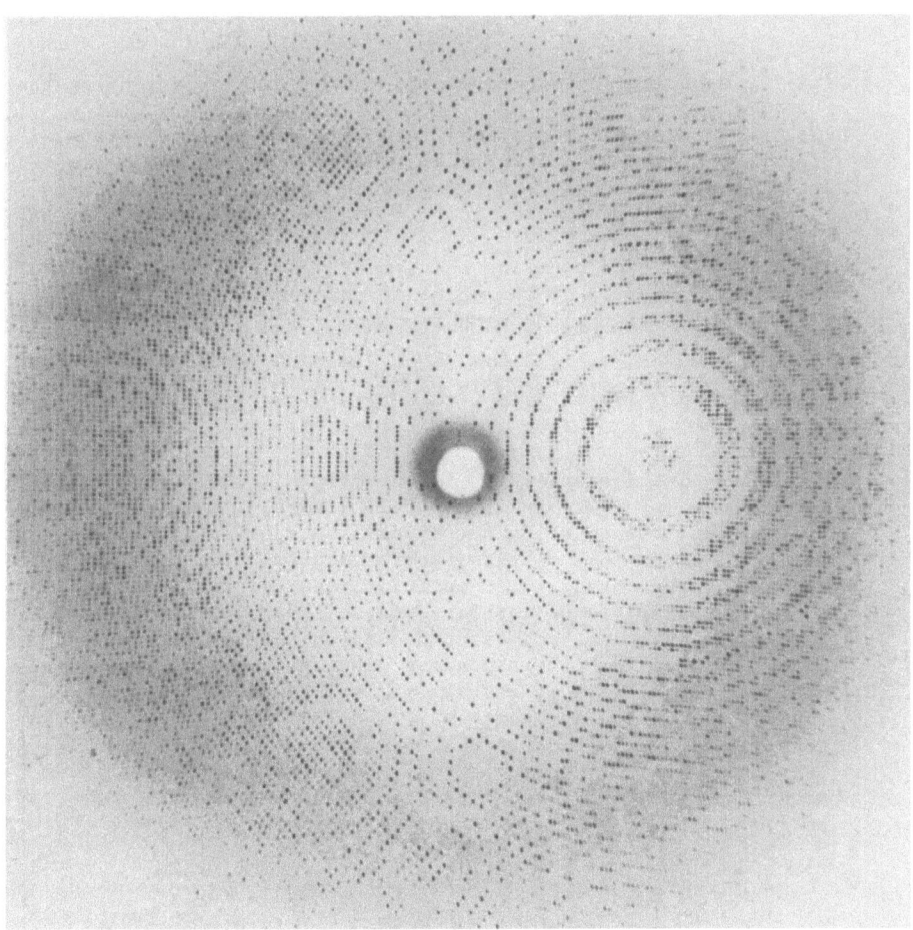

Fig. 4. An oscillation photograph obtained from a BPMV crystal at CHESS. An 0.4° oscillation angle was used with an exposure time of 2'34". The photograph contains 5148 whole reflections and 9890 partial reflections to 2.8 Å resolution. The entire BPMV data set consists of 182 photographs of this quality.

147

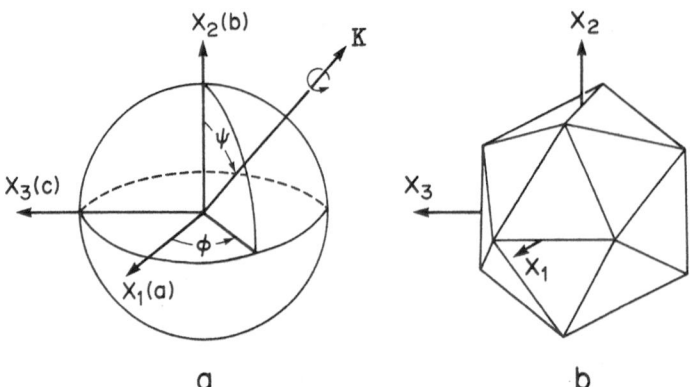

a   b

Fig. 5. (a) The spherical polar coordinate convention for orienting the
rotation vector in the rotation function (Rossmann and Blow,
1962). To test the reciprocal lattice for n fold noncrystallographic symmetry in a particular direction, the vector is oriented by setting the ψ and ϕ angles for the direction and the
reciprocal lattice is rotated by $K=\frac{2\pi}{n}$ about the vector. If the

reciprocal lattice contains the symmetry tested, there will be an
approximate overlap of the original and rotated patterns leading to
a large value of the rotation function which is computed using the
expression

$$R(\psi, \phi, \kappa) = \sum_{h} |F_h|^2 \left[\sum_{P} |F_P|^2 G_{hP} \right] \text{ where } |F_P|^2 \text{ is the rotated}$$

reciprocal lattice and $|F_h|^2$ is the original reciprocal lattice.
Since the points $|F_P|^2$ will not fall directly on lattice points
$|F_h|^2$ unless the rotation corresponds to a crystallographic symmetry operation, G_{hP} is the interference function required for
evaluating the intensity of positions in reciprocal space that do
not fall on reciprocal lattice points. For a crystallographic rotation, the two lattices exactly overlap and the second series be-
comes a single term corresponding exactly to $|F_P|^2$. Here the
rotation function is at its maximum value which is the sum of the
squares of all the intensities.
(b) The orientation of the CPMV particle in the BPMV cell for $\psi=0$;
$\phi=0$; K=0. The CPMV particle corresponds to the BPMV particle orientation when it is rotated by 8.4° about the X_1 axis (see Figs. 7
and 8).

An approximate solution to the translation problem was found by considering the packing of the particles in the unit cell. The only region along the a axis that allows reasonable interparticle contact distances is near x=1/4. The position was refined (Fig. 8) by using structure factors calculated with the Cα coordinates of residues in the cowpea mosaic virus structure (Stauffacher et al., 1987) positioned in the BPMV cell.

The high resolution electron density map was computed by extending the phases from 10 Å resolution, where they were calculated assuming the CPMV Cα positions, to 3Å resolution. Each extension of the phases to higher resolution added a volume of the reciprocal lattice equal to a shell of approximately two crystallographic indices in each direction. For example, the electron density map at 10 Å resolution was averaged over the thirtyfold noncrystallographic symmetry and an envelope was applied that had the

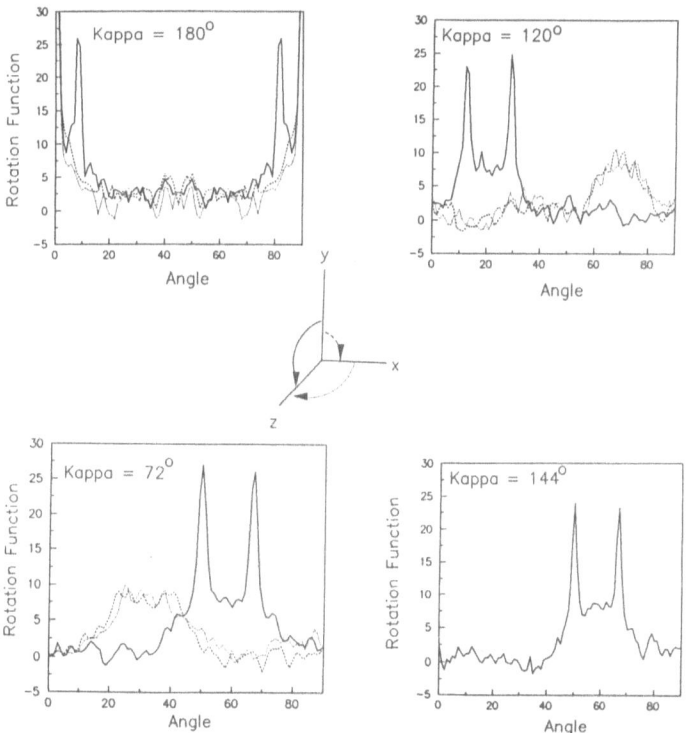

Fig. 6. Rotation functions computed between 12 and 8 Å resolution exploring the reciprocal lattice for symmetry axes expected for an icosahedral particle transform. For each rotation angle K, the rotation vector (Fig. 5a) was set perpendicular to each of the three crystallographic axes and sampled at 1° intervals through 90°. The solid line is perpendicular to x (ϕ=-90°; ψ=0-90°); the fine dashed line is perpendicular to y (ϕ=0-90°; ψ=90°) and the coarse dashed line is perpendicular to z (ϕ=0°; ψ=0-90°). In each case large peaks are found only perpendicular to x. This confirms that the particle twofold axis is coincident with the lattice twofold axis along x. The peaks for different values of K correlate perfectly with an 8.4° rotation of the particle about the X_1 axis shown in Fig. 5b.

exterior shape of the CPMV model (roughly that of an icosahedron) and a 90 Å internal spherical radius. All density not within this envelope was set to zero. The resulting electron density map was Fourier transformed to produce

F^{calc} and α^{calc}. The α^{calc}, F^{obs} and weights, $\omega_F = e^{-\frac{(F^{obs} - F^{calc})}{F^{obs}}}$ were

used to compute a new electron density map. The process was cycled at 10 Å resolution until the average phase change was less than 5° from the previous cycle. The electron density map was then back transformed at 9.5 Å resolution. Calculated amplitudes in the shell between 9.5 and 10Å resolution were very small and would have been zero if the map were not perturbed by averaging and imposition of the envelope. A new electron density map was then computed at 9.5 Å resolution using F^{obs}, α^{calc} and ω_F, and the process cycled to convergence. Normally the phases converged after two or three cycles. The process was continued to 3 Å resolution using a total of 40 extension steps and 121 cycles of averaging and Fourier transformation. At

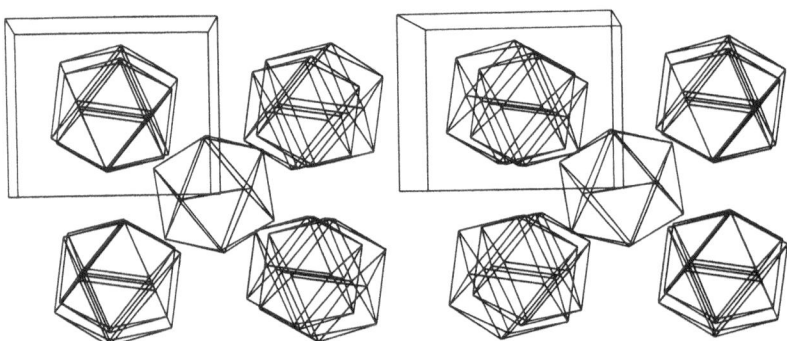

Fig. 7. Particle packing in the BPMV unit cell. The particle orientation was determined from the rotation function (Fig. 6) and the position was determined from the R-factor search using the CPMV Cα coordinates as a model. The packing is pseudo body centered cubic. The crystal axes are oriented as shown in Fig. 5a.

3 Å resolution each cycle executed for 75 minutes on a Cyber 205 computer. Fig. 9 shows the distribution of correlation coefficients as a function of resolution. Fig. 10 shows a typical region of the electron density map at 3.8 Å resolution. The polypeptide chains were easily followed at this resolution, and some of the larger side chains were clearly visible.

The procedure described has also been used to solve Mengo virus using human rhinovirus 14 as a low resolution phasing model (Luo et al., 1987). Phases calculated at 5 Å resolution or lower using multiple isomorphous replacement have been used as a starting point to extend phases to 3.0 Å resolution for human rhinovirus 14 (Rossmann et al., 1985), poliovirus (Hogle et al., 1985), and black beetle virus (Hosur et al., 1987). Experiments in which phases are extended from an electron microscopy model at 30 Å resolution (Olson et al., 1987) are now underway in our laboratory. If these experiments are successful, high resolution structures of viruses may be very straightforward to determine.

Quaternary Structures

Genetic economy dictates that capsids of small RNA viruses must be built from multiple copies of a few subunit types. The quaternary structures found in these shells are based on the symmetry of the icosahedron (Crick and Watson, 1956; Caspar and Klug, 1962). Fig. 1 shows schematically the relationship between shells formed by small RNA viruses. The T=1 shell contains 60 identical subunits, the T=3 shell contains 180 identical subunits, and the picornavirus shell, which is pseudo T=3, contains 60 copies each of three different subunits. Although the primary structures of the three protein types in picornaviruses are not similar, the tertiary

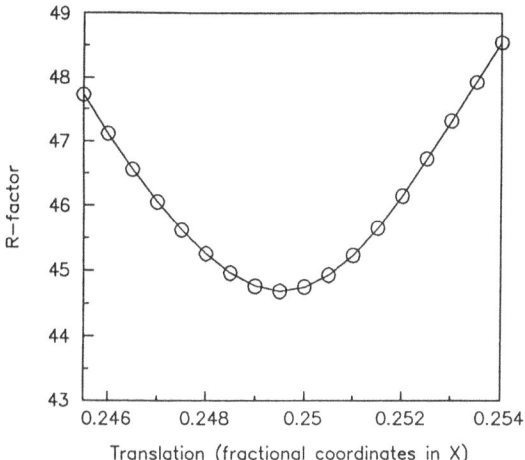

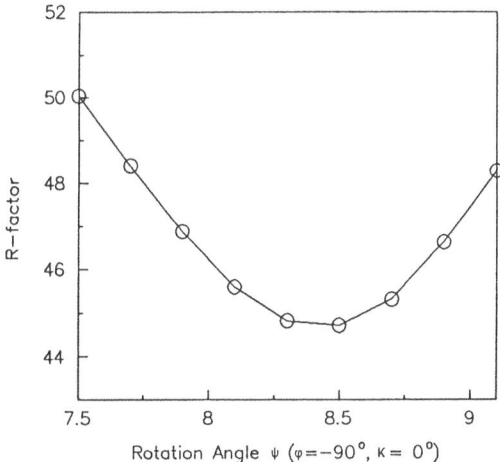

Fig. 8. R-factor searches with data between 12 and 8 Å resolution comparing structure factors calculated for the CPMV Cα coordinates placed in the BPMV cell with the BPMV observed structure factors. The orientation of the particle obtained from the rotation function was used for the translation search (top) and this was confirmed by an R-factor search for orientation (bottom) with the particle center positioned a x=0.2495, y=0, z=0.

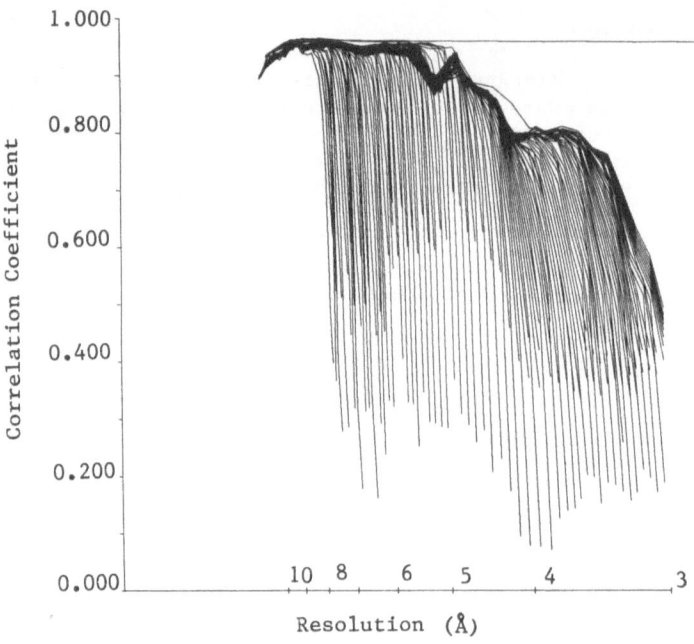

Fig. 9.   Resolution extension of BPMV phases by real space averaging

structures are similar. At low resolution the capsid appears to be a T=3 structure. The comovirus capsid is closely similar to the picornavirus shell, except VP2 and VP3 are covalently linked together as a single poly-peptide, thus there are two protein types (L and S) but three domains. Fig. 11 shows schematically the organization of the T=4 shell based on the low resolution structure of nudaurelia capensis β virus (Finch et al., 1974; Olson et al., 1987). In this virus the icosahedral asymmetric unit contains four identical subunits.

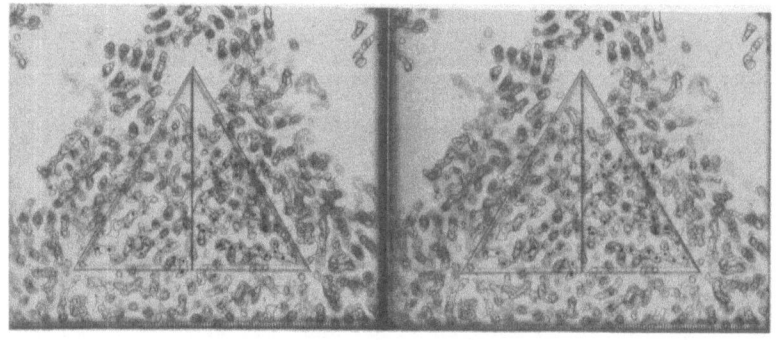

Fig. 10.   3.8 Å electron density map

The quasi-equivalent shells demonstrate a strategy that has evolved for altering similar (quasi-equivalent) contacts between the same subunits to produce different structures. Inspection of Fig. 1 shows that in the T=3 shell contacts between C subunits at the icosahedral twofold axes (bottom of the central triangle) are similar to A/B contacts with neighboring triangles on the left and right of the central triangle. The shape of T=3 shells is due, in part, to the differences in the angle formed at the base (dihedral 180° angle) and the angles formed at the sides (dihedral angle 136°) between the triangles. The chemical interactions at the two joint types are very different. The flat contact (C/C) cannot bend because a polypeptide segment is inserted between two hydrophobic surfaces just inside a line of contact between subunits at the bottom of the triangle. The polypeptide is not present at the A/B interactions, and the subunits pivot about the line of contact which allows the hydrophobic surfaces to contact directly with each other. The mechanism for achieving quasi-equivalent interaction is similar for all the T=3 structures determined, however, the strategies used by plant viruses and insect viruses are different. These will be discussed under tertiary structures.

The low resolution structure of the T=4 capsid shows a quaternary organization that can be explained in terms of trimer interactions similar to those observed in the T=3 structures. All contacts between trimers are either flat like the C/C contacts in the T=3 capsids, or bent like the A/B contacts. The mechanism for achieving the two types of joints may be similar for T=4 structures and this will be determined from the high resolution structure.

A dramatically different type of particle is formed by ilarviruses (Francki, 1985). The type member of this group is tobacco streak virus (TSV). The subunits in these viruses "coat" the RNA, forming asymmetric particles proportional to the size of the nucleic acid encapsulated. The protein-protein interactions in TSV capsids are minimal. Subunits treated with trypsin lose the portion of the protein interacting with RNA (residues 1-87) and the remaining portion of the subunit (residues 88-237) behaves like a globular protein that crystallizes without assembling (Sehnke and Johnson, 1988).

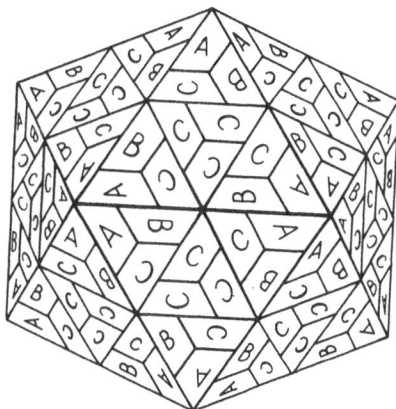

Fig. 11. The organization of a T=4 icosahedral shell. There are 240 identical subunits in the shell, labeled A, B and C for comparison to the T=3 shell (Fig. 1).

Tertiary Structures of Viral Capsids

Table I lists functions of viral capsid proteins. Although many of these functions are common to all viruses it is clear from biochemical and structural studies that different viruses often employ different tactics to perform the same function. The subunit tertiary structure identified with some of these functions are described.

Assembly

The polypeptide chains of small spherical RNA virus subunits are folded into anti-parallel β-barrels containing at least eight strands. The barrel has the topology of the "jelly roll" (Richardson, 1979). The highly conserved secondary structures and the shapes of the subunits (roughly that of the trapezoids representing subunits in Fig. 1) suggest their importance for assembly. The details of the quaternary structures of all the viruses are not the same, but the narrow ends of the wedge are always in contact about the five and quasi-sixfold axes. The chemical surfaces of the subunits are complimentary and lead to an assembling system. The nature of the intersubunit stabilizing forces is different for different viruses, and in many viruses subunit interactions with the RNA is an essential feature of assembly (Erickson and Rossmann, 1982). Both the tertiary and primary structures are exceedingly important for proper assembly, but the tertiary structure is far more conserved than the primary structures of viral subunits. An important difference has been found in the strategy for assembling T=3 shells in insect viruses (Hosur et al., 1987) and the T=3 plant virus structures studied (Abad-Zapatero et al., 1980; Harrison et al., 1978; Hogle et al., 1986). The architecture of the black beetle virus (BBV) strongly resembles that of the plant viruses, and the presence of a protein segment, often called an "arm", in the icosahedral (C/C contact) but not in the quasi-twofold joints (A/B contact), is a feature common to all four viruses (vide supra), but the details in BBV are different. In the case of the plant viruses, the arm is contributed by the C subunits and constitutes an additional strand (βA), which is connected directly to βB in the β-barrel; this strand is invisible in the plant virus A and B subunits. In BBV, however, the arm in the occupied twofold groove is disordered at both termini and is not connected to the βB strand of the C subunit. The insect virus subunit undergoes an autocatalytic cleavage near the carboxy terminus (residue 363) of the polypeptide during particle morphogenesis (Hosur et al., 1987). This cleavage is required to convert the unstable provirion to the stable mature particle (Gallagher and Rueckert, 1988) and the inserted polypeptide arm is almost certainly part of the subunit carboxy terminus. Evidence from the structure and from sequence homologies between related viruses (Kaesberg, personal communication) suggests that the active site for the autocatalytic cleavage is located following βI in a portion of the polypeptide that lies below the β-barrel (Fig. 2). The presence of histidine, asparatic acid and serine residues in the region of the putative active site suggests that the cleavage mechanism in the virus may be similar to that of a serine protease (Kraut, 1977). Helical regions in the polypeptide following the β-barrel fold and within the shell have not been observed in any other viruses, adding support to the hypothesis that this addition to the tertiary structure of the subunit in these insect viruses has evolved as part of a unique assembly mechanism.

The catalytic site for proteolysis and structures for other functions of viral capsid proteins appear to be added to the β-barrel fold suggesting that the viral subunit structures possess a constant portion (the β-barrel) and variable portions that are added at the amino and carboxy termini of the subunits or inserted between strands of the sheet structure.

154

RNA Packaging

Protein-RNA interactions in viruses had not been visualized until recently. The RNA is a linear molecule, not possessing the symmetry of the capsid and the electron density for this portion of the virus is normally not visible. Many plant viruses and some insect viruses possess a basic, histone-like, amino terminus (Savithri and Erickson, 1983; Dasmahaptra et al., 1985) and this portion of the subunit interacts with the RNA in these viruses. The amino termini of the subunits are not visible in the electron density map because they adapt to various conformations of the RNA and thus are also disordered. The mammalian picornaviruses and plant comoviruses contain polyamines in their capsids (Koch and Koch, 1985; Nickerson and Lane, 1977) and the subunits do not have basic amino termini (van Weezenbeck et al., 1983; Callahan et al., 1985). The RNA in picornaviruses is disordered, just as in the other viruses studied (Rossmann et al., 1985), but nearly 20% of the RNA in comoviruses studied conforms to the symmetry of the capsid and these regions are well ordered (Chen et al., 1988). The interactions with the protein involve only a few direct hydrogen bonds to the polynucleotide. The surface of the single-strand polynucleotide forms a complimentary van der Waal's contact with the inner portion of the N terminal β-barrel of the large subunit that lies in the C position of the pseudo T=3 shell (Fig. 2). Most of the tertiary structure of the protein interacting with the RNA is part of the β-barrel, thus the adjustment in the subunit for binding RNA is at the primary structure level. The amino terminus of the large subunit extends toward its twofold related counterpart and also binds to the RNA in this region. This type of amino terminal structure has not been visualized in other virus subunits and may be part of the structure specifically designed for RNA packaging in comoviruses.

Receptor Binding

Rossmann and his colleagues (Rossmann and Rueckert, 1987) proposed that a "canyon" on the surface of rhinovirus 14 functions as a receptor binding site. The bottom of the canyon cannot be an epitope for antibody binding because this area is sterically inaccessible to the F_{ab} portion of an immunoglobulin. Residues in this part of the structure are highly conserved and are under no immune pressure for mutagenesis like much of the virus surface. The canyon is formed both by the quaternary structure of the capsid and insertions between stands of the β-sheet that form the barrel. The canyon is not observed in plant comoviruses. Although the capsid quaternary structure is nearly identical to the picornaviruses, the insertions between strands of the β-sheet are not present, thus the area corresponding to the canyon floor would be accessible to antibodies (Stauffacher et al., 1987). Plant viruses apparently do not bind specifically to cell surfaces (Shaw, 1985) and there is no circulating immune system.

Insect viruses bind to receptors, but there is no circulating immune system in insects and hence, no need to protect a receptor binding site from immune pressure. The structure of black beetle virus shows no protected canyon or pit as observed in mammalian viruses (Hosur et al., 1987). The BBV subunit contains a protrusion formed by three large insertions between strands of the β-barrel. Insertions from subunits related by a quasi-three-fold axis form an intricate terraced pyramid that may be the receptor binding site in this insect virus. Sequence analysis of different nodavirus subunits and biological studies of receptor binding to different cell types will resolve the questions regarding receptor binding in these viruses.

Mutable Regions for Avoiding Immune Recognition

The epitopes for antibody binding on mammalian viruses occur on loops on the outer surface of the particle. These regions are formed by extended

insertions between strands of the barrel and they play no apparent structural role. Amino acids in such regions are readily altered as evidenced by
the variety of escape mutants that contain amino acid changes in these
regions, and hence, avoid neutralization by particular monoclonal antibodies
(Sherry et al., 1986). In most cases the mutated virus is otherwise indistinguishable from the native virus. It is likely that these loops have
evolved to permit mammalian viruses to form different serotypes and hence
avoid immune recognition. Such an hypothesis is supported by the virtual
absence of such loops in the plant and insect viruses, where they would have
no function.

The picornavirus capsid, containing three subunit types, is better
suited to mutate rapidly than a T=3 shell. A mutation in the coat protein
of a T=3 virus introduces changes in three locations in the asymmetric
structure unit; however, a similar mutation in a picornavirus shell (pseudo
T=3) makes an alteration in only one of the three subunit positions. This
permits pseudo T=3 viruses to tolerate mutations that otherwise might be
lethal in a T=3 capsid and thereby allows greater potential for evolution.
The pseudo T=3 shell of picornaviruses may provide an important key to their
success in coping with immune surveillance; the number of known picornavirus
serotypes exceeds 200. In plants and insects that possess no immune system
most spherical viruses have a T=3 capsid and there are no loops corresponding to the epitopes observed in the tertiary structure of mammalian viruses.

Subunit Role in RNA Replication

Capsid subunits are required for RNA replication in two groups of plant
viruses, ilarviruses (TSV is the type member) and in alfalfa mosaic virus.
The actual role played by the subunit has not been determined, but in both
cases the amino terminus of the subunit appears to be the functional component of the polypeptide. The amino terminus of the TSV capsid protein contains a zinc binding amino acid sequence motif (Sehnke et al., 1988b) that
is present in transcription factor IIIA (Miller et al., 1985) and in other
proteins associated with the regulation of gene expression (Berg, 1986).
The inability of the TSV subunit to form symmetric shells (vide supra) may
be related to its role in gene expression. Viruses with more specialized
roles for gene products may have capsid formation associated with one protein and gene regulation associated with another. If this is the case, the
TSV subunit is either a primitive capsid protein where these functions have
not yet separated, or the subunit is a unique protein formed by a recombination of separate genes. We are investigating the TSV subunit using both
biochemical techniques and crystallography in an attempt to answer these
questions (Sehnke and Johnson, 1988).

CONCLUSIONS

The evidence for an evolutionary relationship among small RNA viruses
is strong. Although sequence similarities in unrelated viral capsid subunits is not obvious, the quaternary structures of different viruses are
remarkably similar to each other. Tertiary structures of subunits from
different viruses show similar folds with identical connectivity. The
unique functional roles required of the capsid protein in different viruses,
however, lead to specialized structures for performing these functions.

Nonstructural proteins of RNA viruses show remarkable sequence homologies that are independent of the virus family. Ahlquist et al. (1985)
showed that proteins required for RNA replication in different virus groups
have regions of highly conserved amino acids and they suggest that all RNA
viruses may have evolved from a primitive protovirus. They point out that
in adaptation to new hosts or under exposure to animal immune systems, virus

capsid protein mutations would be readily accepted while the mechanism for RNA replication would be slow in changing and intolerant of alterations in amino acids of the enzyme catalyzing this process. The results obtained from virus crystallography show that while amino acid changes in the capsid proteins are readily accomodated, the subunit fold has been highly conserved. Indeed the spherical viral subunit is an ideal "laboratory" for understanding the relationship between primary and tertiary structures of proteins. In addition, they provide a system that contains a constant portion of highly conserved tertiary structure with variable positions of structure added for specific functions. In this sense viral subunit structures resemble immunoglobulins which also contain constant and variable domains. As the number of subunit primary and tertiary structures increase it will be possible to better establish the relation between structure and function in viral subunits and the evolutionary relationships among viruses and possibly their origin from cellular proteins (Hosur et al., 1987).

REFERENCES

Abad-Zapatero, C., Abdel-Meguid, S. S., Johnson, J. E., Leslie, A. G. W., Rayment, I., Rossmann, M. G., Suck, D., and Tsukihara, T., 1980, Structure of southern bean mosaic virus at 28 Å resolution, Nature (London), 286:33.

Ahlquist, P., Strauss, E. G., Rice, C. M., Strauss, J. H., Haseloff, J., and Zimmern, D., 1985, Sind bis virus proteins ns P1 and ns P2 contain homology to nonstructural proteins from several RNA plant viruses, J. Virol., 53:536.

Argos, P., and Johnson, J. E., 1984, Chemical stability in simple spherical plant viruses, in: "Biological Macromolecules and Assemblies," F. A. Jurnak and A. McPherson, eds., Wiley & Sons, New York.

Argos, P., and Rossmann, M. G., 1980, Molecular replacement method, in: "Theory and Practice of Direct Methods in Crystallography," M. F. C. Ladd and R. A. Palmer, eds., Plenum Press, New York.

Bancroft, J. B., 1962, Purification and properties of bean pod mottle virus and associated centrifugal and electrophoretic components, Virology, 16:419.

Berg, J., 1986, Potential metal-binding domains in nucleic acid binding proteins, Science, 232:485.

Callahan, P. L., Mizutani, S., and Colonno, R. J., 1985, Molecular cloning and complete sequence determination of RNA genome of human rhinovirus type 14. Proc. Natl. Acad. Sci. U.S.A., 82:732.

Caspar, D. L. D., and Klug, A., 1962, Physical principles in the construction of regular viruses, Cold Spring Harbor Symp. Quant. Biol., 27:1.

Chen, Z., Stauffacher, C., Li, Y., Schmidt, T., Kamer, G., Shanks, M., Lomonossoff, G., and Johnson, J. E., 1988, Protein-nucleic acid interactions in a spherical virus: The structure of beanpod mottle virus at 3.0 Å resolution, Science, in preparation.

Crick, F. H. C., and Watson, J. D., 1956, Structure of small viruses, Nature (London), 177:473.

Dasmahapatra, B., Dasgupta, R., Ghosh, A., and Kaesberg, P., 1985, Structure of the black beetle virus genome and its functional implications, J. Mol.Biol., 182:183.

Erickson, J., and Rossmann, M. G., 1982, Assembly and crystallization of a T=1 icosahedral particle from trypsinized southern bear mosaic virus coat protein, Virology, 116:128.

Erickson, J. W., Silva, A. M., Murthy, M. N. R., Fita, I., and Rossmann, M. G., 1985, The Structure of a T=1 icosahedral empty particle from southern bean mosaic virus, Science, 229:625.

Finch, J. T., Crowther, R. A., Hendry, D. A., and Struthers, J. K., 1974, The structure of nudaureha capersis virus: the first example of a capsid with icosahedral surface symmetry T=4, J. Gen. Virol., 24:191.

Francki, R., 1985, The viruses and their taxonomy, in: "The Plant Viruses, Vol. 1, Polyhedral Virions with Tripartite Genomes," R. Francki, H. Fraenkel-Conrat, and R. Wagner, eds., Plenum Press, New York.

Fukuyama, K., Abdel-Meguid, S. S., Johnson, J. E., and Rossmann, M. G., 1983, Structure of a T=1 aggregate of alfalfa mosaic virus coat protein sent at 4.5 Å resolution, J. Mol. Biol., 167:873.

Gallagher, T., and Rueckert, R. R., 1988, Assembly-dependent maturation cleavage in provirions of a small icosahedral insect ribovirus, J. Gen. Virol., in press.

Harrison, S. C., Olson, A. J., Schutt, C. E., Winkler, F. K., and Bricogne, G., 1978, Tomato bushy stunt virus at 2.9 Å resolution, Nature (London), 276:368.

Hogle, J. M., Chow, M., and Filman, D. J., 1985, Three-dimensional structure of poliovirus at 2.9 Å resolution, Science, 229:1358.

Hogle, J. M., Maeda, A., and Harrison, S. C., 1986, Structure and assembly of turnip crinkle virus. I. X-ray crystallographic structure analysis at 3.2 Å resolution, J. Mol. Biol., 191:625.

Hosur, M. V., Schmidt, T., Tucker, R. C., Johnson, J. E., Gallagher, T. M., Selling, B. H., and Rueckert, R. R., 1987, Structure of an insect virus at 3.0 Å resolution, Proteins, 2:167.

Jones, T. A. and Liljas, L., 1984, Structure of satellite tobacco necrosis virus after crystallographic refinement at 2.5 Å resolution, J. Mol. Biol., 177:735.

Kaper, J. M., 1975, "The Chemical Bases of Virus Structure, Disassociation and Reassembly," American Elsevier, New York.

Koch, F., and Koch G., 1985, "The Molecular Biology of Poliovirus," Springer-Verlag, New York.

Kraut, J., 1977, Serine protease: structure and mechanism of catalysis, in: "Annual Review of Biochemistry," Academic Press, New York.

Luo, M., Vriend, G., Kamer, G., Minor, I., Arnold, E., Rossmann, M. G., Boege, U., Scraba, D. G., Duke, G. M., and Palmenberg, A. C., 1987, The atomic structure of mengo virus at 3.0 Å resolution, Science, 235:182.

Martelli, G. P., and Russo, M., 1977, Plant virus inclusion bodies, Adv. Virus Res., 21:175.

McPherson, A., 1982, "Preparation and Analysis of Protein Crystals," Wiley & Sons, New York.

Miller, J., McLachlan, A., and Klug, A., 1985, Repetitive zinc-binding domains in the protein transcription factor. III. A from Xenopus oocytes, EMBO J., 4:1609.

Nickerson, K. W., and Lane, L. C., 1977, Polyamine content of several RNA plant viruses, Virology, 81:455.

Olson, A. J., Bricogne, and Harrison, S. C., 1983, Structure of tomato bushy stunt virus. IV. The virus particle at 2.0 Å resolution, J. Mol. Biol., 171:61.

Olson, N. H., Baker, T. S., Bomu, W., Johnson, J. E., and Hendry, D. A., 1987, The three-dimensional structure of frozen-hydrated nudaurelia capersis virus, in: "Proc. 45th Ann. Meet. Elec. Microscopy Soc. of Am.," San Francisco Press, California.

Rayment, I. A., Baker, T. S., Caspar, D. L. D., and Murakami, W., 1982, Polyoma virus capsid structure at 22.5 Å resolution, Nature, 295:110.

Richardson, J. S., 1979, The anatomy and taxonomy of protein structure, Adv. Prot. Chem., 34:167.

Roberts, M. M., White, J. L., Grutter, and Burnett, R. M., 1986, Three-dimensional structure of the adenovirus major coat protein hexon, Science, 232:1148.

Robinson, I. K., and Harrison, S. C., 1982, Structure of the expanded state of tomato bushy stunt virus, Nature, 297:563.

Rossmann, M. G., 1979, Processing oscillation diffraction data for very large unit cells with an automatic convolution technique and profile fitting, J. Appl. Crystallogr., 12:225.

Rossmann, M. G., and Blow, D. M., 1962, The detection of sub-units within the crystallographic asymmetric unit, Acta Crystallogr., 15:24.

Rossmann, M. G., and Rueckert, R. R., 1987, What does the molecular structure of viruses tell us about viral functions? Microbiol. Sci., 4:206.

Rossmann, M. G., Leslie, A. G. W., Abdel-Meguid, S. S., and Tsukihara, T., 1979, Processing and post-refinement of oscillation camera data, J. Appl. Crystallogr., 12:570.

Rossmann, M. G., Arnold, E., Erickson, J. W., Frankenberger, E. A., Griffith, J. P., Hecht, H. J., Johnson, J. E., Kamer, G., Luo, M., Mosser, A. G., Rueckert, R. R., Sherry, B., and Vriend, G.,1985, Structure of a human common cold virus and functional relationship to other picornaviruses, Nature (London), 317:145.

Savithri, H. S., and Erickson, J. W., 1983, The self-assembly of the cowpea strain of southern bean mosaic virus: formation of T=1 and T=3 nucleo protein particles, Virology, 126:328.

Sehnke, P. C., and Johnson, J. E., 1988, Crystallization of a proteolytically modified subunut of tobacco streak virus, Virology, in preparation.

Sehnke, P. C., Harrington, M., Hosur, M. V., Li, Y., Usha, R., Tucker, R. C., Bomu, W., Stauffacher, C. V., and Johnson, J. E., 1988a, Crystallization of viruses and virus proteins, J. Crystal Growth, in press.

Sehnke, P. C., Mason, A., Hood, S. J., Lister, R. M., and Johnson, J. E., 1988b, A zinc-finger type binding domain in tobacco streak virus coat protein, unpublished.

Shaw, J. G., 1985, Early events in plant virus infections, in: "Molecular Plant Virology, Vol. II," J. W. Davies, ed., CRC Press, Boca Raton, Florida.

Sherry, B., Mosser, A. G., Colonno, R. J., and Rueckert, R. R., 1986, Use of monoclonal antibodies to identify four neutralization immunogens on a common cold picornavirus, human rhinovirus 14, J. Virol., 57:246.

Silva, A. M., and Rossmann, M. G., 1987, Refined structure of southern bean mosaic virus at 2.9 Å resolution, J. Mol. Biol., 197:69.

Stauffacher, C. V., Usha, R., Harrington, M., Schmidt, T., Hosur, M. V., and Johnson, J. E., 1987, The structure of cowpea mosaic virus at 3.5 Å resolution, Crystallogr. Mol. Biol., 126:293.

van Wezenbeek, P., Verver, J., Harmsen, J., Vos, P., and van Kammen, A., 1983, Primary structure and gene organization of the middle component RNA of cowpea mosaic virus, EMBO J., 2:941.

Varghese, J. N., Laver, W. G., and Colman, P. M., 1983, Structure of the influenza virus glycoprotein antigen neuraninidase at 2.9 Å resolution, Nature, 303:35.

Wilson, I. A., Skehel, J. J., and Wiley, D. C., 1981, Structure of the lae magglutinin membrane glycoprotein of influenza virus at 3 Å resolution, Nature, 289:366.

IRON CORE FORMATION IN FERRITINS

Elizabeth C. Theil[1] and Dale E. Sayers[2]

Departments of Biochemistry[1] and Physics[2]
North Carolina State University
Raleigh, NC 27695

INTRODUCTION

Iron is the most abundant transition metal in biology. Proteins containing iron are important for DNA synthesis, respiration, photosynthesis,
nitrogen fixation and the transport and activation of dioxygen. Two properties of Fe(III) in aqueous solution lead to the formation of large, in-
soluble aggregates of iron and oxygen: first, the pKa of a proton in water
coordinated to Fe(III) is ca 3; second, after proton loss, the conjugate
bases form stable oxygen bridges to each other with the elimination of
water. Thus unless ligands other than water are coordinated to Fe(III), the
solubility of Fe(III) is very low under physiological conditions (ca 10^{-18} M)
(Biedermann and Schindler, 1957). Animals, plants and microorganisms use
ferritin, an iron protein complex, to accumulate reserves of iron suffi-
cently high for the synthesis of proteins that have iron at the active
center. Ferritin also provides a site for the detoxification of excess iron
which enters an organism or cell when the normal barriers to controlled iron
uptake are breached.

Iron reserves in ferritin are in the form of polynuclear iron aggregates similar to those that form spontaneously in water at pH 7. However,
the iron aggregate is inside a coat of protein which is constructed from 24
subunits, ca 20 kDa each, which are of similar or identical sequence. The
iron core varies in size (up to 4500 iron atoms per molecule of protein),
phosphate content, and both long and short range order. Differences in the
structure of ferritin iron cores are related in an unknown way to the source
of ferritin (structure of the protein coat) and the cellular environment
(Theil, 1983 and 1987). However, the coincidence of variations in iron core
structure, protein coat structure and the function of the iron stored in
ferritin (intracellular use, extracellular use, or detoxification) suggests
an interdependence. Ferritin provides control over the intracellular utilization of iron by the cell through its influence on rates of iron deposition into or released from the iron core.

A major role of the protein coat in the formation of the ferritin iron
core appears to be the spatial orientation, of the first Fe(III)-oxo
clusters inside the protein coat. With the correct orientation of the initial nucleus, subsequent spontaneous additions of Fe(III) atoms to the grow-
ing core will automatically fill the hollow center of the protein coat. The
process begins with the binding of Fe(II) to the protein coat followed by

oxidation and polynuclear complex formation. A variety of spectroscopic techniques has been used to study the early stages of iron core formation including UV-vis (Macara et al., 1973; Treffry and Harrison, 1984; Crichton et al., 1980) and EPR (Chasteen and Theil, 1982; Wardeska, 1986). However, the application of X-ray absorption spectroscopy (XAS), using synchrotron radiation as a source to examine the properties of iron in ferritin directly under physiological conditions and in mononuclear or polynuclear arrays has provided novel insights both about ferritin iron core formation (Sayers et al., 1983; Yang et al., 1987; and Rohrer et al., 1988) and also about the structure of the iron core and models for the core (Theil et al., 1979; Mansour et al., 1985; and Yang et al., 1986). The results of XAS studies during the early stages of iron core formation will be reviewed in terms of current ideas about the process.

RESULTS AND DISCUSSION

Fe(III)-Protein Complexes

Ferritin can be reconstituted in aqueous, buffered solutions of the protein coats and Fe(II). When the full core is present, the numbers of iron atoms are so large that the average environment, detected by XAS, is that of the inorganic core phase. To probe the environment of iron attached to the protein, complexes were formed by oxidizing a small number of Fe(II) atoms below the number which appeared to saturate the protein; earlier studies had shown that the maximum number of Fe(II) atoms bound by the protein coat appeared to be 8-12 (Chasteen and Theil, 1982; Wardeska, et al., 1986). It is important to note that XAS measurements were made 24 hours after the admission of air to the Fe(II)-protein complex, since migration of iron atoms and reorganization of environments appear to occur during that period (Theil and Chasteen, 1982; Treffry and Harrison, 1986) and, as will be discussed below, Fe(II) can persist inside the protein coat for as long as 16-24 hours (Rohrer et al., 1987).

EXAFS analysis of the environment of iron formed when 10 Fe(II) atoms were added to the protein coats of horse spleen ferritin and allowed to oxidize indicated that each Fe(III) was attached to a protein ligand (carboxylate-like)(Sayers, Theil and Rennick, 1983; Yang et al., 1987). However, the Fe(III) atoms were not solitary, in contrast to Fe(II), but were bridged to ca 2 other Fe(III) atoms which were, in turn, attached to the protein. The existence of such a cluster explains the regeneration of vanadyl(IV) sites when ferritin protein coats saturated with Fe(II) were exposed to vanadyl (IV) after the oxidation of Fe(II).

Analysis of the Fe(III)-protein complex by Mossbauer spectroscopy, when [57]Fe was used, confirmed the existence of the cluster. The blocking temperature was much lower than for full ferritin iron cores (8 K rather than 38 K) indicating a smaller magnetic domain. Detection of magnetic coupling by Mossbauer spectroscopy when only 2 Fe neighbors were detected by EXAFS analysis suggests that the cluster may be very asymmetric. The location of the Fe(III) oxo cluster(s) in the protein coats of ferritin is not known. However, the presence of three carboxylate ligands in each subunit at the dimer interface in all ferritins for which sequence information is available (mammals, birds and amphibia reviewed by Theil, 1987), the presence of a flexible groove on the interior surface of the molecule at that site (Rice et al., 1983) and the effect of crosslinks between subunit pairs on iron core formation (Mertz and Theil, 1982) make the dimer interface an attractive candidate for the site. Moreover, the groove at the dimer interface appears to provide the correct spatial orientation of the cluster for core growth toward the hollow center of the protein coat. Experiments to test the hypothesis with protein from genetically altered cloned DNA are currently in progress (Sreedharan, Rohrer, and Theil, unpublished results).

Oxidation of Fe(II) During Formation of Ferritin Iron Core

Before the availability of XAS to study the formation of the ferritin iron core, the main property of the reconstitution that was studied was the appearance of an amber color and the associated broad absorbance spectrum when Fe(II) was mixed with protein coats of ferritin. Changes in absorbance were most often monitored at 310 or 420 nm (Macara et al., 1973; Crichton, et al., 1980; and Mertz and Theil, 1983); the reaction was rapid (only minutes were required for completion). Since the availability of Fe(II) to form complexes with o-phenanthrolene decreased coincidently with the increase in the absorbance, it seemed reasonable to assume that oxidation of Fe(II) had occurred; the idea that the protein coat catalyzed the oxidation of Fe(II) during iron core formation was developed. More recently, the protein coats of ferritin were shown to increase the rate at which Fe(III)-transferrin forms from Fe(II) suggesting again, albeit indirectly, that ferritin protein coats have "ferroxidase" activity. None of the analyses described could analyze the oxidation of Fe directly or in the absence strong chelators for Fe(III) such as apotransferrin. Moreover, the conclusions rested on the assumption that all the Fe (II) atoms mixed with the protein coats of ferritin followed the same pathway to the core, beginning with oxidation on the surface (transferrin is unlikely to acquire Fe(III) from inside the protein coat), and migration to form clusters and polynuclear complexes inside the protein coat; a corollary was that if iron in the core were converted to Fe(II), it would find its way back to the outside of the protein. In fact, protein coats were made from ferritin by the reduction of Fe(III) in the core. The general ideas were widely accepted. However, somewhat puzzling was the length of time required to remove iron from ferritin protein coats: the exhaustive dialysis in the presence of reductant and the necessity for the absence of dioxygen to remove the last traces of iron, particularly at pH 7 (Chasteen and Theil, 1982). Moreover, the use of Mossbauer spectroscopy and gel filtration had shown that Fe(II) could be retained by the protein coat after reduction (Watt et al., 1985; Watt et al., 1986; and Frankel et al., 1987).

We decided to determine the rates of oxidation of Fe(II) during formation of the ferritin iron core using XAS in the dispersive mode and exploiting the differences in the spectra of Fe(II) and Fe(III) in the region of the near edge(XANES). Protein coat of ferritin from horse spleen were mixed with 480 Fe(II)/molecule at pH 7 in Hepes buffer. Comparisons were made to the rate of the absorbance change (420 nm) and the availability of Fe(II) to form pink complexes with o-phenanthrolene. In confirmation of the earlier studies, the absorbance change at 420 nm and the decrease in the availability of Fe(II) to react with o-phenanthrolene were very rapid; changes occurred within minutes of mixing Fe(II) with the protein coats. However, the change in the XAS was very slow with no change detectable at two hours; complete conversion to the Fe(III) spectrum of ferritin iron cores took 16-24 hours (Rohrer et al., 1987) (Fig. 1). The results indicated that the bulk of the Fe(II) was rapidly taken inside the protein coat (inaccessibility to o-phenanthrolene) where oxidation was slow, possibly because the Fe(II) was in a solid phase (ferrous hydroxide?). [Note that the concentration of Fe(II) in the hollow center of the protein coat is high enough to form hydroxide precipitates at pH 7, which also form immediately in the buffer without protein coats.] The colored complex, and the Fe(III) available for apotransferrin may in fact represent only a small fraction of the Fe(II), which oxidizes on the outer surface of the ferritin protein coat, and may account for the EPR signal of solitary Fe(III) observed even when completed iron cores are present (Rosenberg and Chasteen, 1982). Apotransferrin in the experiment by Bakker and Boyer (1986) would sequester the Fe(III) as it formed, driving all the Fe(II) down the same pathway (surface oxidation) until all the apotransferrin was saturated.

The observed stabilization of Fe(II) inside ferritin protein coats was so unexpected in light of existing ideas that subsequent experiments have been undertaken to answer the following questions: Is there an artificial source of electrons that reduced the Fe(III)? [Hepes is known to form such a radical under the experimental conditions but Mops, e.g., does not (N. Chasteen, personal communication)]. If the Fe(II) is inside the protein for long periods of time can it be recovered? Is the XAS spectrum of Fe(II) in the protein really a form of Fe(III) with an unusual XAS?

An effect of a stable radical was sought using dispersive XAS, in collaboration with A. Fontaine, by comparing the oxidation rate of Fe(II) mixed with ferritin protein coats the presence of Hepes to that in Mops at pH 7.

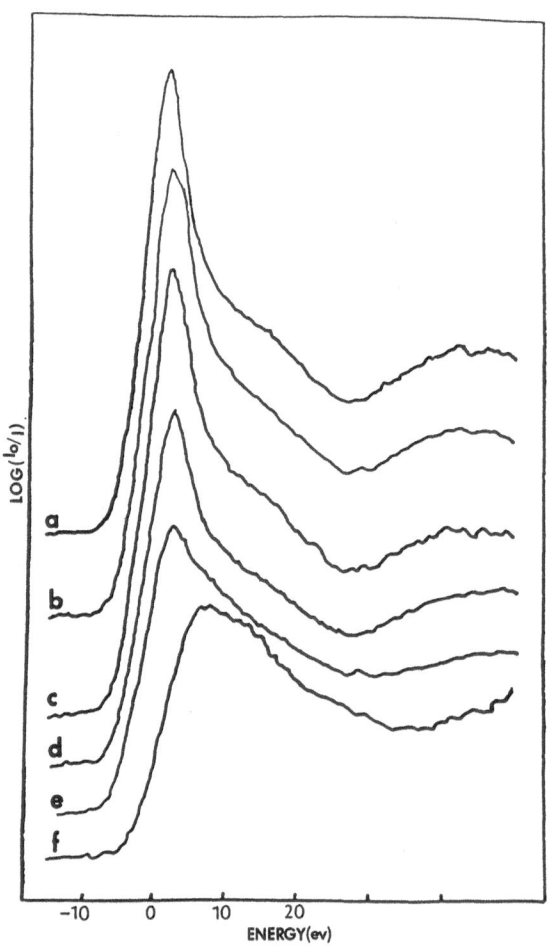

Fig. 1. X-ray absorption near edge spectra of Fe(II) after mixing with the protein coats of horse spleen ferritin. The reaction occurred at pH7 in 0.15 M Hepes at room temperature. Continuous purging with air had no effect. Note the stability of Fe(II) in the presence of protein. The data are taken from Rohrer et al., 1987. Top to bottom: a) Fe(II) in 0.1 NHNO$_3$; b-e) Fe(II) mixed with protein after 0.5 hours, 2 hours, 4 hours, and 12.5 hours; f) Fe(III) in 0.1N HNO$_3$.

Table I. The Amount of Fe(II) Two Hours After Mixing with
          The Protein Coats of Horse Spleen Ferritin

Analysis	% Fe(II)
X-Ray Absorption Spectroscopy(XANES)	> 80
Mossbauer Spectroscopy	68
Fe(II)-o-phenanthrolene complex (Denatured protein, 1 N HCl, 100 C, 30 min.)	77
Fe(II)-o-phenanthrolene complex (Native protein coats)	33
Rate of absorbance change at 420 nm	0

Both oxidation rates are slow, and to a first approximation, the same. Moreover, the rate of oxidation appeared to be dependent upon pH, suggesting that there are no radiation-induced reductants.

If Fe(II) is sequestered inside ferritin protein coats during the early stages of the formation of ferritin iron core, and is therefore inaccessible to o-phenanthrolene, it should be possible to destroy the protein and re-cover the Fe(II). Two hours after mixing Fe(II) with protein coats of fer-ritin, when 75-80% of the Fe(II) did not react with o-phenanthrolene and the XAS was that of Fe(II), the mixture was boiled for 30 minutes in 1 N HCL; 75-85% of the Fe(II) added formed a pink complex with o-phenanthrolene show-ing that the Fe had been inaccessible, not oxidized.

Mossbauer spectroscopy also can be used to distingush between Fe(II) and Fe(III) in ferritin (Watt et al., 1985). In collaboration with R. Frankel and G. Papafythymiou, preliminary analysis has been made of the Mos-sbauer spectra obtained on samples of ^{57}Fe and protein coats of ferritin frozen at various intervals after mixing, under the same conditions used for XAS, UV-vis spectroscopy and measurement of availability to o-phenanthrolene. At two hours only a fraction of the Fe(II) is oxidized. More is oxidized by 12 hours (66%), but the rate appears to be similar to that measured by XAS. A comparison of the oxidation rates at 2 hours measured by XAS and Mossbauer spectroscopy, and the effect of destroying the protein coat on the avail-ability of Fe(II) for complex formation with o-phenanthrolene is presented in Table 1.

Clearly Fe(II) can be stabilized for substantial amounts of time inside the protein coat of ferritin. In cells such as macrophages, which recycle iron through ferritin in short periods of time, some of the iron may never be oxidized, minimizing the massive electron flux previously thought to be associated with the deposition and utilization of ferritin iron. Moreover, the original role of ferritin may have been stabilizing and sequestering Fe(II), not Fe(III). If so, ferritin may not be an evolutionary response of cells to dioxygen but may in fact be among the older proteins in iron-using organisms.

The advantages of XAS studies of ferritin core formation are the abil-ity to examine directly the environment of the iron, under a variety of con-ditions of pH, concentration, buffer anions, and electron acceptors which are within the range of physiology, to understand more completely the steps by which organisms stabilize and sequester the iron needed for life. XAS, using dispersive and standard measurement modes with analysis of both the EXAFS and XANES spectral regions, has already produced important results: the characterization of the Fe-protein interaction in a putative nucleation complex and the reevaluation of the pathway of core formation, the role of oxidation in the macrophage recycling of red blood cell iron, as well as the role and evolutionary age of the protein. Many of the results of ferritin studies are important not only for medicine and agriculture but also for biomineralization and biocorrosion. With the development of new sources, beam lines and analytical approaches, the future of XAS studies of ferritin iron cores at all stages of formation and dissolution looks very bright.

ACKNOWLEDGMENT

Supported in part by the National Institutes of Health grants DK20251 and GM 34675 (to E.C.T) and the Division of Materials Science of the Department of Energy under contract DE-AS05-80-ER10742(D.E.S) for Beam Line X-11 at the National Synchrotron Light Source. X-ray absorption facilities were used at the Stanford Synchrotron Radiation Laboratory, supported by the Department of Energy Office of Basic Energy Services and the National Institutes of Health Biotechnology Research program, Division of Research Resources, and at the Laboratoire pour l'ulitisation du Rayonnement Electromagnetique Scientifique, which is partly supported by the Centre National de la recherche Scientific as well as at beam line X-11 at the National Synchrotron Light Source. This paper is a contribution from the Department of Biochemistry, the College of Agriculture and Life Sciences and the College of Physical and Mathematical Sciences.

REFERENCES

Bakker, G. R., and Boyer, R. F., 1986, Iron incorporation into apoferritin, J. Biol. Chem., 261:13182.
Biedermann, G., and Schindler, P., 1957, On the solubility product of precipitated Iron(III) hydroxide, Acta. Chem. Scand., 11:731.
Chasteen, N. D., and Theil, E. C., 1982, Iron binding by horse spleen apoferritin: a vanadyl (IV) EPR spin probe study, J. Biol. Chem. 257:7672.
Crichton, R. R., Roman, F., Roland, F. Paques, E., and Vandamme, E., 1980, Ferritin Iron Deposition and Mobilisation, J. Mol. Catal. 7:267.
Frankel, R. B., Papaefthymiou, G. C., and Watt, G. D., 1987, Binding of Fe(II) by mammalian ferritin, Hyperfine Interactions, 33:233.
Macara, I. G., Hoy, T. G., and Harrison, P. M., 1973, The formation of ferritin from apoferritin, Biochem. J., 135:343.
Mansour, A. N., Thompson, C., Theil, E. C., Chasteen, N. D., and Sayers, D. E., 1985, Fe(III) ATP complexes: models for ferritin and other polynuclear iron complexes with phosphate, J. Biol. Chem., 260:7975.
Mertz, J. R., and Theil, E. C., 1983, Subunit Dimers in sheep spleen apoferritin, J. Biol. Chem., 258:11719.
Rice, D. W., Ford, G. C., White, J. L., Smith, J. M. A., and Harrison, P. M., 1983, The Spatial structure of horse spleen ferritin, Adv. in Inorganic Biochem., 5:39.
Rohrer, J. S., Joo, M. -S., Dartyge, E., Sayers, D. E., Fontaine, E., and Theil, E. C., 1987, Stabilization of iron in a ferrous form by ferritin, J. Biol. Chem., 262:13385.
Rosenberg, L. P., and Chasteen, N. D., 1982, Initial Iron Binding to Horse Spleen Apoferritin, in "Proteins of Iron Metabolism," J. Hegenauer and P. Saltmen, eds., Elsevier, Amsterdam.

Sayers, D. E., Theil, E. C., and Rennick, F. J., 1983, A distinct environment for Iron(III) in the complex of horse spleen apoferritin observed by X-ray absorption spectroscopy, J. Biol. Chem., 258:14076.

Theil, E. C. , 1983, Ferritin structure, function and regulation, Adv. In Inorg. Biochem., 5:1-38.

Theil, E. C., 1987, Ferritin: structure, gene regulation and cellular function in animals, plants and microorganisms, Ann. Rev. Biochem. 56:289-315.

Theil, E. C., Sayers, D. E., and Brown, M. A., 1979, Similarity of the structure of ferritin and iron-dextran (Imferon) determined by X-ray absorption fine structure analysis, J. Biol. Chem., 254:8132.

Treffry, A., and Harrison, P. M., 1984, Spectroscopic studies on the binding of iron, terbium and zinc by apoferritin, J. Inorg. Biochem., 28:9.

Wardeska, J. G., Viglione, B., and Chasteen, N. D., 1986, Metal ion complexes of apoferritin: evidence for initial binding in the hydrophilic channels, J. Biol. Chem., 261:6677.

Watt, G. D., Frankel, R. B., and Papaefythymiou, G. C., 1985, Reduction of mammalian ferritin, Proc. Nat'l. Acad. of Sci. USA, 82:3640.

Watt, G. D., Frankel, R. B., and Papaefthymiou, G. C., Spartalian, K., and Stiefel, E. I., 1986, Redox properties and Mossbauer spectroscopy of Azotobacter vinelandii bacterioferritin, Biochem., 25:4330.

Yang, C. -Y., Bryan, A. M., Theil, E. C., Sayers, D. E., and Bowen, L. H., 1986, Structural variatons in soluble iron complexes of models for ferritin: an X-ray absorption spectroscopy comparison of horse spleen ferritin to Blutal (iron-chondroitin sulfate) and Imferon (iron-dextran), J. Inorg. Biochem., 28:393.

Yang, C. -Y., Meagher, A., Huynh, B. H., Sayers, D. E., and Theil, E. C., 1987, Iron(III) clusters bound to horse spleen apoferritin: An X-ray absorption and Mossbauer spectroscopy study that shows iron nuclei can form on the protein, Biochem., 26:497.

X-RAY ABSORPTION SPECTROSCOPIC STUDIES OF VANADO-ENZYMES:

NITROGENASE AND BROMOPEROXIDASE

C. David Garner and Judith M. Arber

The Department of Chemistry
Manchester University
Manchester M13 9PL, U.K.

S. Samar Hasnain and Barry R. Dobson

The Daresbury Laboratory
Daresbury
Warrington, WA4 4AD, U.K.

Robert R. Eady and Barry E. Smith

AFRC, IPSR Nitrogen Fixation Laboratory
University of Sussex, Brighton
BN1 9PQ, U.K.

Eize de Boer and Ron Wever

Laboratory of Biochemistry
University of Amsterdam
P. O. Box 20150, 1000HD Amsterdam
The Netherlands

Tadashi Matsushita and Masaharu Nomura

The Photon Factory
National Laboratory for High Energy Physics
Oho-machi, Tsukuba-gun, Ibaraki-ken 305
Japan

INTRODUCTION

Vanadium is well recognised as an essential trace element, is widely
distributed throughout the lithosphere and biosphere, and is present in all
mammalian tissues at concentrations of $\leq$ 10µM (Chasteen, 1983). However,
only in a few instances can a precise biological role be assigned to this
element. It is a potent inhibitor of phosphatases, Na,K-ATPases, and a variety of other important enzymes, doubtless because of the similarity of
$AO_4{}^{3-}$ (for A=V or P) and their derivatives. Amavadin, an oxovanadium(IV)
complex with two molecules of N-(1-carboxyethyl)-N-hydroxyalanine, has been
isolated from the mushroom Amanita muscaria (Kneifel and Bayer, 1986) and
may function as an electron-transfer catalyst (Nawi and Riechel, 1987). Two
enzyme systems were shown to be dependent upon vanadium: nitrogenases from

Azotobacter (Robson et al., 1986) and bromoperoxidases from marine algae (de Boer et al., 1986; de Boer et al., 1986). Herein we describe the results of vanadium K-edge X-ray absorption spectroscopic studies for a member of each of these two types of vanadium-dependent enzymes.

Vanadium Nitrogenase

The free-living, nitrogen-fixing bacterium Azotobacter chrooccocum has been shown to contain two nitrogenase systems. One of these is the well-characterised molybdenum-containing enzyme and the other, which is encoded by different genes, contains vanadium. Both enzyme systems consist of two proteins: an iron protein and a molybdenum- or vanadium-containing protein which also contains iron. The vanadium-containing nitrogenase protein, Ac1*, has a relative molecular mass of ca. 210,000 and contains 2 vanadium atoms, 23 iron atoms, and 20 acid-labile sulfide ions per mole (Robson et al., 1986).

We have obtained vanadium K-edge X-ray absorption data for three forms of Ac1*: thionine-oxidized, dithionite reduced, and the further reduced form generated during enzyme turnover.

Ac1* was isolated and assayed as previously reported (Robson et al., 1986). Thionine-oxidised protein was prepared without significant loss of activity by incubation with excess solid thionine for 5 min. before freezing. To prepare samples of Ac1* under turnover conditions, 0.4 cm^{-3} of Ac1* was cooled to 6° and added to a freeze-dried reaction mixture (0.2 cm^{-3}) containing MgATP and an ATP-regenerating system (Eady et al., 1972). After thorough mixing, 20 μmol of $S_2O_4^{2-}$ was added. Enzyme turnover was initiated by the addition of Ac2* (0.2 cm^{-3}, pre-cooled to 6°, containing 30 mg of protein). The reaction mixture was transferred to a pre-cooled EPR tube and an EXAFS sample cell, and frozen in petroleum ether at -50°, 2.75 min. after the addition of Ac2*. EPR measurements indicated that Ac1* was > 75% 'super-reduced' under these conditions. Samples were stored in liquid nitrogen after being loaded anaerobically into cells.

X-ray absorption spectra were recorded in fluorescence mode on EXAFS station 7C at the Photon Factory operating at 2.5GeV and an average current of 200mA; ca. 10 scans were recorded and averaged for each sample. A sagittal focussing, double crystal Si(111) monochromator was used together with a 200mm long fused quartz mirror placed at an angle of ca. 8mrad to reduce harmonic contamination. The incident intensity was measured using a windowless ionisation chamber whilst a Lytle cell, filled with 100% argon and masked with a titanium foil, monitored the fluorescent intensity. During data collection, an average sample temperature of 80K was maintained by use of a liquid nitrogen cryostat.

Data analysis was accomplished via the single-scattering curved-wave method of EXAFS calculation and phaseshifts were derived from ab initio calculations as described previously (Gurman et al., 1984; Perutz et al., 1982).

Fig. 1 shows the vanadium K-absorption edge and XANES of both the thionine-oxidised and dithionite-reduced forms of Ac1*, together with a single scan for the protein during enzyme turnover in the presence of MgATP and Ac2*. The position of the pre-edge feature and the edge, and the structure of the edge and XANES are similar for all samples, namely a weak pre-edge feature and edge inflection and a 'doublet' in the XANES. The positions of the pre-edge feature and edge are the same within the estimated experimental errors (±1eV). This overall similarity implies that no major changes are occurring in the local environment of the vanadium.

Table 1. Parameters Used To Simulate The EXAFS Associated
          With The Vanadium K-edge of Dithionite-reduced and
          Thionine-oxidised Ac1* [a]

System	Atom	N	R/Å	$2\sigma^2/Å^{2}$ [b]
Dithionite-	O [c]	3.0(1.0)	2.13(3)	0.001(3)
reduced Ac1*	S	3.0(1.0)	2.34(3)	0.018(4)
	Fe	3.0(1.0)	2.70(3) [d]	0.018(3)
Thionine-	O [c]	3.0(1.0)	2.13(3)	0.008(4)
oxidised Ac1*	S	3.0(1.0)	2.33(3)	0.024(5)
	Fe	3.0(1.0)	2.70(3) [d]	0.020(4)

[a] E_{0}=18.27eV; figures in parentheses indicate estimated errors
[b] Debye-Waller parameter
[c] O, N, C are all possible first shell ligands
[d] This value is considered to be underestimated by 0.05Å

Fig. 2 shows the vanadium K-edge EXAFS data for dithionite-reduced and
thionine-oxidised Ac1*, together with their Fourier transforms and simulations employing the parameters presented in Table 1. The two sets of data
resemble each other closely and are also very similar to the previously published EXAFS data for the dithionite-reduced protein (Arber et al., 1987).

The major differences between these simulations and that previously
reported involve amplitude effects. Thus, for both the dithionite-reduced

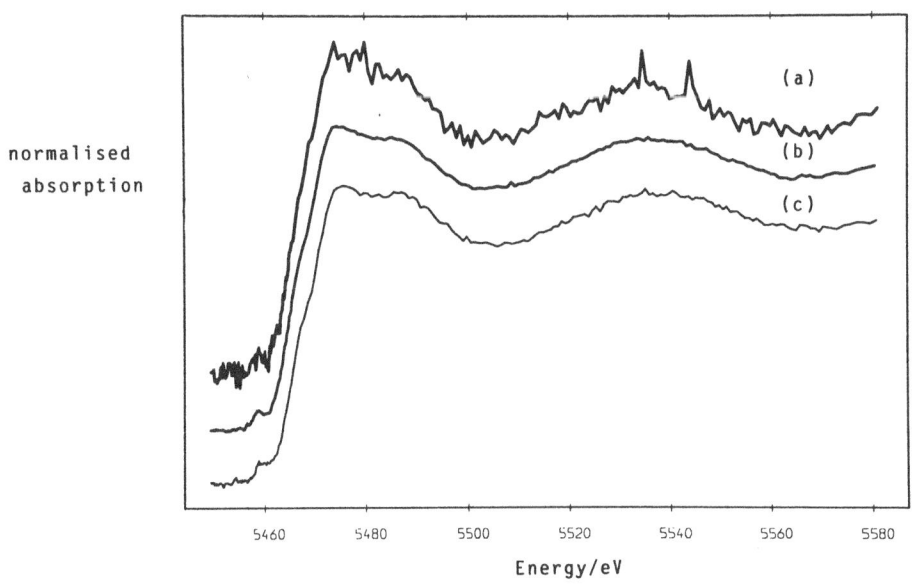

normalised
absorption

Fig. 1. Vanadium K-edge and XANES of (a) super-reduced Ac1*, (b) dithionite
         reduced Ac1* and (c) thionine-oxidised Ac1*.

and thionine-oxidised forms the data were best simulated with 3±1 oxygen and 3±1 sulfur atoms, compared to 4±1 and 2±1, respectively, previously suggested (Arber et al., 1987) for the dithionite-reduced protein. Thus, the edge, XANES, and EXAFS data presented here for the dithionite-reduced and thionine-oxidised forms of Acl* are consistent with vanadium being octahedrally coordinated in a similar environment to that proposed previously for the reduced protein and present in the synthetic compound [NMe$_4$][VFe$_3$S$_4$Cl$_3$ (dmf)$_3$] (Kovacs and Holm, 1986). This situation is analogous to that of the MoFe-protein of nitrogenases, where X-ray absorption studies of reduced and dye-oxidised samples showed that molybdenum is present as part of an MoFeS cluster, and that no appreciable differences in the molybdenum K-edge EXAFS occur upon a change in the oxidation level of the protein (Cramer et al., 1978; Cramer et al., 1978).

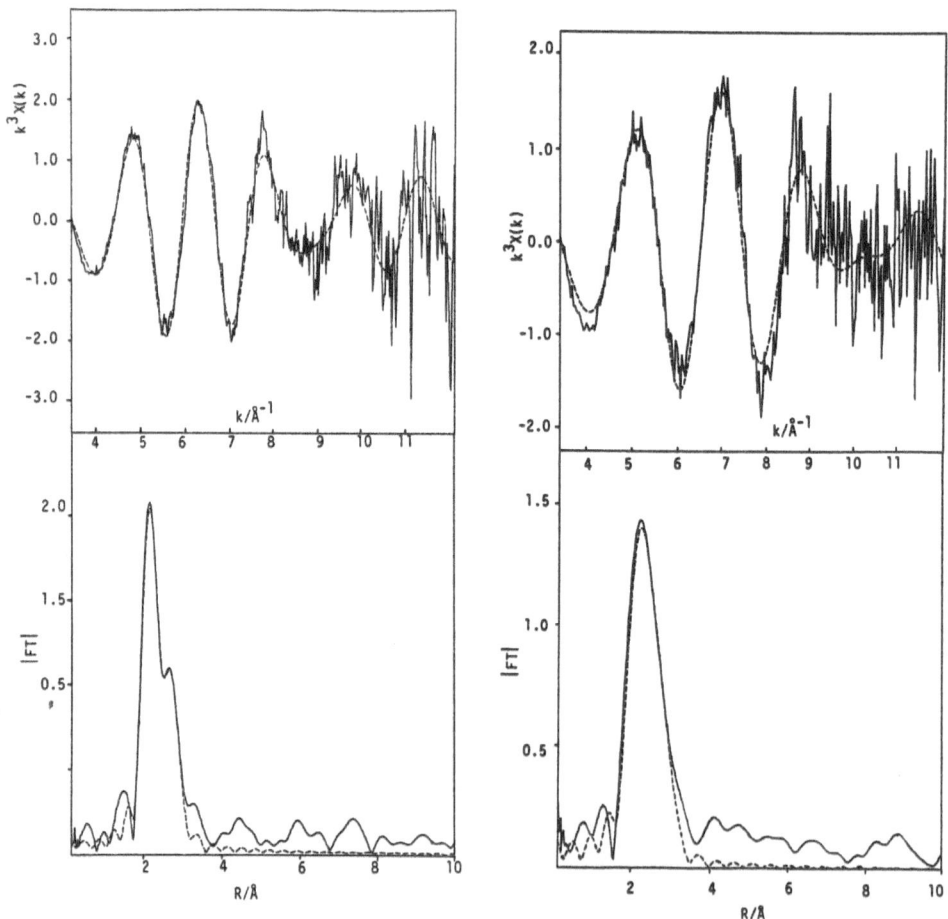

Fig. 2. Vanadium K-edge EXAFS (xk^3) and Fourier transform of dithionite-reduced Acl* (left-hand Figure) and thionine oxidised Acl* (right-hand Figure). Solid line represents experimental data; broken line theory using parameters presented in Table 1.

Vanadium Bromoperoxidase

Many organisms are able to synthesise halogenated metabolites. These
halometabolites are formed upon oxidation of halides by haloperoxidases and
hydrogen peroxide, in the presence of a nucleophilic reagent. Halometabolites appear to be involved in chemical defence mechanisms since they display
considerable antibacterial and anticellular activities. Most haloperoxidases contain haem at the active site, however, vanadium(V) is essential for
the brominating activity of bromoperoxidases purified from several marine
algae including Ascophyllum nodosum (de Boer, 1986, v.869) and Laminaria
saccharina (de Boer et al., 1986).

Bromoperoxidase from A. nodosum was purified as described previously
(de Boer et al., 1986) and samples of the reduced and native enzyme, and the
native enzyme in the presence of bromide and hydrogen peroxide, were loaded
into sealed cells, stored and studied at ca. 80K.

X-ray absorption spectra were recorded in fluorescence mode on EXAFS
station 8.1 at the Daresbury Synchrotron Radiation Source operating at 2GeV
and an average current of 130mA, a single NaI scintillation detector with a
titanium foil being used to record the fluorescent intensity. A slit-less
double crystal Si(111) monochromator was employed and crystal glitches limited the data range to approximately 400eV beyond the edge. For each bromoperoxidase sample ca. 15 scans were recorded and the data were averaged.
Data analysis was accomplished as described previously (Gurman et al., 1984;
Perutz et al., 1982).

Fig. 3 shows the vanadium K-edge and XANES for the four samples studied. The profile and position of the edge and XANES for the native,

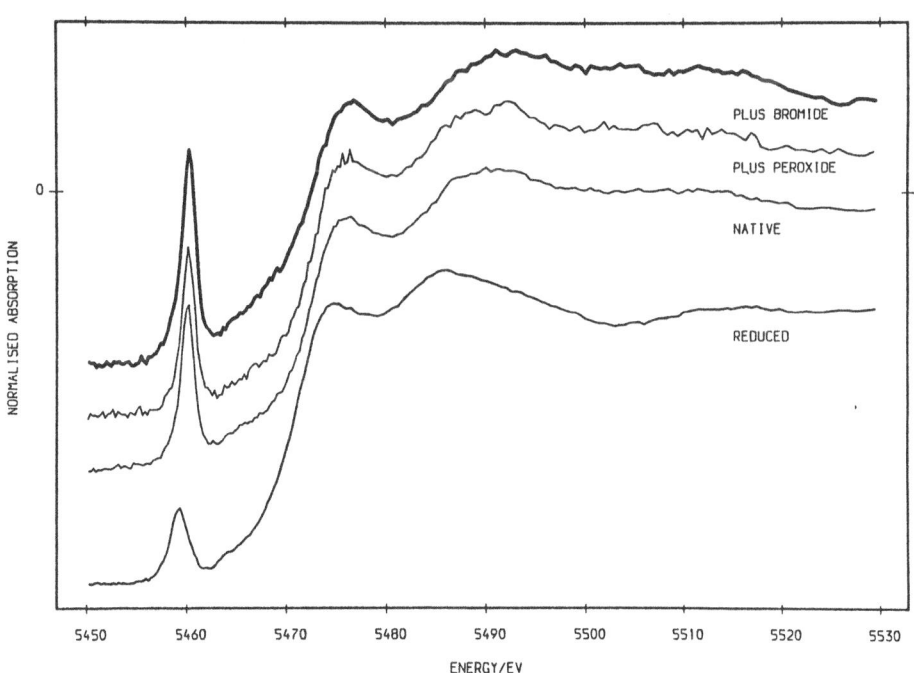

Fig. 3. Vanadium K-edge and XANES of bromoperoxidase samples.

bromide, and peroxide samples are very similar; the data for the reduced enzyme differs considerably. This implies that the environment of the vanadium is altered significantly on reduction, but not on addition of bromide or peroxide to the native enzyme.

The changes observed in the edge and XANES regions on reduction of the native enzyme are reflected in the very different EXAFS of the two samples. The EXAFS amplitude is decreased and the frequency is increased on reduction of the native enzyme, implying a considerable change in the vanadium environment. In the reduced sample the dominant contribution is from light atom backscattering at ca. 2Å and only a weak contribution is made by shorter bonds. Other light-atom backscattering contributions occur at ca. 3.0 and 4.2Å. The data can be successfully interpreted by backscattering from two imidazole groups at 1.95(3)Å and substantially weaker contributions from one oxygen at 1.68(5)Å and two oxygens at 1.98(5)Å, suggesting a distorted square pyramidal geometry in reduced bromoperoxidase. This environment is typical of vanadium(IV) compounds. The EXAFS of native bromoperoxidase is dominated by a short backscattering contribution 1.75(3)Å that lowers the frequency of the EXAFS oscillations compared to reduced bromoperoxidase. Successful interpretations of the data suggest that the dominant backscattering contributions could arise from oxygens of tyrosine groups, with an additional oxygen atom at 1.59(3)Å and two imidazole groups at 2.07(3)Å.

CONCLUSIONS

These studies have provided the first structural information concerning the nature of the vanadium sites in nitrogenases and bromoperoxidases and demonstrated clear chemical differences between the two systems. Also, whereas the vanadium centre of bromoperoxidase undergoes a profound structural change upon a change in the oxidation level of the protein this is not the case for the vanadium centre of nitrogenase.

ACKNOWLEDGMENTS

We thank the SERC (U.K.), AFRC (U.K.), and DSM N.V. (Netherlands) for financial support and the Directors of the Daresbury Laboratory and the Photon Factory for provision of facilities.

REFERENCES

Arber, J. M., Dobson, B. R., Eady, R. R., Stevens, P., Hasnain, S. S., Garner, C. D., and Smith, B. E., 1987, Vanadium K-edge, X-ray absorption spectrum of the VFe-protein of the vanadium nitrogenase of Azotobacter chrooccocum, Nature, 325:372.

Chasteen, N. D., 1983, The biochemistry of vanadium, Structure and Bonding, 53:107.

Cramer, S. P., Gillum, W. O., Hodgson, K. O., Mortenson, L. E., Stiefel, E. I., Chisnell, J. R., Brill, W. I., and Shah, V. K., 1978, The molybdenum site of molybdenum cofactor by X-ray absorption spectroscopy, J. Amer. Chem. Soc., 100:3814.

Cramer, S. P., Hodgson, K. O., Gillum, W. O., and Mortenson, L. E., 1978, The molybdenum state of nitrogenase. Preliminary structural evidence from X-ray absorption spectroscopy, J. Amer. Chem. Soc., 100:3398.

de Boer, E., Van Kooyk, Y., Tromp, M. G. M., Plat, H., and Wever, R., 1986, Bromoperoxidase from Ascophyllum nodosum: a novel class of enzymes containing vanadium as a prosthetic group, Biochim. Biophys. Acta, 869:48.

de Boer, E., Tromp, M. G. M., Plat, H., Krenn, G. E., and Wever, R., 1986, Vanadium(V) as an essential element for haloperoxidase activity in marine brown algae: purification and characterization of a vanadium(V)- containing bromoperoxidase from <u>Laminaria saccharina</u>, <u>Biochim. Biophys. Acta</u>, 872:104.

Eady, R. R., Smith, B. E., Cook, K. A., and Postgate, J. R., 1972, Nitrogenase of <u>Klebsiella pneumoniae</u>. Purification and properties of the component proteins, <u>Biochem. J.</u>, 128:655.

Gurman, S. J., Binsted, N., and Ross, I., 1984, A rapid, exact curved-wave theory for EXAFS calculations, <u>J. Phys. C: Solid State Phys.</u>, 17:143.

Kneifel, H., and Bayer, E., 1986, Stereochemistry with total synthesis of Amavadin, the naturally occurring vanadium compound of <u>Amanita muscaria</u>, <u>J. Amer. Chem. Soc.</u>, 108:3075.

Kovacs, J. A., and Holm, R. H., 1986, Assembly of vanadium- iron-sulphur cubane clusters from mononuclear and linear trinuclear reactants, <u>J. Amer. Chem. Soc.</u>, 108:340.

Nawi, M. A., and Riechel, J. L., 1987, The electrochemistry of Amavadine, a vanadium natural product, <u>Inorg. Chim. Acta</u>, 136:33.

Neidleman, S. L., and Geigert, J., 1986, "Biohalogenation: principles, basic roles and applications" Ellis Horwood Ltd., Chichester.

Perutz, M. F., Hasnain, S. S., Duke, P. J., Sessler, J. L., and Hahn, J. E., 1982, Stereochemistry of iron in deoxyhaemoglobin, <u>Nature</u>, 295:535.

Robson, R. L., Eady, R. R., Richardson, T. H., Miller, R. W., Hawkins, M., and Postgate, J. R., 1986, The alternative nitrogenase of <u>Azotobacter chrooccocum</u> is a vanadium enzyme, <u>Nature</u>, 322:388.

X-RAY ABSORPTION SPECTROSCOPY OF

PSEUDOMONAS CEPACIA PHTHALATE DIOXYGENASE

James E. Penner-Hahn

Department of Chemistry
University of Michigan
Ann Arbor, MI 48109

INTRODUCTION

Bacterial degradation of aromatic compounds generally proceeds by a series of oxygenations. In Pseudomonas cepacia, the first step in phthalate metabolism is dihydroxylation to give phthalate 4,5-dihydrodiol. This reaction is catalyzed by a novel non-heme iron oxygenase. The phthalate dioxygenase from P. cepacia is a large (ca. 192,000 kDa) tetrameric protein containing one "Rieske-like" 2Fe/2S cluster and one mononuclear Fe site per monomer (Batie et al, 1987). We used x-ray absorption spectroscopy (XAS) to determine the structures of the metal sites in phthalate dioxygenase.

Iron-sulfur clusters

Although their existence has been known for less than 40 years, iron sulfur (Fe/S) proteins are now recognized as ubiquitous constitutents of biological systems (Beinert, 1973). A wide variety of biochemical, synthetic, and spectroscopic methods has been used in studies of Fe-S proteins, and detailed structures are now known for biological iron-sulfur centers of several types. These include Fe(SR)4, Fe2S2(SR)4, and Fe4S4(SR)4 clusters, found in, respectively, rubredoxins, plant-type ferredoxins, and high potential iron-sulfur proteins and clostridial ferredoxins (Palmer, 1973; Stout, 1982).

In addition to these well characterized clusters, there exists an additional cluster type, the so-called Rieske-like Fe-S cluster. This site, first discovered in the mitochrondrial ubiquinol-cytochrome c reductase system (Rieske et al., 1964), has since been found in photosynthetic electron transfer proteins (Nelson, 1972) and in several bacterial oxygenases (Geary et al., 1984; Axcell et al., 1975; Gibson et al., 1970; Sauber et al., 1977). The Fe/S cluster in the P. cepacia phthalate dioxygenase is spectroscopically indistinguishable from other Rieske-like clusters.

Rieske-like clusters are distinguished, empirically, by their unusual EPR spectra (g=2.025, 1.89, 1.8) and by their redox potentials (Em > +150 mV) (Kuila and Fee, 1986), which are unusually high for [2Fe-2S] centers. The Rieske-like Fe/S clusters are known to contain two Fe atoms and two acid-labile sulfides and are generally believed to have a [2Fe-2S] core similar to that found in other 2Fe Fe/S clusters. However, the amino acid composition of the Rieske Fe/S protein from Thermus thermophilus demonstrates that each cluster has no more than 2 thiolate (cysteine) ligands (Fee et al., 1984). Thus, the Rieske-like centers apparently differ from plant-type ferredoxins in that at least two of the thiolates have been re-placed by other ligands. On the basis of EPR and ENDOR measurements (Cline et al., 1985), at least one of these new ligands appears to be an imidazole nitrogen.

Mononuclear Fe(II) site

Ferrous iron must normally be added to phthalate dioxygenase (and to the other Rieske dioxygenases) in order to activate the enzyme fully. Batie et al. (1987) showed that if phthalate dioxygenase is isolated in the presence of phthalate, no additional iron is required for full activity, whereas if the enzyme is dialyzed against EDTA, one equivalent of Fe must be added to activate the enzyme. They find that three equivalents of Fe are required per monomer for maximal activity. Other divalent metals will bind in the place of Fe, however the resulting enzyme is catalytically inactive.

Very little is known about the mononuclear Fe site, since it is not amenable to conventional spectroscopic probes. This iron is tightly bound ($K_d < 1$ micromolar); however, in the absence of phthalate it is readily re-moved by chelating agents. In the presence of phthalate, the mononuclear Fe is not removed, even by overnight dialysis against 10 mM EDTA. The mono-nuclear iron is believed to be in the Fe(II) oxidation state, at least in the resting form of the enzyme, and has been proposed as the site of oxygen activation (Batie et al., 1987).

An important advance in the characterization of the mononuclear Fe site came with the discovery that this site can be readily removed and recon-stituted (C.J. Batie and D.P. Ballou, unpublished). Of particular impor-tance in the present context is the fact that the Fe(II) can be replaced with a variety of divalent metal ions, including Zn(II) and Co(II). Al-though these mixed metal derivatives are not catalytically competent, they do appear to have structures similar to those of the native protein. The evidence for this comes from the fact that the UV-visible and EPR spectra of the Rieske site are slightly perturbed when the mononuclear Fe is removed. Addition of one equivalent of either Co(II) or Zn(II) restores these spectra to those observed for the native protein (Batie and Ballou, unpublished).

We used extended x-ray absorption fine structure (EXAFS) and x-ray ab-sorption near edge structure (XANES) to study the metal sites in phthalate dioxygenase. To study selectively the Rieske-like site, we measured data for phthalate dioxygenase in which the mononuclear Fe had been removed. To study selectively the mononuclear site, we measured Zn and Co EXAFS for phthalate dioxygenase samples in which the mononuclear site had been recon-stituted with Zn or Co. The samples that we studied are summarized in Table 1.

EXPERIMENTAL METHODS

All XAS spectra were measured at the Stanford Synchrotron Radiation Laboratory. All measurements were made at 10K using an Oxford Instruments Cryostat. Typical protein concentrations were 3-5 mM. Data were measured as fluorescence excitation spectra, using a large solid-angle ion chamber to detect the x-ray fluorescence and the Z-1 filters to reject scattered x-rays. EXAFS data reduction followed standard procedures (Penner-Hahn et al., 1986). Curve-fitting analysis used both empirical and ab initio para-meters (Teo and Lee, 1979) to define the EXAFS amplitude and phase func-tions. Identical results were obtained using either parameters.

Table 1. Samples studied using XAS

Rieske-cluster oxidation	Phthalate	Metal in mononuclear site	EXAFS+ XANES	XANES only
Oxidized	+	Fe	Fe	
Oxidized	+	Zn	Fe,Zn	
Oxidized	+	Co	Co	
Oxidized	+	--	Fe	
Oxidized	-	Zn	Fe,Zn	
Oxidized	-	Co	Co	
Reduced	+	--	Fe	
Reduced	+	Zn	Fe,Zn	
Reduced	+	Co	Fe,	Co
Reduced	-	Co	Fe,	Co

In order to make quantitative XANES measurements we developed a new procedure for the normalizing of XANES data (G.S. Waldo and J.E. Penner-Hahn, unpublished). In brief, this procedure involves adjustment of the experimental data to agree (above and below the absorption edge) with tabulated x-ray absorption cross-sections. Adjustments are limited to subtraction of a single low-order polynomial (fitted over the entire data range) and multiplication by a single constant. In tests with model compounds, we find that this procedure allows us to normalize XANES spectra with a precision of ca. 1%.

RESULTS AND DISCUSSION

Rieske-like cluster

The EXAFS spectra for the Rieske-like site are shown in Fig. 1. The obvious beat pattern in these data reflects the fact that there are at least two shells of scatterers making significant contributions to the overall EXAFS. It is obvious from a simple comparison of the EXAFS spectra for the oxidized and reduced Rieske clusters that this cluster must undergo some sort of structural rearrangement on reduction (this is especially obvious in the k=6-9 Å^{-1} region).

The Fourier transforms of the EXAFS data (Fig. 2) show the expected contributions from Fe-S and Fe-Fe scattering. The quantitative curve fitting results for these data are given in Table 2. In general, the cluster has structural parameters consistent with those seen for other Fe/S clusters (Teo et al., 1979; Teo and Shulman, 1982). This finding, while not unexpected, is the first direct evidence that the unusually high redox potential of Rieske-like clusters does not arise from gross structural changes in the Fe/S core. Reduction of the cluster results in a 0.04 Å increase in the Fe-S distance and no detectable change in the Fe-Fe distance. The increase in Fe-S bond length on reduction is somewhat larger than that observed for ferredoxin-like clusters, possibly reflecting the less polarizable nature of the ligands in the present case.

It is interesting to note that, within the precision of EXAFS, the structure of the Rieske cluster is not affected by either the metal in the mononuclear site or the binding of substrate. This is somewhat surprising in view of the fact that the EPR spectrum of the Rieske-site _is_ perturbed if the metal is removed from the mononuclear site. Our results mean that the

179

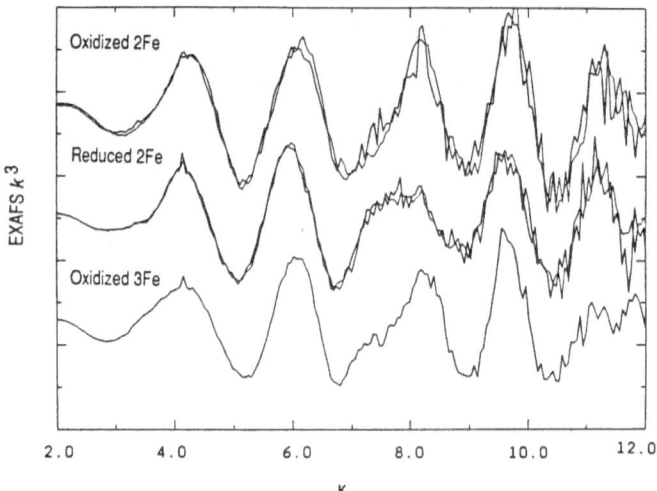

Fig. 1. EXAFS spectra for oxidized and reduced Rieske clusters in phthalate
dioxygenase. The top 2 sets are for samples having either Zn or
nothing in the mononuclear site. The lower curve is for the native
enzyme and includes contributions from the mononuclear site. The
same EXAFS structure is observed for samples having Co rather than
Zn in the mononuclear site. Spectra measured with or without sub-
(not shown) show no changes.

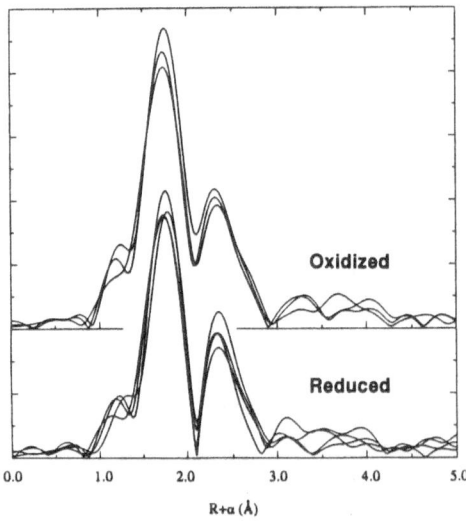

Fig. 2. Fourier transform for Fe EXAFS for all 2Fe phthalate dioxygenase
samples studied. Two peaks correspond (primarily) to Fe-S and
Fe-Fe scattering. In addition to small shift in Fe-S distance,
there is a characteristic change in the second shell between
oxidized and reduced centers.

Table 2. Curve-Fitting Results for Rieske-like Cluster

Sample	Fe-S		Fe-N	Fe-Fe	
	R	σ	R	R	σ
Oxidized					
+ phthalate	2.20	0.072	2.05	2.69	0.031
+ Zn + phthalate	2.21	0.067	2.00	2.70	0.031
+ Zn - phthalate	2.20	0.058	2.03	2.70	0.021
Reduced					
+ phthalate	2.24	0.083	2.07	2.7	0.048
+ Zn + phthalate	2.25	0.082	2.06	2.69	0.038
+ Co + phthalate	2.24	0.078	2.08	2.70	0.038
+ Co - phthalate	2.25	0.074	2.08	2.71	0.020
Models (Teo et al., 1979)					
Fe(SR)4 - oxidized	2.26	0.048			
- reduced	2.32	0.057			
Fe2S2(SR)4 - oxidized	2.23	0.063	2.70	0.07	
- reduced	2.24	0.059	2.72	0.08	

EPR changes must arise from conformational rearrangements and not from changes in the overall coordination.

Perhaps the most interesting aspect of the Rieske clusters is their nitrogen coordination. It is therefore somewhat disappointing that the Fe-N distance is only poorly defined by the EXAFS data. The reason for this effect lies in the low scattering power of one nitrogen in comparison with three sulfurs or one iron. Indeed, if there were not independent evidence for the presence of nitrogen ligation, it would be difficult to prove, using EXAFS alone, that nitrogen is bound as a ligand.

Fortunately, we know a great deal about the coordination environment of the iron atoms and can make use of this information to restrict the number of variable parameters in the curve fitting analysis. If we restrict ourselves to solutions having integer or half-integer coordination numbers, we find that 3 S + 1 N gives the optimal fit for the oxidized Rieske cluster. Interestingl,, we find that for the reduced cluster a simulation using 3 S + 1.5 N consistently gives a better fit than a simulation using 3 S + 1 N. The uncertainty in coordination number determination by EXAFS is generally estimated as ±25%, thus the distinction between average Fe-N coordination numbers of 1 and 1.5 N is far from conclusive. Nevertheless, the observation that this difference is consistently observed for seven different data sets measured under different conditions suggests that there may be a real change in Fe coordination number between oxidized and reduced Rieske clusters.

The 1s→3d transition can be used as an independent test of this hypothesis. It is well known that the intensity of the 1s→3d transition is dependent on the geometry of a site, with the very weak transitions being observed for centrosymmetric metals and relatively strong transitions being characteristic of tetrahedral coordination. Roe et al. (1985) showed that it is possible to correlate the Fe(III) 1s→3d intensity with coordination number. The normalized areas (in units of 10^{-2} eV) are 6-9 for 6-coordinate, 12-19 for 5-coordinate and 23-25 for 4-coordinate Fe(III).

The 1s→3d transition for the Rieske cluster (Fig. 3) shows a significant decrease in intensity when the cluster is reduced. The normalized

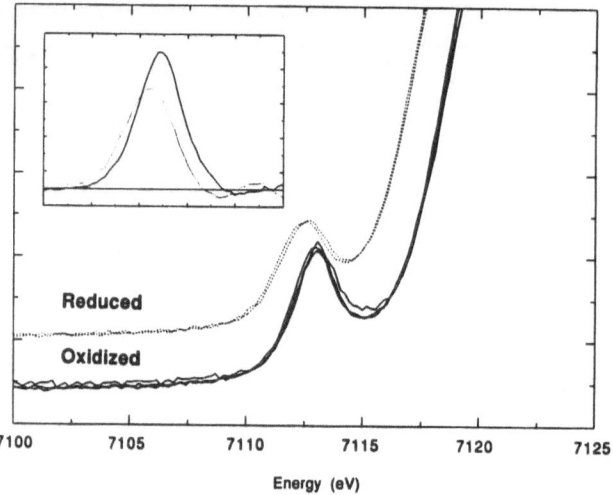

Fig. 3.   XANES structure for oxidized and reduced Rieske clusters.   Inset shows isolated 1s→3d transition for oxidized (solid) and reduced (dashed) clusters.

1s→3d intensity for the oxidized cluster is the same (26) as that found by Roe (1985) for 4-coordinate Fe centers, while the intensity for the reduced cluster decreases to 18.   A 10% decrease in 1s→3d intensity is expected as a result of the decrease in d-vacancies between Fe(III) and Fe(II) (Roe, 1985 and personal communication).   The remaining decrease is consistent with a model in which one iron has an intensity of 26 while the second iron has an intensity of 14.   This again suggests that there is a change in coordination number when the Rieske cluster is reduced, although further work will be necessary to confirm this proposal.   Of particular value will be studies of models for the reduced Rieske cluster.

Mononuclear site

The Fourier transforms of the EXAFS data for Co- and Zn-substituted phthalate dioxygenase samples are shown in Fig. 4.   These spectra are characterized by a single peak at a position corresponding to an absorber scatterer distance of ca. 2 Å.   There is no evidence for the outer shell peaks that are characteristic of imidazole coordination (compare Feiters et al., 1987).   Quantitative curve fitting analysis of these data (see Table 3) show that they can be modelled by a single shell of oxygen or nitrogen scatterers.   There is no evidence for contributions from heavier scatterers.

The coordination number determined by EXAFS is 6±1.   Additional insight into the coordination can be gained from an examination of the known coordination chemistry of Zn (as reflected by structures included in the Cambridge Crystallographic Data Base).   Average Zn-(O,N) bond lengths vary from 2.03 Å for ZnO_5 to 2.18 Å for ZnN_6.   Only 5-coordinate Zn complexes having predominantly oxygen ligation show bond lengths close to those found experimentally for the Zn site in phthalate dioxygenase.   From this we conclude that the mononuclear site has predominantly oxygen ligation.   This is consistent with the lack of any detectable second shell contributions to the Zn EXAFS.   When Zn is in the mononuclear site, this site appears to be 5-coordinate, at least for the samples we have studied.

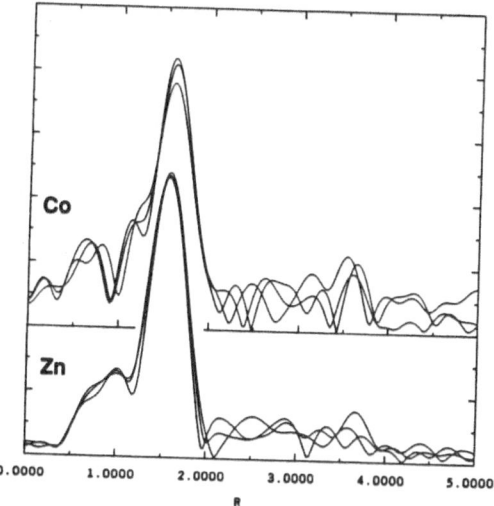

Fig. 4. Comparison of Fourier transforms for Co and Zn EXAFS for
substituted phthalate dioxygenase samples.

The Co EXAFS spectra are qualitatively similar to those for Zn. The
0.03 Å increase in bond length between Zn and Co (reduced Rieske cluster +
phthalate) is only marginally significant. The observed Co-O bond length
(2.04 Å) is still consistent with known high spin 5-coordinate Co complexes.
On the other hand, the 0.05 Å increase in Co-O bond length when phthalate is

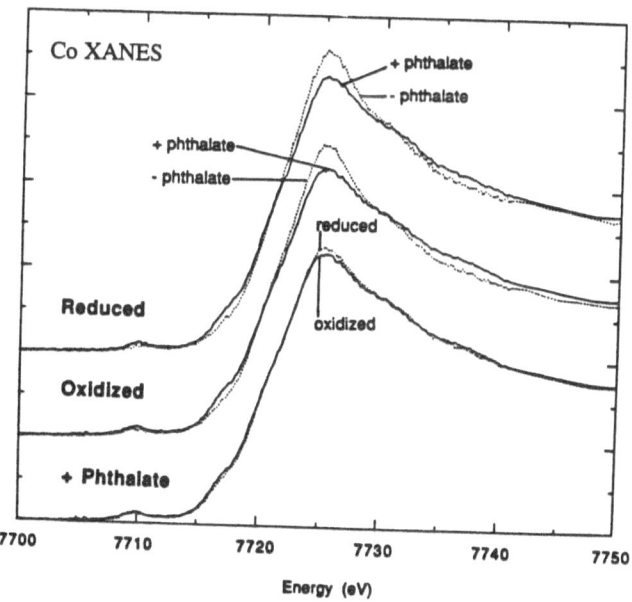

Fig. 5. Co XANES structure for Co-substituted phthalate dioxygenase.
Identical spectral changes are observed for the Zn-substituted
protein.

183

Table 3. XAS results for the mononuclear site

Rieske cluster	Phthalate	Bond-length Zn	Co	1s→3d area
Oxidized	+	2.01		10.9
	−	2.01		8.9
Reduced	+	2.01	2.04	7.7
	−		2.09	5.8

absent and is outside the limits of EXAFS uncertainty, suggesting that the coordination number increases from 5 to 6 when phthalate is removed.

The XANES structure for both Co and Zn show that the mononuclear site is sensitive to the oxidation state of the Rieske cluster and to the presence/absence of phthalate. The Co XANES spectra are shown in Fig. 5; identical effects are observed for Zn. It is difficult to interpret these changes structurally due to the complexity of XANES spectra. These data do serve, however, as the first direct evidence that the mononuclear site is affected by the substrate. This adds support to the belief that the mononuclear site is the locus of substrate oxygenation.

An advantage of Co XANES is that these spectra have a 1s→3d transition that can be used to monitor the geometry of the absorbing site. The isolated 1s→3d transitions for the four different Co substituted proteins are shown in Fig. 6. The decrease in 1s→3d intensity as the Rieske site is reduced and as phthalate is removed probably reflects a shift in equilibrium between two limiting forms, rather than the existence of four discrete structures for the Co site. The oxidized+phthalate enzyme has a 1s→3d intensity consistent with 5-coordination, while the reduced−phthalate intensity is consistent with 6-coordination.

Combining the EXAFS and XANES results for Co and Zn, we conclude that the mononuclear site has predominantly oxygen ligation and can assume either 5- or 6-coordinate geometries. Binding of phthalate and/or oxidation of the Rieske cluster appear to favor the 5-coordinate form for Co. We have seen no evidence for a 6-coordinate form for Zn, however, this may exist when the Rieske cluster is reduced and phthalate is removed. The similarity between the Co and Zn data suggest that these data are relevant to the native Fe-containing site, and not mere artifacts of substitution. Consistent with this, Fe EXAFS data for the mononuclear site (calculated by difference) are similar to the Co and Zn data, albeit with a higher level of noise.

CONCLUSIONS

The Rieske-like cluster has a structure very similar to that found for other Fe/S clusters. This site appears to bind an additional low-Z ligand when reduced (as judged by both EXAFS and XANES). In addition to stabilizing the reduced form, this structural change offers a mechanism for communicating the oxidation state of the Rieske-site to the mononuclear site.

The mononuclear site has predominantly oxygen ligation with a coordination number of 5-6. There is no apparent (by EXAFS/XANES) communication from the mononuclear site to the Rieske, however the mononuclear site is sensitive to both the oxidation state of the Rieske cluster and the presence of phthalate. Oxidation of the Rieske cluster and/or binding of phthalate

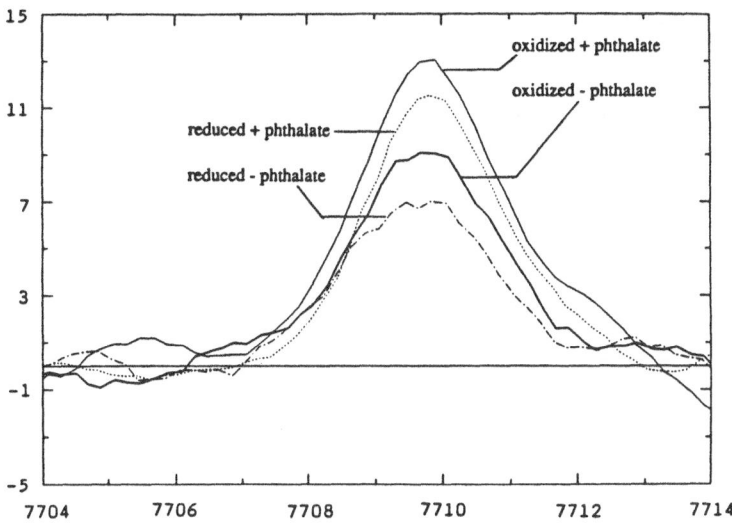

Fig. 6. Isolated 1s}3d transitions for Co substituted phthalate dioxygenase samples.

seem to favor a 5-coordinate form of the mononuclear site, while reduction of the Rieske and/or removal of phthalate appear to favor a 6-coordinate form of the mononuclear site. This is the first direct evidence for an interaction between phthalate and the mononuclear Fe. Future experiments will seek to characterize further the nature of the mononuclear site.

ACKNOWLEDGMENTS

I thank H.T. Tsang and G.S. Waldo for experimental assistance. Phthalate dioxygenase samples were prepared by C. J. Batie and D. P. Ballou. This work was supported in part by the Camille and Henry Dreyfus Foundation and the National Institutes of Health (GM-38047). X-ray absorption measurements were made at the Stanford Synchrotron Radiation Laboratory, which is supported by the U. S. Department of Energy and the National Institutes of Health.

REFERENCES

Antonio, M. R., Averill, B. A., Moura, I., Moura, J. J. G., Orme-Johnson, W. H., Teo, B. K., and Xavier, A. V., 1982, Core dimensions in the 3Fe cluster of Desulfovibrio gigas ferredoxin II by extended x-ray absorption fine structure spectroscopy, J. Biol. Chem., 257:6646.
Axcell, B. C., and Geary, P. J., 1975, Purification and properties of a soluble benzene-oxidizing system from a strain of Pseudomonas, Biochem. J., 146:173.
Batie, C. J., LaHaie, E., and Ballou, D. P., 1987, Purification and characterization of phthalate oxygenase and phthalate oxygenase reductase from Pseudomonas cepacia, J. Biol. Chem., 262:1510.

Beinert, H., 1973, Development of the field and the nomenclature, in "Iron-Sulfur Proteins," Lovenberg, W., ed., Academic Press, New York.
Cline, J. F., Hoffman, B. M., LaHaie, E., Ballou, D. P., and Fee, J. A., Evidence for nitrogen coordination to iron in the [2Fe-2S] clusters of Thermus Rieske protein and phthalate dioxygenase from Pseudomonas, 1985, J. Biol. Chem., 260:3251.

Fee, J. A., Findling, K. L., Yoshida, T., Hille, R., Tarr, G. E., Hearshen, D. O., Dunham, W. R., Day, E. P., Kent, T. A., and Munck, E., 1984, Purification and characterization of the Rieske iron-sulfur protein from Thermus thermophilus. Evidence for a [2Fe-2S] cluster having non-cysteine ligands, J. Biol. Chem., 259:124.

Feiters, M. C., Al-Hakim, M., Navaratnam, S., Allen, J. C., Veldink, G. A., and Vliegenthart, J. F. G., 1987, Extended X-ray absorption fine structure (EXAFS) study of iron in native soybean lipoxygenase-1, Recl. Trav. Chim. Pays-Bas, 106:227.

Geary, P. J., Saboowalla, F., Patil, D., and Cammack, R., 1984, An investigation of the iron-sulfur proteins of benzene dioxygenase from Pseudomonas putida by electron spin resonance spectroscopy, Biochem. J., 217:667.

Gibson, D. T., Hensley, M., Yoshioka, H., and Mabry, T. J., 1970, Oxidative degradation of aromatic hydrocarbons by microorganisms. III. Formation of (+)-cis-2,3-dihydroxy-1-methyl-4,6-cyclohexadiene from toluene by Pseudomonas putida, Biochemistry, 9:1626.

Kuila, D., and Fee, J. A., 1986, Evidence for a redox-linked ionizable group associated with the 2-iron-2sulfur cluster of Thermus Rieske protein, J. Biol. Chem., 261:2768.

Nelson, N., and Neumann, J., 1972, Isolation of a cytochrome b6-f particle from chloroplasts, J. Biol. Chem., 247:1817.

Palmer, G., 1973, Current insights into the active center of spinach ferredoxin and other iron-sulfur proteins, in: "Iron-Sulfur Proteins," Lovenberg, W., ed., Academic Press, New York.

Penner-Hahn, J. E., Eble, K. S., McMurry, T. J., Renner, M., Balch, A. L., Groves, J. T., Dawson, J. H., and Hodgson, K. O., 1986, Structural Characterization of Horseradish Peroxidase Using EXAFS spectroscopy. Evidence for Fe=O litation in compounds I and II, J. Am. Chem. Soc., 108:7819.

Rieske, J. S., Hansen, R. E., and Zaugg, W. S., 1964, Studies of the electron transfer system. LVIII. Properties of a new oxidation-reduction component of the respiratory chain as studied by electron paramagnetic resonance spectroscopy, J. Biol. Chem., 239:3017.

Roe, A. L., Hodgson, K. O., Reem, R. C., Solomon, E. I., and Whittaker, J. W., 1985, SSRL Activity Report, X-ray absorption studies of oxygen:non-heme iron interactions, Proposal No. 932B.

Roe, A. L., Scheider, D. J., Mayer, R. J., Pyrz, J. W., Widom, J., and Que, L. Jr., 1985, X-ray absorption spectroscopy of iron-tyrosinate proteins, J. Am. Chem. Soc., 106:1676.

Sauber, K., Frohner, C., Rosenberg, G., Eberspacher, J., and Lingens, F., 1977, Purification and properties of pyrazon dioxygenase from pyrazon-degrading bacteria., Eur. J. Biochem., 74:89.

Stout, C. D., 1982, Iron-sulfur protein crystallography, Met. Ions Biol., 4:97.

Teo, B. K., and Lee, P. A., 1979, Ab initio calculations of amplitude and phase functions for extended x-ray absorption fine structure spectroscopy, J. Am. Chem. Soc., 101:2815.

Teo, B. K., Shulman, R. G., Brown, G. S., and Meixner, A. E., 1979, EXZFS studies of proteins and model compounds containing dimeric and tetrameric iron-sulfur clusters, J. Am. Chem. Soc., 101:5624.

Teo, B. K., and Shulman, R. G., 1982, X-ray absorption studies of iron-sulfur proteins and related compounds, Met. Ions Biol., 4:343.

FLUORESCENCE SPECTRAL CHARACTERISTICS AND FLUORESCENCE DECAY

PROFILES OF COVALENT POLYCYCLIC AROMATIC CARCINOGEN-DNA ADDUCTS

Seog K. Kim[1], Nicholas E. Geacintov[1], David
Zinger[2], and John C. Sutherland[2]

[1]Chemistry Department
New York University
New York, NY 10003

[2]Biology Department
Brookhaven National Laboratory
Upton, NY 11973

INTRODUCTION

Many polycyclic aromatic hydrocarbons (PAH) are known to be mutagenic
and carcinogenic. The relationships between structure, physico-chemical
properties, and differences in biological activities of PAH compounds, have
long been a source of fascination to chemists (Lehr et al., 1985; Harvey and
Geacintov, 1988). In living cells, the sparingly soluble and relatively
inert PAH molecules are metabolically converted to a variety of oxygenated
derivatives (reviewed by Conney, 1982); results obtained in many different
laboratories indicate that the ultimate mutagenic and tumorigenic forms of
PAH compounds are diol epoxide derivatives.

Benzo[a]pyrene is one of the most widely studied compounds of this
class of carcinogens, and the structures of two of the four stereoisomers of
the diol epoxide benzo[a]pyrene-trans-7,8-dihydrodiol-anti-9,10-epoxide
(BPDE) are shown in Fig. 1. It is widely believed that the covalent binding
of these and other highly reactive metabolites to cellular DNA constitutes
the critical step in mutagenesis and the initiation of the complex phenomenon of carcinogenesis. The two enantiomers of BPDE are merely mirror
images of one another, yet their biological activities are remarkably different; the (+) enantiomer is a potent tumorigen in mice, while the (-)
enantiomer is at best a weak carcinogen (Conney, 1982). Brookes and Osborne
(1982) have shown that the mutation efficiency (induced mutation frequency,
normalized to the number of surviving cells, per DNA adduct) of (+)BPDE is
4-5 times greater than that of (-)BPDE in Chinese hamster V79 cells. In
both cases, the dominant adduct involves the binding of BPDE to the exo-
cyclic amino group of deoxyguanine (dG); however, in the case of the (-)
enantiomer, significant binding to deoxyadenine (dA) and the O6 position of
dG is also observed. Brookes and Osborne speculated that the spatial conformation of the covalently bound (+)BPDE molecules is different from that
of the (-)BPDE molecules, and that this difference influences the template
proporties of the modified DNA regions towards replication and/or enzymatic
repair. Utilizing a variety of spectroscopic techniques we have shown that
there are indeed striking differences in the conformations of covalent

adducts derived from the binding of tumorigenic and non-tumorigenic PAH diol epoxides to DNA (Geacintov, 1985). In this paper, we discuss the differences in the characteristics of the adducts derived from the binding of the two enantiomers of BPDE to native DNA and to selected synthetic polynucleotides of defined base composition and sequence. Recent results obtained by fluorescence techniques are emphasized. There are several motivating factors for studying the fluorescence properties of PAH diol epoxide-DNA adducts. For example, fluorescence spectroscopy can be used to determine the type of PAH metabolite bound to cellular DNA (Vigny and Duquesne, 1979), and the overall level of binding of these carcinogens to DNA; utilizing synchronous fluorescence scanning techniques, Vahakangas et al. (1985) were able to detect one BPDE adduct per 10^6-10^7 DNA bases. Furthermore, fluorescence is probably the most sensitive spectroscopic technique for distinguishing different types of BPDE-DNA adducts. While different spectroscopic tools (Geacintov, 1985) can be readily utilized to study the properties of the adducts at higher levels of binding (one adduct/100-1000 bases, only the fluorescence methods are useful at the low levels of binding to DNA encountered in cells. In particular, it is shown here that adducts in which the aromatic chromophores are involved in stacking interactions with the DNA

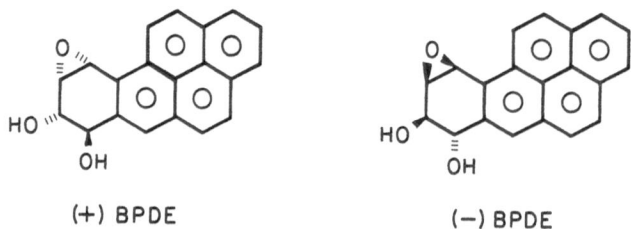

(+) BPDE   (−) BPDE

Fig. 1. Structures of the BPDE enantiomers

bases, and those which are not, can be clearly distinguished from one another by fluorescence methods. Techniques for distinguishing these two types of adducts may be of importance because the solvent-exposed, non-intercalative adducts are formed from the covalent binding of highly tumorigenic and mutagenic PAH diol epoxide derivatives; biologically less active epoxides on the other hand, give rise predominantly to quasi-intercalative, carcinogen-DNA base stacked conformations (Geacintov, 1985; Kolubayev et al., 1987; Harvey and Geacintov, 1988).

EXPERIMENTAL PROCEDURES

The enantiomers (+)BPDE and (−)BPDE were obtained from the National Cancer Institute Chemical Carcinogen Reference Standard Repository. The nucleic acids were purchased from Pharmacia (Piscataway, NJ). The preparation of the covalent adducts derived from the chemical reactions of (+)BPDE and (−)BPDE with DNA has been previously described (Geacintov et al., 1984a,b). BPDE-DNA adducts are known to be unstable and to dissociate to the tetraols 7,8,9,10-tetrahydrotetrahydroxybenzo[a]pyrene (BPT), particularly in the presence of light (Geacintov et al., 1987; Zinger et al., 1987). Therefore, special precautions were taken to minimize the

contributions of the highly fluorescent BPT molecules to the overall fluorescence of solutions of BPDE-nucleic acid adducts. In order to remove the
BPT molecules, the BPDE-nucleic acid reaction mixtures were extracted 12
times with ether, followed by five cycles of exhaustive dialysis against
buffer solutions (volume ratio 1:100). Control experiments in which known
amounts of BPT were added to DNA solutions, indicate that this treatment is
sufficient to remove all tetraols, detectable by fluorescence, from the DNA
solutions.

All of the experimental results described here were obtained with the
adducts dissolved in aqueous buffer solutions (5 mM sodium cacodylate
buffer, pH 7.0) at 23±1°, unless otherwise noted. Only 2-20% of the BPDE
molecules, depending on the enantiomer and on the base composition and base
sequence of the nucleic acids, bind covalently to either dG or dA. The
remaining fraction of BPDE molecules is hydrolyzed to BPT. The (+) enantiomer of BPDE binds 4-5 times more extensively to DNA than (-)BPDE; in the
case of the alternating polymer poly(dG-dC), the (+)BPDE/(-)BPDE binding
ratio is about 10. In the case of the alternating polynucleotide
poly(dA-dT), only the (+) enantiomer gives rise to any significant binding.

Absorption spectra were determined utilizing a Perkin Elmer 320 UV
spectrophotometer (Perkin Elmer Corporation, Norwalk, CT), while the fluorescence excitation and emission spectra were obtained with a single
photon counting fluorometer (Spex Industries, Edison, NJ). The fluorescence decay profiles were determined at the fluorometer at port U9B of
the National Synchrotron Light Source at Brookhaven National Laboratory;
this instrument and the non-linear least squares deconvolution computer
program for the analysis of the fluorescence decay curves, have been described by Laws and Sutherland (1986).

The orientations of the long axes of the pyrenyl residues bound to the
DNA were determined utilizing a flow dichroism method (Geacintov et al.,
1984b); in this technique, which is schematically outlined in Fig. 2, a
Couette cell (consisting of a stationary outer cylinder and a rotating inner
cylinder) is utilized to obtain oriented DNA molecules (the long axes of the
double-helices tend to align themselves along the flow direction, tangent to
the surfaces of the cylinders). The orientation of the transition moments
of the nucleic acid bases and the pyrenyl residues is then probed utilizing
linearly polarized light. The absorbances $A_{//}$ and $A_\perp$, measured with the
directions of polarization oriented either parallel or vertical with respect
to the flow lines, define the linear dichroism LD according to the relation:

$$LD = A_{//} - A_\perp \tag{1}$$

Transition moments, which tend to be oriented parallel to the flow
lines, give rise to a positive LD signal, while those which tend to be oriented along the perpendicular direction are characterized by a negative LD
signal. For example, because the planes of the DNA bases tend to be aligned
perpendicular to the flow lines, and because the transition moments are
oriented within these planes, the LD signal within the DNA absorption band
below 300 nm is negative in sign.

RESULTS AND DISCUSSION

Classification of types of BPDE-DNA adducts

The absorption spectrum of the pyrenyl residue is very sensitive to its
microenvironment. This property can be utilized to distinguish between different types of adducts. As a reference, we use the absorption spectrum of
the tetraol BPT dissolved in aqueous solution. This spectrum (Fig. 3A,

dashed line) is characterized by relatively sharp vibronic-electronic absorption bands (due to the So ----> S2 transition) at 313, 327 and 343 nm. The absorption spectra of the covalent (+)BPDE-poly(dG-dC), (-)BPDE-poly(dG-dC), and (+)BPDE-poly(dA-dT) adducts are quite different from the BPT absorption spectrum, as well as from one another (Fig. 3).

In general, two types of conformations derived from the binding of BPDE and similar compounds to DNA have been identified using UV absorption and linear dichroism spectroscopic techniques (Geacintov, 1985). One of these, called site I, appears to involve significant aromatic carcinogen-nucleic

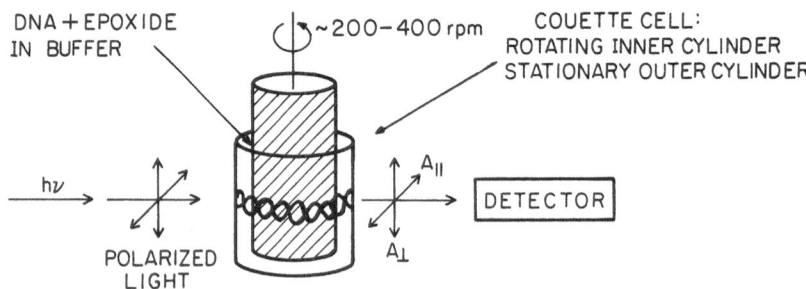

LINEAR DICHROISM:
● LD = A$_{||}$ − A$_{\perp}$ (Differences in Absorbances)

Fig. 2.   The flow linear dichroism apparatus for determining the orientation of the polycyclic aromatic residues bound to DNA or synthetic polynucleotides.

acid base stacking interactions and a ~10-11 nm red shift in the absorption spectrum of the bound pyrenyl residues; such characteristics are consistent with intercalative conformations in which the planar aromatic portions of the carcinogens are inserted between neighboring base pairs (LeBreton, 1985). As expected, since the planes of the aromatic residues tend to be parallel to those of the DNA bases, the LD spectra are negative in sign. In contrast to site I, site II is characterized by only a small red shift and a positive LD spectrum, suggesting that the long axis of the pyrene residue is tilted closer to the normals of the planes of the DNA bases.

Characteristics of Adducts Derived From the Binding of BPDE to Synthetic Polynucleotides

The covalent binding of the enantiomers of BPDE to native DNA can, in principle, give rise to significant conformational heterogeneities of adduct conformations. All four nucleotide building blocks are present in native DNA and therefore, in principle, many different base sequences containing the BPDE-dG adducts are possible. In an attempt to reduce the degree of heterogeneity of binding sites, we utilized the synthetic alternating purine/pyrimidine polynucleotides poly(dG-dC) and poly(dA-dT).

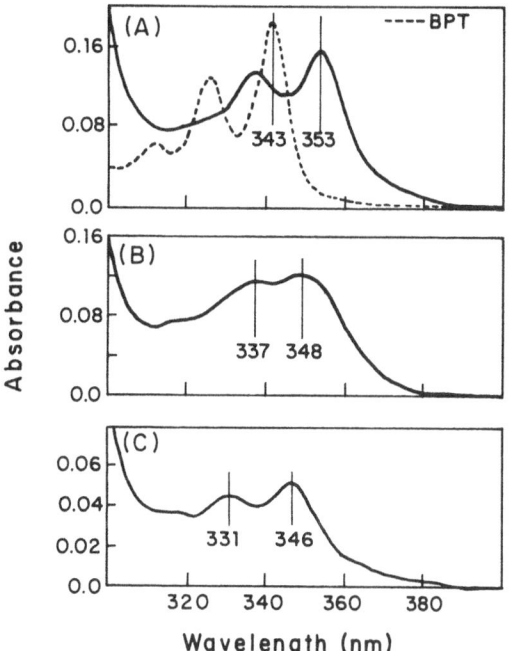

Fig. 3. Absorption spectra of adducts.
(A) Solid line: (+)BPDE-poly(dA-dT) adduct, 1.4% of bases modi-
fied. Dashed line: BPT (6×10^{-6} M) in aqueous buffer solution.
(B) (−)BPDE-poly(dG-dC) adduct, 1.4% of bases modified.
(C) (+)BPDE-poly(dG-dC) adduct, 0.8% of bases modified.

The absorption spectrum of the pyrenyl chromophore in the poly(dA-dT) adducts is clearly of the site I type, since the absorption spectrum is red-shifted by 10 nm relative to the BPT spectrum, and the maxima occur at 337 and 354 nm (Fig. 3A). In the (+)BPDE-poly(dG-dC) adducts, the maxima occur at 330 and 346 nm, and the red shift is only 3 nm (Fig. 3C). These characteristics are consistent with site II conformations. The absorption spectrum of the (-)BPDE-poly(dG-dC) adducts is significantly broader than in the other two polynucleotides and displays maxima at 337 and 348 nm (Fig. 3B). These features suggest a greater heterogeneity of adduct conformations; linear dichroism spectra appear to be indeed superpositions of site I and site II adducts (Roche, 1987). These characteristics are quite similar to those of (-)BPDE-DNA adducts (Zinger et al., 1987).

While significant binding of (+)BPDE to poly(dA-dT) is observed, the (-) enantiomer is much less reactive and thus we were not able to investigate the spectroscopic properties of (-)BPDE-poly(dA-dT) adducts. While poly(dA-dT) does not become as well oriented in the flow gradient of the Couette cell as poly(dG-dC) and native DNA, a significant LD spectrum due to the partially oriented dA-dT base pairs is observed; however, in contrast to the BPDE adducts of the other polymers, the LD signal above 310 nm is zero (data not shown). This implies that the pyrenyl residues in (+)BPDE-poly(dA-dT) adducts are not oriented. This lack of a linear dichroism signal is attributed to a significant disruption of the double helical structure of poly(dA-dT) in the immediate vicinity of the BPDE binding sites. The undisturbed polynucleotide regions contribute to the LD spectrum, while the disrupted regions and the pyrene residues in these regions do not.

In summary, the absorption spectra in Fig. 3 and the LD data (Roche, 1987), suggest that (+)BPDE-poly(dG-dC) adducts are predominantly of the site II type, while the (-)BPDE-poly(dG-dC) adducts are predominantly of the site I type. In (+)BPDE-poly(dA-dT) adducts, the red-shifted pyrenyl absorption spectra suggest considerable base-stacking interactions of the site I type; however, the pyrenyl moieties appear to reside in locally disordered regions of the polymer.

Fluorescence Excitation and Emission Characteristics of BPDE-Poly(dG-dC) Adducts at Room Temperature

The fluorescence excitation and emission spectra of (+)BPDE-poly(dG-dC) adducts at room temperature are depicted in Figs. 5A and A' those of (-)BPDE-poly(dG-dC) in Figs. 5B and B'. The fluorescence is clearly heterogeneous in nature. If the emission is monitored at 400 nm, the excitation spectrum of both the (+) adducts (Fig. 5A) and the (-) adducts (Fig. 5B) resemble, but are not quite identical to the corresponding absorption spectra (Fig. 3). The emission maxima at 380 and 400 nm are due to the normal, monomer-like emission of the pyrenyl chromophore. A broad emission band with a maximum at 475 nm is also observed and coincides with the known emission spectrum of pyrene excimers (Winnik et al., 1987). Similar, though somewhat less pronounced spectra are also observed with BPDE-DNA adducts, and a broad emission above 400 nm in BPDE-DNA adducts has also been reported recently by Vahakangas et al. (1985) and Eriksson et al. (1988). The latter workers ascribed this emission to the formation of pyrene excimers attributed to the interactions of an excited pyrenyl chromophore with a ground state pyrene residue, both located at highly flexible, nearby binding sites; the concept of conformationally mobile microenvironments of the pyrene residues is supported by the very low degrees of fluorescence polarization observed by Eriksson et al., 1988. Consistent with a bimolecular origin of the 475 nm emission, we observe that the intensity of this emission relative to the monomer emission at 380 nm tends to generally increase with increasing levels of covalent BPDE adducts per nucleotide (data not shown).

Excimer fluorescence emission spectra have also been recently observed in pyrene-labeled (Hydroxypropyl)cellulose polymers with similar levels of covalent attachment of pyrene residues (Winnik et al., 1987); the fluorescence excitation spectrum of the pyrene excimer emission should resemble

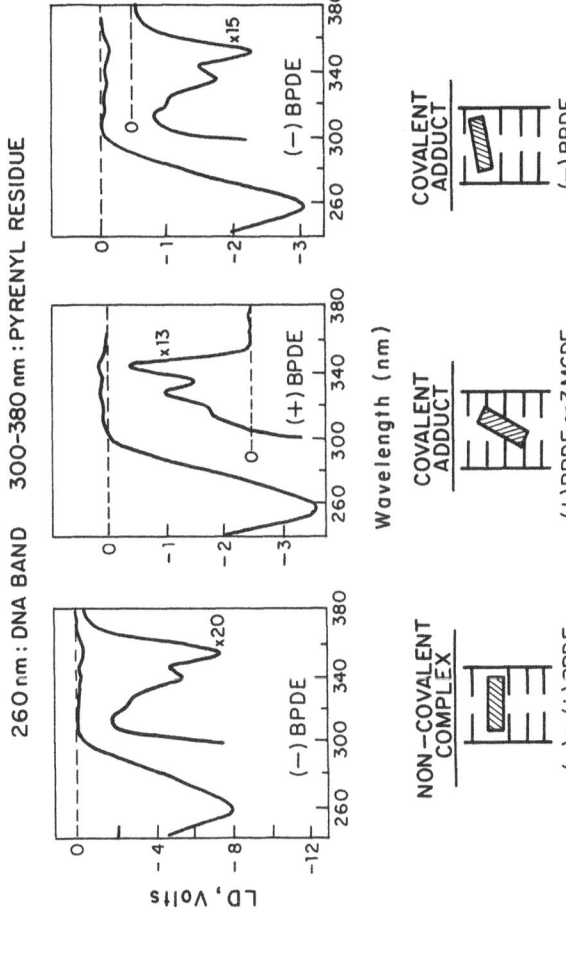

Fig. 4. Characteristic LD spectra. Left: noncovalent (−)BPDE-DNA complex, site II. Middle: covalent (+)BPDE-DNA complex, site I. Right: covalent (−)BPDE-DNA complex, site I. Another potent tumorigenic diol epoxide derived from 3-methylcholanthrene (3MCDE) gives rise to DNA adducts of similar conformation as (+)BPDE (data not shown).

the monomeric pyrene absorption spectrum, and this was indeed observed by Winnik et al. (1987). In the case of the BPDE-poly(dG-dC) adducts however, the excitation spectra of the fluorescence monitored at 475 nm are broad and

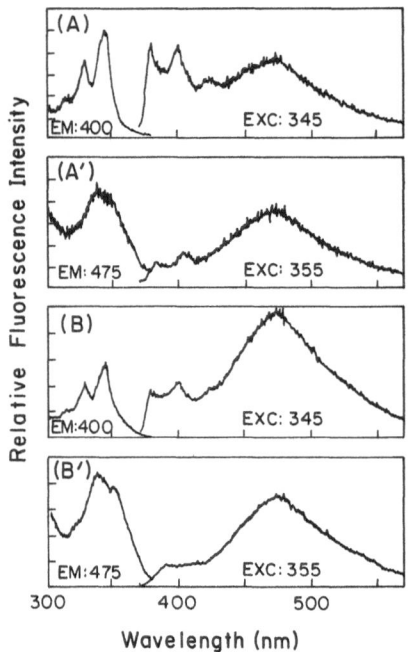

Fig. 5. Room temperature fluorescence excitation and emission spectra of (+)BPDE- and (-)BPDE-poly(dG-dC) adducts, (A,A') and (B,B'), respectively (the adducts are the same as those described in the legend corresponding to Fig. 3).

structureless with a maximum at 340 nm (Figs. 4A' and 4B'). Thus there are significant differences in the excitation spectra when the emission is moni- tored either at 400 nm or at 475 nm. This suggests that the 475 nm emission is due to a subset of molecules which are in a different, heterogeneous mi- croenvironment, which alters and broadens the vibrational spectrum of the pyrenyl residue. This emission, as in the case of acetylaminofluorene-dG adducts in DNA (Van Houte et al., 1987), could involve a significant charge transfer component between the excited states of the pyrene residues and one of the neighboring nucleic acid bases (exciplex-type emission). The fact that the excitation spectrum does not resemble a monomeric pyrene-like spectrum seems to argue against the excimer mechanism and in favor of an exciplex origin of the 475 nm emission. Perhaps both mechanisms are operative simultaneously. In summary, the exact origin of this long wavelength emission has not yet been clarified and is still under investigation.

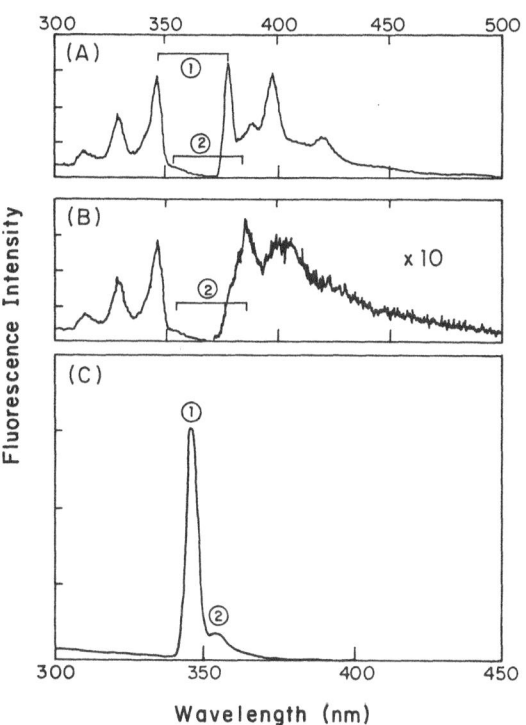

Fig. 6. Fluorescence characteristics of (−)BPDE-poly(dG-dC) adduct at 77 K. (A) and (B) fluorescence and excitation spectra. (C) Synchronous fluorescence scan spectrum with Δλ = 31 nm. The (+)BPDE-poly(dG-dC) adducts at 77 K are characterized by the Synchronous flourescence spectrophotometry (SFS) peak "1" and a weaker contribution of "2" as compared to the (−)BPDE-poly(dG-dC) spectrum shown. Same adducts as in Fig. 5.

There is a significant difference between the excitation spectra of the (+)BPDE- and (−)BPDE-poly(dG-dC) adducts when the fluorescence due to mono-meric pyrenyl residues is viewed at 400 nm (Figs. 5A and 5B). In the case of the (−) adducts, there is a significant shoulder in the excitation spectrum above 350 nm, suggesting that site I adducts are characterized by a non-zero fluorescence yield as described by Chen (1986). While a similar

shoulder is observable in the (+) adducts as well, it is much less pronounced. These observations are consistent with a greater contribution of
site I binding sites in the (-)BPDE-poly(dG-dC) adducts than in
(+)BPDE-poly(dG-dC) adducts, as already deduced from their absorption and
linear dichroism properties.

Fluorescence Characteristics of (-)BPDE-poly(dG-dC) Adducts at 77 K

Characteristic fluorescence excitation and emission spectra are shown
in Fig. 6A and Fig. 6B. At 77 K, the bands in both the emission and
excitation spectra are considerably sharpened in comparison to the room temperature spectra. While the excitation maxima are about the same at 77 k
and at 296 K, the emission maxima are blue shifted by about 3 nm at low
temperatures when the fluorescence is excited at 345 nm (Fig. 6A); these
fluorescence characteristics are attributed to site II type adducts. When

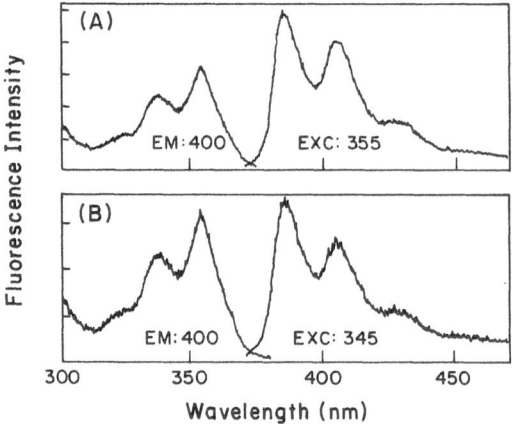

Fig. 7. Fluorescence excitation and emission spectra of (+)BPDE-poly(dA-dT)
adducts in the presence of 0.7 M acrylamide to suppress the tetraol
(BPT) fluorescence. Same adduct as described in the legend corresponding to Fig. 3.

the excitation wavelength is shifted to 353 nm, corresponding to site I absorption, the site II emission peaks are no longer visible, but are responsible for the 378 nm shoulder and structure at 398 nm (Fig. 6B); the emission band at 386 nm and the structure at 406 nm are attributed to site I
emission.

The disappearance of the broad exciplex/excimer emission above 440 nm
at low temperatures is noteworthy. This result is consistent with the notion that at room temperature this broad emission involves a conformational
mobility of some of the bound pyrene residues, which is frozen out at low
temperatures.

Vahakangas et al. (1985) utilized synchronous fluorescence spectrophotometry (SFS) methods to quantitate the level of covalent binding of BPDE
to DNA. In the SFS method, both the excitation wavelength monochromator and
the emission wavelength monochromator are scanned simultaneously with the
wavelength interval between the two monochromators remaining constant during
the scan. A maximum in the fluorescence signal as a function of the excitation wavelength is observed when the excitation wavelength corresponds to
the excitation and emission maxima as shown by the horizontal bar, 31 nm
wide, labeled "1" in Fig. 6A. As expected, when the SFS signal is plotted
as a function of the excitation wavelength, a maximum due to site II adducts
is observed at 346 nm (Fig. 6C). A small additional peak in the SFS spectrum, labeled "2", is also observed when the excitation wavelength is 354 nm
(Fig. 6C); as shown in Fig. 6B, this occurs when the (excitation wavelength
+ 31 nm) = 385 nm, the site I emission maximum.

In summary, at low temperatures the broad structureless fluorescence
emission observed at room temperature and attributed to excimers and/or
exciplexes, is no longer apparent. The fluorescence yield is enhanced
significantly, and the emission and excitation peaks are sharper, thus permitting a better resolution of the site I and site II adducts; the individual contributions of each of these two types of binding sites are clearly
apparent in the SFS spectra.

Fluorescence Characteristics of (+)BPDE-poly(dA-dT) Adducts

Characteristic fluorescence and excitation spectra of these adducts at
room temperature are shown in Fig. 7. In contrast to the BPDE-poly(dG-dC)
adducts, there is no change in the emission spectra when the fluorescence is
excited either at 345 or 355 nm. These spectra were determined in the presence of 0.7 M acrylamide, a known fluorescence quencher, in order to minimize the contributions of free tetraols; the (+)BPDE-poly(dA-dT) adducts
are sufficiently unstable so that the BPT dissociation products interfere
significantly with an accurate recording of the (+)BPDE-poly(dA-dT) spectra. At this particular acrylamide concentration, the fluorescence yields
of free tetraols in air-saturated solutions are known to be quenched by a
factor of 70 (Geacintov et al., 1987; Zinger et al., 1987).

The excitation maxima occur at 338 and 354 nm, while the emission
maxima are situated at 385 and 405 nm. The fluorescence excitation maxima
coincide with the site I type, red shifted absorption maxima (Fig. 3C). The
positions of the emission peaks are similar to those of the site I emission
maxima of (-)BPDE-poly(dG-dC) adducts at 77 K (Fig. 6B). Thus, there is no
apparent site II type emission in (+)BPDE-poly(dA-dT) adducts.

Finally, it is interesting to note that there is no exciplex/excimer
emission in this polymer. The broad, structureless emission observed in
poly(dG-dC) and in native DNA thus appears to be associated with deoxyguanine residues.

Fluorescence Lifetimes and Relative Amplitudes of the Decay Components

A knowledge of the fluorescence decay properties of ligand-macromole-
cule complexes or adducts is crucial for a thorough understanding of the
fluorescence properties of such systems. For example, it is important to
ascertain that the fluorescence emanates from a major fraction of the chromophores, and that these are not quenched statically. If the observed fluorescence is due to only a minor fraction of the ligand molecules in the
sample, then fluorescence studies of such systems are not worthwhile. This
issue can be resolved by comparing the absolute fluorescence quantum yields
with the fluorescence decay parameters (Geacintov, 1987). Previous results
obtained with BPDE-DNA complexes have shown that the fluorescence of the

pyrene residues is quenched dynamically by interaction with the DNA environment, and that the fraction of non-fluorescent (statically quenched) chromophores is negligibly small (Undeman, 1983; Jernstrom et al., 1984; Geacintov et al., 1987).

Typical lifetime data obtained with racemic BPDE-DNA adducts and excited with synchrotron radiation pulses have been reported (Geacintov et al., 1987). In air-saturated solutions, an ubiquitous 130 ns fluorescence component of variable amplitude was attributed to tetraol dissociation products (Geacintov et al., 1987); the fluorescence decay profiles of the ad-

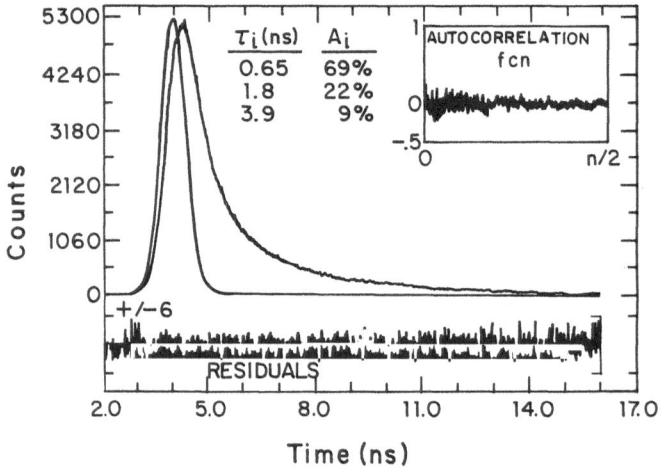

Fig. 8. Fluorescence decay profile of (+)BPDE-poly(dG-dC) adducts in the presence of 0.8 M acrylamide. The response function of the system is also shown. Excitation wavelength: 346 nm; emission viewing wavelength: 400 nm (predominantly monomeric pyrenyl residue fluorescence).

ducts were accounted for in terms of two lifetime components of about 1-1.5 ns (relative amplitude of about 2/3), and a longer-lived component of about 5-7 ns (amplitude of about 1/3); it was concluded that these three components can account reasonably well for the shapes of the fluorescence decay profiles, but that the presence of a larger number of fluorescence decay components could not be ruled out.

In this work, typical fluorescence decay parameters obtained with adducts derived from the covalent binding of (+)BPDE and (-)BPDE to calf thymus DNA and to poly(dG-dC), and of (+)BPDE to poly(dA-dT), are discussed. The

fluorescence decay profiles, F(t), can be analyzed in terms of a 2-4
component multi-exponential decay model:

$$F(t) = \sum_i a_i \exp[-t/\tau i] \qquad (2)$$

where a_i and τi are the relative amplitudes and lifetimes, respectively, of
each of the components. The lifetime of the fourth, 130 ns tetraol component, was always fixed, while its amplitude was treated as a variable
parameter (Geacintov et al., 1987). The fluorescence decay profiles were
then analyzed (Laws and Sutherland, 1986) in terms of either 2 or 3 additional components, depending on whether a three-component analysis gave a
better "chi square" value than a two-component analysis; the value of this
parameter was generally in the range of 0.7 - 1.5, which probably means that
the kinetics are not exactly describable in terms of eq. 2 and the number of
decay components chosen. Therefore, these analyses are only approximate in
nature, and the kinetics underlying the fluorescence decay, particularly if
time-dependent exciplex or excimer interactions are involved, could be significantly more complex than suggested by this simple model (eq. 2).

A typical fluorescence decay profile of (+)poly(dG-dC) adducts (346 nm
excitation, 400 nm emission), the values of the lifetimes and associated amplitudes, and the residuals and autocorrelation of residuals (Laws and
Sutherland, 1986) are shown in Fig. 8. This particular decay profile, and
the set of amplitudes and lifetimes shown in the Figure, were determined in
the presence of 0.8 M acrylamide in order to supress the fluorescence of tetraol dissociation products. When the excitation wavelength is 340 nm and
the emission wavelength is 475 nm, the decay profiles in BPDE-poly(dG-dC)
and BPDE-DNA adducts can be accounted for in terms of two components. This
wavelength combination corresponds to the excimer/exciplex emission. A typical fluorescence decay profile obtained with (+)BPDE-DNA adducts is depicted in Fig. 9. The lifetimes of these two components of roughly equal
amplitudes are about 1.8 and 14 ns. Neither the fluorescence yield, nor the
lifetimes measured at 475 nm were sensitive to acrylamide (up to 0.8 M).

Since the exciplex/excimer fluorescence is characterized by significant amplitudes in the excitation spectrum at 346 nm and the emission
spectrum at 400 nm (Fig. 5), it becomes important to take this emission into
account in analyzing the decay profiles at this set of wavelengths. It thus
became necessary to include a total of three parameters, rather than two
(besides the 130 ns component, Geacintov et al., 1987) in the analysis of
the fluorescence decay profiles. As before, only the lifetime of the 130 ns
BPT component was fixed, while all of the other parameters were free-
running. Typical results of such analyses are summarized in Tables 1 and 2.

Results for (+)BPDE-DNA and (-)BPDE-DNA adducts are summarized in
Table 1. When the emission wavelength is fixed at 400 nm, and the excitation wavelengths are fixed at either 346 or 352 nm, a 15 - 22 ns component
is recovered; these values are close to the 13 -17 ns component observed in
DNA and poly(dG-dC) adducts when the emission wavelength is fixed at 475 nm;
in the BPDE-poly(dG-dC) and (-)BPDE-DNA adducts, the lifetimes of the third
component are below 10 ns, in apparent disagreement with the longest lifetime obtained at 475 nm. These differences could be due to inadequacies of
the simple multi-exponential fluorescence decay model and the relatively
large number of parameters used in the analysis of the experimental decay
curves. Because the characteristics of the fluorescence decay profiles of
the enantiomeric BPDE-polynucleotide adducts appear just as complex as in
the native BPDE-DNA adducts, efforts to analyze these profiles in greater
detail were abandoned. Nevertheless, it can be concluded that the relative
amplitude of the ~15-22 ns component is less than 15% when excited at 346 nm
and viewed at 400 nm, which is reasonably consistent with the spectral data

in Fig. 5. Thus, while the excimer/ exciplex emission can account for a significant fraction of the overall emission, depending on the excitation and emission wavelengths, the relatively small amplitude at 346 nm excitation suggests that this long-lived emission arises from a minor fraction of bound pyrenyl residues.

In view of the foregoing results, only the first two decay components listed in Tables 1 and 2 can be attributed to covalently bound, monomeric pyrenyl residues when the emission is viewed at 400 nm. In the case of 346 nm excitation, these two fluorescence lifetimes (1.4 and 5.0 ns) are somewhat longer in (+)BPDE-DNA than in (-)BPDE-DNA (0.8 and 3.1 ns), and close to the values previously reported for racemic BPDE-DNA adducts (Geacintov et

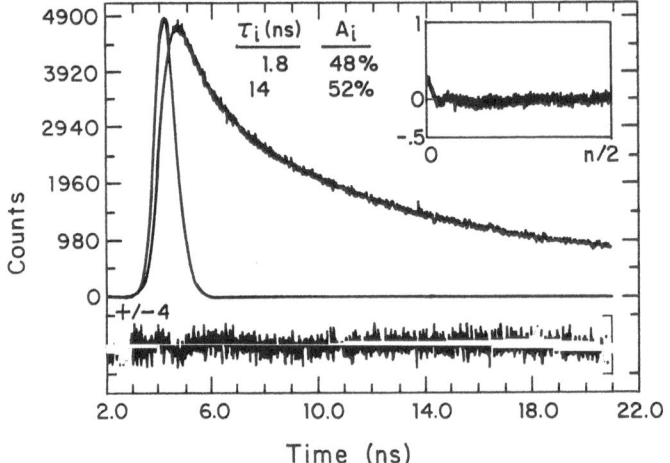

Fig. 9. Fluorescence decay profile of (+)BPDE-DNA adduct (1.1 % of bases modified). Excitation wavelength: 355 nm; emission viewing wavelength: 475 nm (exciplex/excimer emission is dominant under these conditions).

al., 1987). In the case of 352 nm excitation, the lifetimes and amplitudes of the (+) and (-)BPDE-DNA adducts are similar in value (about 0.8 and 3.0 ns).

The first two fluorescence decay times and amplitudes of the (+) and (-)BPDE-poly(dG-dC) adducts are similar to one another (about 0.5 - 0.6 and 1.4 - 1.6 ns), and are shorter than the analogous values in the DNA adducts; the lifetime of the second component is slightly longer (2.4 - 2.6 ns) at the 352 nm excitation wavelength than at 346 nm.

In the case of the (+)BPDE-poly(dA-dT) adducts, it was found that two lifetime components (in addition to the 130 ns BPT component) were adequate for describing the observed fluorescence decay profile; three-component fits

Table 1. Fluorescence decay times (ns) and relative amplitudes (in parentheses) of adducts derived from the covalent binding of (+)BPDE or
(-)BPDE to double-stranded calf thymus DNA. Excitation wavelength
either 346 or 352 (±1) nm with emission viewing wavelength of
400 nm, or excitation at 340 and viewing at 475 nm.

	346/400	352/400	340/475
(+)BPDE-DNA	1.4(0.45)	0.8(0.44)	1.8(0.48)
	5.0(0.37)	2.7(0.41)	14 (0.52)
	19 (0.13)	15 (0.14)	
	130(0.05)		
(-)BPDE-DNA	0.8(0.69)	0.8(0.37)	1.1(0.69)
	3.1(0.19)	3.1(0.44)	9.2(0.31)
	22 (0.11)	15 (0.19)	
	130(0.01)		

did not lead to better chi-square values than two-component fits. About
half of the fluorescence is due to a 1 ns component, while the remainder can
be attributed to a longer, 8-13 ns component. As mentioned previously, no
excimer/exciplex emission is evident in this particular polynucleotide. In
general, the highest mean lifetimes are obtained in the dA-dT polymer (1.0
and 8.5 ns), with intermediate lifetime values in native DNA (0.8 - 5.0
ns), while the shortest decay times (0.5 - 0.6 and 1.4 - 1.6 ns) are observed in poly(dG-dC). It can thus be concluded that dG-dC base pairs are
most effective in quenching the fluorescence of covalently bound pyrenyl
residues, while dA-dT base pairs exert a weaker quenching effect.

Fluorescence Quenching by Acrylamide and Solvent Accessibility of Pyrenyl Residues

Bimolecular collisions between excited fluorophores and quencher molecules result in a decrease in the fluorescence yield. If only one type of
fluorophore is present, the intensity of the fluorescence in the absence and
presence of quencher, denoted by F_0 and F, respectively, is given by the
Stern-Volmer expression:

$$F_0/F = 1 + \tau K[Q] \qquad (3)$$

where K is the bimolecular collisional encounter rate constant and [Q] is
the quencher concentration.

The magnitude of the bimolecular collisional rate constant K is a
measure of the solvent exposure of the fluorophores bound to DNA. The
efficiency of acrylamide as a quencher of the fluorescence of various fluorophores bound to DNA in relation to their mode of binding, has been recently
discussed (Zinger and Geacintov, 1988). Acrylamide fluorescence quenching
studies of (+)BPDE-DNA and (-)BPDE-DNA adducts (Zinger et al., 1987) and of
adducts derived from the binding of (+)BPDE or (-)BPDE to poly(dG-dC) (data
not shown) are consistent with the notion that the fluorescence of pyrene
residues at type II binding sites is sensitive to acrylamide, while site I
residues which involve pyrenyl-base stacking interactions are not (Zinger et
al., 1987). These conclusions are supported by results obtained by optical
detection of magnetic resonance (ODMR) techniques (Kolubayev et al., 1907).

TABLE 2. Fluorescence decay times (ns) and relative amplitudes of
covalent adducts derived from the binding of (+)BPDE or (-)BPDE
to synthetic polynucleotides. Excitation wavelengths either 346
or 352 ($\pm$1) nm, emission viewing wavelength: 400 nm.

	346/400	354/400
(+)BPDE-poly(dG-dC)	0.5(0.58)	0.7(0.67)
	1.4(0.32)	2.4(0.29)
	4.4(0.07)	17 (0.04)
	130(0.03)	
(-)BPDE-poly(dG-dC)	0.6(0.61)	0.8(0.73)
	1.6(0.24)	2.6(0.21)
	6.1(0.10)	16 (0.06)
	130(0.05)	
(+)BPDE-poly(dA-dT)	1.0(0.40)	1.1(0.44)
	8.5(0.48)	13 (0.56)
	130(0.12)	

Acrylamide Stern-Volmer quenching plots obtained with
(+)BPDE-poly(dA-dT) adducts are shown in Fig. 10. The quenching curve
obtained with free BPT in aqueous solution is shown for comparison (solid
line); the quenching constant for free BPT is $7.5.10^8$ $M^{-1}s^{-1}$. Utilizing an
excitation wavelength of 345 nm, it appears that tetraol dissociation
products are contributing to the fluorescence since the F_o/F plot for the
(+)BPDE-poly(dA-dT) adduct seems to follow the BPT line until the acrylamide
concentration reaches about 0.05 M. As the quencher concentration is

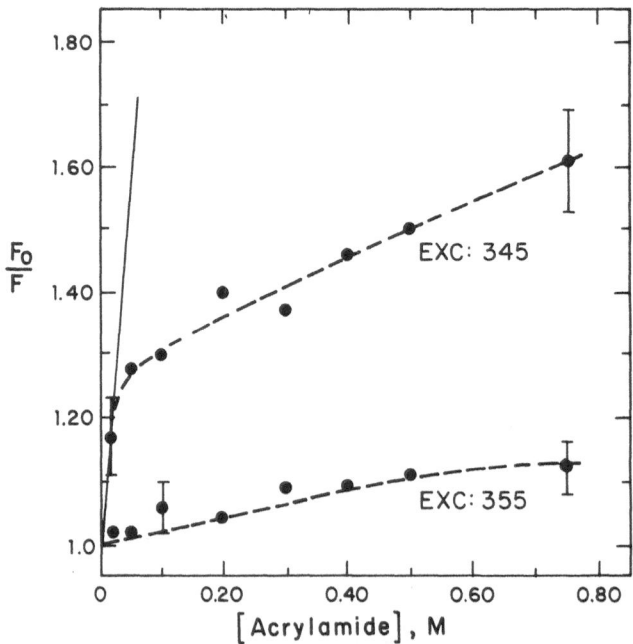

Fig. 10. Stern-Volmer acrylamide fluorescence quenching plots. Solid
lines: free BPT in aqueous buffer solution. Dashed lines:
(+)BPDE-poly(dA-dT) adducts, which excitation at two different
wavelengths.

increased beyond this value, the F_o/F plot is still linear, but with a greatly diminished slope. At the 355 nm excitation wavelength, tetrols are no longer contributing to the fluorescence and an apparent straight-line portion with an intercept of 1.0 can be visualized, at least at acrylamide concentrations below 0.5 M. A value of $K\tau$ of 0.20 ± 0.03 can be estimated from this slope. Utilizing the two fluorescence lifetimes of 1.1 and 13 ns (Table 2), values of K of $\sim 1.8 \cdot 10^8$ and $\sim 1.5 \cdot 10^7$ $M^{-1}s^{-1}$ can be calculated, respectively. These values are about 5-50 times lower than the value of K for free tetrols. This suggests that the site I type residues in (+)BPDE-poly(dA-dT) adducts are only slightly accessible to acrylamide; this is expected for adducts in which base-stacking interactions tend to sterically inhibit the interactions of acrylamide with the photoexcited chromophores. According to the linear dichroism results discussed earlier, these pyrene residues appear to be located in locally disordered regions of the poly(dA-dT) polymer.

CONCLUDING REMARKS

Both time-resolved and steady-state fluorescence methods can provide significant information on the characteristics of polycyclic aromatic carcinogen - nucleic acid adducts. Fluorescence techniques, particularly at low temperatures, are valuable for distinguishing site I and site II BPDE-nucleic acid polymer adducts. Site II conformations involve external and solvent-exposed binding modes, and are dominant in adducts derived from the chemical reactions of the more tumorigenic PAH diol epoxide derivatives with DNA; the formation of site I adducts, which involve carcinogen-base stacking interactions, is generally associated with less tumorigenic or non-tumorigenic PAH epoxide derivatives (Geacintov, 1985; Harvey and Geacintov, 1988). Ivanovic et al. (1978) have shown that fluorescence methods can be used to trace the evolution of adducts formed from the binding of BPDE metabolites to DNA and RNA in hamster embryo cell cultures; a further refinement of such techniques might be used to follow more specifically the kinetics of appearance and disappearance of site I and site II adducts in cells exposed to BP or BPDE.

ACKNOWLEDGMENTS

At New York University the work was supported by the U.S. Public Health Service, Grant CA 20851 awarded by the National Cancer Institute, and by the Department of Energy (Grants DEFGO2-86ER60405 and DEFGO2-288ER60674), while the research at Brookhaven National Laboratory was supported by a grant from the National Institute of General Medical Sciences (GM 35662) and by the Office of Health and Environmental Research, United States Department of Energy. The National Synchrotron Light Source at Brookhaven National Laboratory is supported by the Office of Basic Energy Sciences, USDOE.

REFERENCES

Brookes, P. and Osborne, M. R., 1982, Mutation in mammalian cells by stereoisomers of anti-benzo[a]pyrene-diol epoxide in relation to the extent and nature of the DNA reaction products, Carcinogenesis, 3:1223.

Chen, F. M., 1986, Binding of enantiomers of trans-7,8-dihydroxy-anti-9, 10-epoxy-7,8,9,10-tetrahydrobenzo[a]pyrene to polynucleotides, J. Biomol. Struct. and Dynamics, 3:401.

Conney, A. H., 1982, Induction of microsomal enzymes by foreign chemicals and carcinogenesis by polycyclic aromatic hydrocarbons, Cancer Res., 42:4875.

Eriksson, M., Norden, B., Jernstrom, B., and Graslund, A., 1988, Binding geometries of benzo[a]pyrene diol epoxide isomers covalently bound to DNA. Orientational distribution, Biochemistry (USA), 27:1213.

Geacintov, N. E., Ibanez, V., Gagliano, A. G., Jacobs, S. A., and Harvey, R. G., 1984a, Stereoselective covalent binding of anti-benzo[a]pyrene diol epoxide to DNA. Conformation of enantiomer adducts, J. Biomol. Struct. Dynamics, 1:1473.

Geacintov, N. E., Yoshida, H., Ibanez, V., Jacobs, S. A., and Harvey, R. G., 1984b, Conformations of adducts and kinetics of binding of the optically pure enantiomers of anti-benzo[a]pyrene diol epoxide, Biochem. Biophys. Res. Commun., 122:33.

Geacintov, N. E., 1985, Mechanisms of interaction of polycyclic aromatic diol epoxides with DNA and structures of the adducts, in: "Polycyclic Hydrocarbons and Carcinogenesis," R. G. Harvey, ed., ACS Symposium No. 83, The American Chemical Society, Washington, D.C.

Geacintov, N. E., 1987, Principles and applications of fluorescence techniques in biophysical chemistry, Photochem. Photobiol., 45:547.

Geacintov, N. E., Zinger, D., Ibanez, V., Santella, R., Grunberger, D., and Harvey, R. G., 1987, Properties of covalent benzo[a]pyrene diol epoxide-DNA adducts investigated by fluorescence techniques, Carcinogenesis, 8:925.

Harvey, R. G. and Geacintov, N. E., 1988, Intercalation and binding of carcinogenic hydrocarbon metabolites to nucleic acids, Acc. Chem. Res., 21:66.

Ivanovic, V., Geacintov, N. E., Yamasaki, H., and Weinstein, I. B., 1978, DNA and RNA adducts formed in hamster embryo cell cultures exposed to benzo[a]pyrene, Biochemistry (USA), 17:1597.

Jernstrom, B., Lycksell, P. O., Graslund, A., and Norden, B., 1984, Spectroscopic studies of DNA complexes formed after reaction with anti-benzo[a]pyrene-7,8-dihydrodiol-9,10-oxide enantiomers of different carcinogenic potency, Carcinogenesis, 5: 1129.

Kolubayev, V., Brenner, H. C., and Geacintov, N. E., 1987, Stereoselective covalent binding of enantiomers of anti-benzo[a]pyrene diol epoxide to DNA as probed by optical detection of magnetic resonance, Biochemistry (USA), 26:2638.

Laws, W. R. and Sutherland, J. C., 1986, The time-resolved photon counting fluorometer at the national synchrotron light source, Photochem. Photobiol., 44:343.

LeBreton, P. R., 1985, The intercalation of benzo[a]pyrene and 7,2-dimethyl-benz[a]anthracene metabolites and metabolite model compounds into DNA, in: "Polycyclic Hydrocarbons and Carcinogenesis," R.G. Harvey, ed., ACS Symposium No. 283, The American Chemical Society, Washington, D.C.

Lehr, R. E., Kumar, S., Levin, W., Wood, A. W., Chang, R. L., Conney, A. H., Yagi, H., Sayer, J. M., and Jerina, D. M., 1985, The bay region theory of polycyclic aromatic hydrocarbon carcinogenesis, in: "Polycyclic Hydrocarbons and Carcinogenesis," R.G. Harvey, ed., ACS Symposium Series No. 283, The American Chemical Society, Washington.

Roche, C. J., 1987, "The physical and covalent interactions of polycyclic aromatic hydrocarbon epoxide derivatives with nucleic acids," Ph.D. dissertation, New York University.

Undeman, O., Lycksell, P. O., Graslund, A., Astlind, T., Ehrenberg, A., Jernstrom, B., Tjerneld, F., and Norden, B., 1983, Covalent complexes of DNA and two stereoisomers of benzo[a]pyrene 7,8-dihydrodiol-9,10-epoxide studied by fluorescence and linear dichroism, Cancer Res., 43:1851.

Vahakangas, K., Haugen, A., and Harris, C. C., 1985, An applied synchronous fluorescence spectrophotometric assay to study benzo[a]pyrene-diol epoxide - DNA adducts, Carcinogenesis, 6:1109.

Van Houte, L. P. A., Bokma, J. T., Lutgerink, J. T., Westra, J. G., Retel, J., Van Grondelle, R., and Blok, J., 1987, An optical study of the conformation of the aminofluorene-DNA complex, Carcinogenesis, 8:759.

Vigny, P. and Duquesne, M., 1979, Luminescence, a tool in hydrocarbon carcinogenesis, J. Luminescence, 18/19:587.

Winnik, F. M., Winnik, M. A., Tazuke, S., and Ober, C. K., 1987, Synthesis and characterization of pyrene-labeled (hydroxypropyl) cellulose and its fluorescence in solution, Macromol., 20:38.

Zinger, D., Geacintov, N. E., and Harvey, R. G., 1987, Conformations and selective photodissociation of benzo[a]pyrene diol epoxide enant-iomer-DNA adducts, Biophys. Chem., 27:131.

Zinger, D. and Geacintov, N. E., 1988, Acrylamide and molecular oxygen fluorescence quenching as a probe of solvent accessibility of aromatic fluorophores complexed with DNA in relation to their conformations: coronene-DNA and other complexes, Photochem. Photobiol., 47: 181.

TIME-RESOLVED FLUORESCENCE IN

STUDIES OF PROTEIN STRUCTURE AND DYNAMICS

William R. Laws[1] and David M. Jameson[2]

[1]Department of Biochemistry
Mount Sinai School of Medicine
New York, NY 10029

[2]Department of Pharmacology
University of Texas Southwestern Medical Center
Dallas, TX 75235

INTRODUCTION

Fluorescence spectroscopy continues to develop as a useful technique for understanding the complex relationships between structure and function of proteins and other biological macromolecules. Fluorescence spectroscopy has several advantages, including sensitivity (only a small amount of sample is needed), the ability to examine processes over a wide time range (picoseconds to seconds), the ability to work under "physiological" conditions, and the ability to measure many different physical parameters (quantum yield, lifetime, emission energies, anisotropy) that can contribute to the understanding of the system.

Time-resolved fluorescence studies involve the investigation of the rate of loss of either the first excited singlet electronic state (intensity decay) or an induced polarization of the excited state (anisotropy decay). Protein structural fluctuations can perturb either the intensity or anisotropy decay; these perturbations may be related to local events near or involving the probe (relatively fast), to "global" processes involving the entire protein (relatively slow), or to both. Since the lifetime of the excited state for most probes used to study proteins is subnanosecond to tens of nanoseconds, protein dynamics which occur on this time scale can be examined.

Fluorescent probes of protein structure and dynamics include the intrinsic aromatic amino acids tryptophan, tyrosine, and phenylalanine, and extrinsic chromophores that can be added to the system. Extrinsic probes can either be covalently attached to the protein through a free amino or sulfhydryl group, or linked through noncovalent interactions as, for example, a fluorescent substrate analogue binding to the active site of an enzyme. This diverse set of probes for protein structure and dynamics studies necessitates diverse spectral and temporal requirements for the excitation source. Synchrotron radiation is capable of providing many of these requirements. Several protein systems have been examined using time-resolved fluorescence techniques by taking advantage of synchrotron radiation.

TIME-RESOLVED FLUORESCENCE

Intensity Decay

Most of the planar aromatic molecules used as probes of proteins can be promoted into their first excited singlet electronic state within an energy range between the near ultraviolet and the near infrared. The absorption of the photon, creating the excited state from the ground state, occurs in less than a femtosecond. Vibrational relaxation processes occurring on the picosecond time scale force the chromophore to attain the lowest vibrational level of the first excited singlet state. It is from this energy level that most planar aromatic molecules will undergo fluorescence in competition with all the nonradiative processes. The deactivation of the excited state (A*) can be expressed in terms of the fluorescence rate constant, k_F, and a combined rate constant for all the nonradiative events, k_{nr}, by the equation

$$-d[A^*]/dt = (k_F + k_{nr}) [A^*]. \qquad (1)$$

On integrating equation 1, the fluorescence intensity decay should follow the simple exponential function

$$I(t) \approx \exp(-t/\tau), \qquad (2)$$

where τ is the lifetime of the excited state and equals the reciprocal of the sum of the two rate constants. This expression holds only if the probe is in a noninteracting environment like a vacuum.

Fluorescent molecules in an interacting system are not likely to have simple photophysics. Complicating processes can occur either in the ground state or the excited state. In the ground state, even though the molecule is chemically pure, interactions with itself, other solutes, or the solvent can result in more than one environment for the probe. This ground-state heterogeneity will lead to more than one excited state; these different excited states may have different fluorescence and nonradiative rate constants, and this will complicate the analysis of the intensity decay function.

In the excited state, the probe is essentially a different chemical entity. The dipole moment of the excited state is different than that of the ground state, and this change can induce photochemistry and/or interactions with the environment. Depending on the relative rates of these processes, the overall kinetic scheme describing the decay of the excited state could become more complex. If the interactions are faster than the resolution of the experiment, then the effect is averaged and a single exponential decay is expected. If the processes are much slower than the fluorescence and nonradiative rates, then fluorescence will have occurred before those events could affect the kinetic scheme. But if the interactions are on the same time scale as the excited-state lifetime, then excited-state reactions are occurring and the kinetic expression for the loss of the excited species can become very complex.

In most cases of complex kinetic decay mechanisms, the overall intensity decay can be expressed by the sum of many first order events, i.e.

$$I(t) = \sum_i \alpha_i \exp(-t/\tau i), \qquad (3)$$

where α's are the pre-exponential weighting factors, or amplitudes. Other kinetic mechanisms can yield different complex rate expressions; for example, a diffusion-dependent system will require a square root of time term.

Anisotropy Decay

The polarization of an excited state, created by the absorption of polarized light, can be lost through rotation of the excited molecule, provided that the rotation occurs on the same time scale or faster than the decay of the excited state. Relationships have been derived relating the dimensions of the rotating unit, sweeping out a volume, to the time dependence of the decay of the initial polarized excited state (Rigler and Ehrenberg, 1973). A symmetrical molecule rotating as a sphere will exhibit only one emission anisotropy decay time constant. If the molecule is an ellipsoid of revolution, theory predicts five rotational correlation times containing the three diffusion constants. In practice, however, only three of the five are "obtainable" since two of the correlation parameters are essentially degenerate with the others (Small and Isenberg, 1977). The preceeding description assumed that the rotating fluorophore itself is a rigid body. A protein probe, either intrinsic or extrinsic and either covalent or noncovalent, could have depolarizing motions other than those caused by the rotation of the protein. The fluorophore may have rotational motion about its linkage independent of the protein. The portion (domain) of the polypeptide chain where the fluorophore is linked may also have independent motion (segmental flexibility). Thus the "simple" case of three exponentials to describe the decay of the emission anisotropy for an ellipsoid of revolution can be complicated by other motions within the protein. For the specific case of a mobile fluorescent probe covalently attached to a spherical protein, the decay of the emission anisotropy is the sum of three exponential terms (Rigler and Ehrenberg, 1973), where one term represents the global motion of the entire unit and the other two terms are related to motions involving the entire unit and the mobile probe.

The anisotropy decay, $r(t)$, is typically evaluated by the sum of exponentials

$$r(t) = \sum_j \beta_j \exp(-t/\phi_j). \tag{4}$$

The ϕ_j terms must then be related to the rotational correlation times of the system. The pre-exponential β terms depend on trigonometric functions of the angles between the absorption and emission dipoles and a rotational symmetry axis (Rigler and Ehrenberg, 1973). The sum of the β terms equals the initial anisotropy of the system, r_0, before any depolarizing motions can occur.

Instrumentation

Our ability to probe biological systems through time-resolved fluorescence techniques has advanced rapidly over the recent years. Advances have come in both instrumentation and data analysis capabilities. Two conceptually different, but complementary methods have been developed and are used extensively. One method works in the time domain and is known as the pulse technique (Badea and Brand, 1979). The other method, known as the phase/modulation technique, works in the frequency domain (Gratton et al., 1984a).

Time Domain. Pulse fluorometers operate by accumulating a histogram of the probability of the decay of the excited state as a function of time. This operation is accomplished by measuring the time between the detection of a short pulse of light used to excite the sample and the detection of the photon emitted by fluorescence. One count is added to a multichannel memory device at the channel that corresponds to the measured time interval. This process is performed under single photon-counting conditions, and repeated thousands of times to collect a decay curve as outlined in Fig. 1.

The pulse of exciting light should be a true delta function with respect to the temporal resolution required by the experiment. In many cases, this can be achieved using current laser technology. The detection electronics, particularily the photomultiplier tube, of the pulse fluorometer, however, establish an instrument response function that can "broaden" a few picosecond-wide delta function into a measured distribution on the order of several hundreds of picoseconds up to more than a nanosecond. The collected decay curve, unfortunately, is therefore convolved with the instrument response function. Although recent advances in photomultiplier technology have significantly reduced the instrument response function, the temporal resolution required to resolve events with rate constants of 10^8 S^{-1} or faster will still result in an appreciable instrument response function.

To obtain the parameters describing the decay, the collected decay curve must be analyzed by curve fitting techniques. Several mathematical algorithms have been employed to perform this "deconvolution," including nonlinear least squares, method of moments, and LaPlace and Fourier transformations (Badea and Brand, 1979). Fig. 2 demonstrates the use of the nonlinear least squares technique to resolve the sum of two exponentials where the lifetimes are short and only separated by a factor of two. The ability to resolve lifetimes on the order of a 10% difference has recently been achieved through the use of "global" data analysis, where many decay curves collected as a function of an independent variable such as wavelength, pH, or temperature are analyzed together for common parameters (Knutson et al., 1983). The ability to resolve complex kinetic decay mechanisms has also been improved by the introduction of the linked-function analysis approach where specific decay parameters can be restrained within certain limits based on information obtained from another physical measurement (Ross et

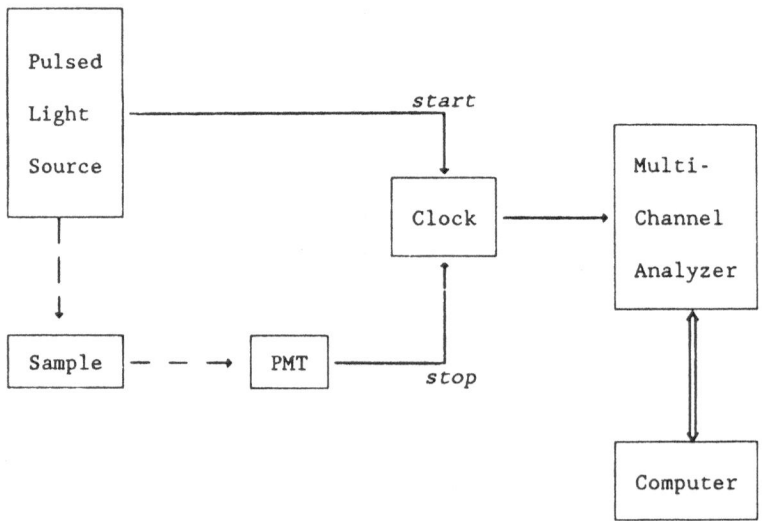

Fig. 1. Block diagram of a pulse fluorometer. Broken arrows represent light while solid arrows represent electrical signals. The start signal indicates when the light source has pulsed and the stop signal from the photomultiplier tube (PMT) is generated by the detection of a photon through fluorescence of the sample.

al., 1986b). By combining the linked-function approach with the global
method, previously unresolvable systems, such as tryptophan fluorescence
from a protein, can begin to be unraveled.

With the pulse method, time-resolved emission anisotropy can be
determined by collecting decay curves through a polarizer oriented parallel
and perpendicular to vertically polarized excitation light, hence obtaining
I_p and I_s respectively. Eq. 5 defines $r(t)$, where $D(t)$ denotes the

$$r(t) = \{[I_p(t) - I_s(t)]/[I_p(t) + 2I_s(t)]\} = D(t)/S(t) \qquad (5)$$

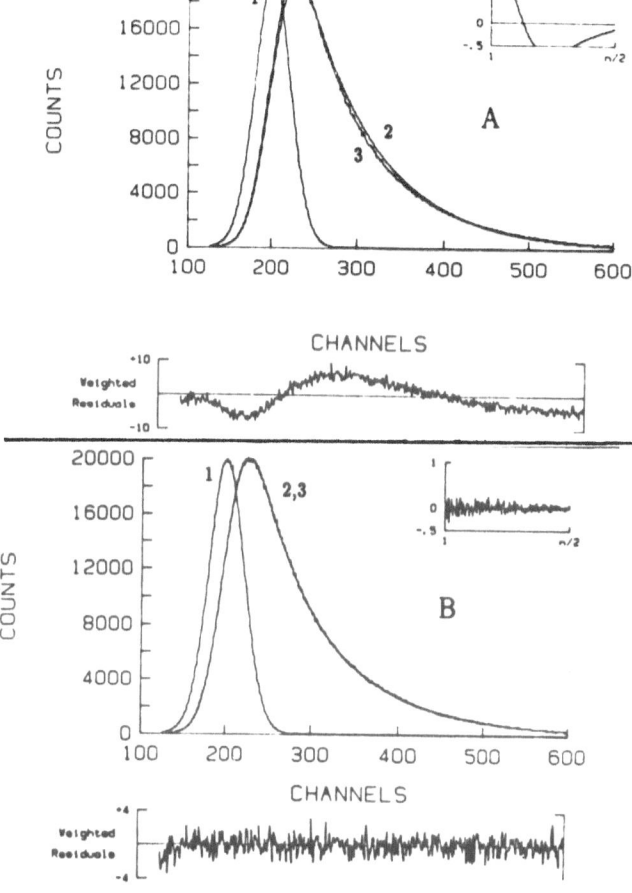

Fig. 2. Analysis of a simulated fluorescence decay curve. Curve 1 is the
instrument response function, curve 2 is the calculated fit, and
curve 3 is the decay curve. The weighted residuals are displayed
below and the autocorrelation function of the residuals is in the
inset. The decay was generated with $\tau_1 = 1.0$ ns, $\tau_2 = 2.0$ ns,
$\alpha_1 = \alpha_2$, and "a" timing calibration of 20 ps/channel. Panel A: a
fit to a single exponential. Panel B: a fit to a sum of two
exponentials; the recovered parameters were within 5% of those used
to generate the decay.

211

difference curve and S(t) is the sum curve which gives the total intensity
decay. The parameters for r(t), as given in equation 4, can be determined
by constructing S(t), obtaining its decay parameters using equation 3, and
then iterating r(t) parameters, using the S(t) values, to fit D(t). With
the recent advent of global data analysis, it is more precise to analyze the
parallel and perpendicular decays together, without constructing D(t). As
shown in equations 6 and 7, these decays depend both on the

$$I_p(t) = 1/3 \{S(t) [1 + 2r(t)]\} \tag{6}$$

$$I_s(t) = 1/3 \{S(t) [1 - r(t)]\} \tag{7}$$

total intensity as well as the anisotropy decay. Again by knowing S(t)
parameters, r(t) parameters can be iterated to fit both polarized decays.

Frequency Domain. In frequency domain fluorescence spectroscopy, the
fluorescent molecules are excited by a light source with an intensity sinusoidally modulated at high angular frequency, ω, typically in the megahertz
range. If the fluorophore is characterized by a single exponential decay
time, τ, then the emitted light will be sinusoidally modulated at the same
frequency but delayed in phase and demodulated with respect to the excitation (Fig. 3). Equations 8 and 9 give the relationships for the phase shift
Φ and relative modulation M,

$$\tan \Phi = \omega\tau \tag{8}$$

$$M = \{(ac/dc)_{EM} / (ac/dc)_{EX}\} = [1 + (\omega\tau)^2]^{-1/2}. \tag{9}$$

For a homogeneous emitting population the phase (τ^P) and modulation
(τ^M) lifetimes will be equal and independent of the modulating frequency.
If the emission is multiexponential or nonexponential, then a composite sinusoidal emission waveform results with frequency ω and a phase delay and de-

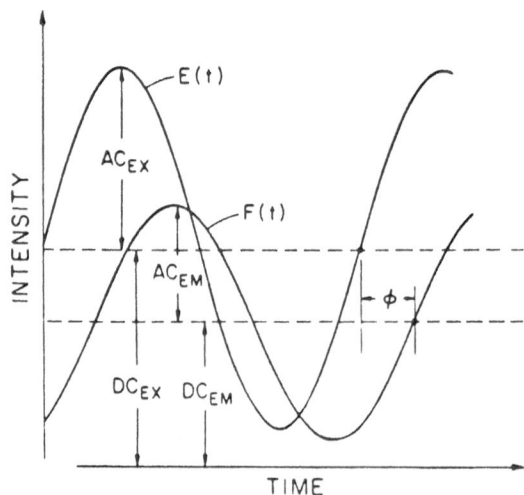

Fig. 3.   Schematic representation of the excitation E(t) and fluorescence
          F(t) waveforms. Fluorescence is delayed by an angle Φ and demod-
          ulated with respect to the excitation.

modulation given by equations 10 and 11, where S and G are defined in equations 12 and 13,

$$\Phi = \tan^{-1} (S/G) \qquad (10)$$

$$M = (S^2 + G^2)^{-1/2} \qquad (11)$$

$$S = \sum_i f_i M_i \sin\Phi_i \qquad (12)$$

$$G = \sum_i f_i M_i \cos\phi_i. \qquad (13)$$

In equations 12 and 13 ϕ_i, M_i, and f_i are the phase delay, demodulation, and fractional intensity associated with each individually emitting species ($\sum f_i = 1$). In such a system, $\tau^P \neq \tau^M$ and both τ^P and τ^M are frequency dependent. The functions S and G are the sine and cosine transforms of the impulse response as shown in equations 14 and 15 (Weber, 1981).

$$S = \frac{\int_0^\infty I(t) \sin(\omega t)\, dt}{\int_0^\infty I(t)\, dt} \qquad (14)$$

$$G = \frac{\int_0^\infty I(t) \cos(\omega t)\, dt}{\int_0^\infty I(t)\, dt} \qquad (15)$$

These equations form the basis for the mathematical equivalence between the time domain and the frequency domain approaches. Consequently, the parameters describing the decay of the fluorescence (equation 3) are obtained by regression algorithms to fit the phase and modulation data as a function of ω. This fitting procedure can be enhanced by taking data as a function of an independent variable and analyzing in a global manner as was done in the pulse technique (Beechem et al., 1983).

Dynamic polarization measurements are the frequency domain equivalent of anisotropy decay measurements. In this method, the fluorescent solution is illuminated by intensity modulated light which is polarized parallel to the vertical laboratory axis. The phase delay, $\Delta\Phi$, between the parallel and perpendicular polarization components of the emission can then be directly determined as well as the ratio of their AC components, Y. Equations 16 and 17 can be derived for an isotropic rotator, where r

$$\Delta\Phi = \tan^{-1} \{(3\omega rR)/[(k^2 + \omega^2)(1 + r - 2r^2) + R(R + 2k + kr)]\} \quad (16)$$

$$Y^2 = \{[k + 6R/(1 - r)]^2 + \omega^2\}/\{[k + 6R/(1 + 2r)]^2 + \omega^2\} \quad (17)$$

is the limiting anisotropy, R is the rotational diffusion coefficient, and k is the radiative decay rate (Gratton et al., 1984a). Examples of simulated dynamic polarization curves are shown in Fig. 4.

The relatively recent appearance of true multifrequency phase and modulation fluorometers, developed in the laboratory of Enrico Gratton (Gratton and Limkeman, 1983), has enormously extended the scope and power of this method. A true multifrequency instrument provides facile selection of arbitrary modulation frequencies over a wide frequency range. Two of the more significant advances in phase fluorometry were the application of cross-correlation techniques (Spencer and Weber, 1969) and the use of electro-optic devices (Pockels cells) for light modulation.

Excitation Light Sources. The pulse technique requires that the exciting light come from a pulsed source. Recently, utilization of intrinsically modulated sources such as synchrotron radiation and mode-locked lasers to collect phase and modulation data has affected a union of the impulse and harmonic response approaches. This union has stemmed from the need to resolve all of the parameters involved in a complex kinetic scheme or to accurately recover events occurring with rates of 10^8 s^{-1} or faster. This goal requires that frequency domain instruments be able to provide modulation frequencies into the gigahertz range, which is presently impossible through electro-optic devices.

High repetition pulsed light sources, such as synchrotron radiation, were suggested as possible light sources for multifrequency phase and modulation fluorometry in an analogous manner to the traditional sinusoidally modulated source (Gratton and Lopez-Delgado, 1980; Munro, 1983; Munro and Schwentner, 1983). As shown in Fig. 5 for the temporal structure of the ADONE storage ring at the National Laboratory in Frascati, Italy, the pulsed nature of the light source corresponds to a set of equally spaced frequencies separated by the inverse of the time between the pulses (Gratton et al., 1984b). This frequency set has a Gaussian envelope (because the synchrotron pulse is Gaussian) with a half width the reciprocal of the half width of the pulse duration. Consequently, the shorter the pulse the higher the frequency distribution will extend into the gigahertz region.

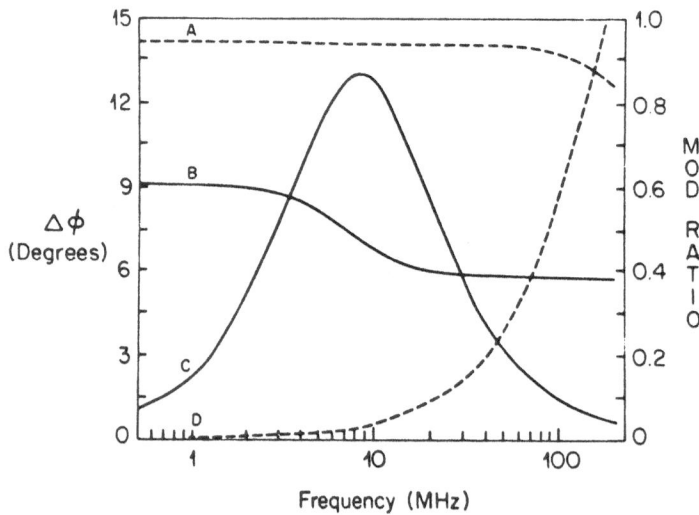

Fig. 4. Simulated dynamic polarization results for two specific cases. Solid lines denote ΔΦ and dashed lines denote Y, or the modulation ratio. Curves B and C were generated with τ = 26 ns and the Debye rotational relaxation time ρ = 87 ns. Curves A and D were generated with τ = 1.9 ns and ρ = 0.54 ns.

An ideal light source for both a pulse and a phase/modulation fluorometer would have: 1) easy selection of a 1 nm-wide band of a continuous wavelength distribution from 200 to 800 nm; 2) high photon flux at all wavelengths; 3) a pulse width less than 10 ps and 4) polarized light. The two instruments have different repetition rate requirements, however. The pulse instrument would be benefitted by having a light source with a selectable pulse rate up to 25 MHz, while the phase/modulation instrument would be best operated at a fixed frequency near 1 MHz to obtain optimal harmonic content.

Synchrotron radiation from storage rings can provide many of these qualifications for an ideal light source for either type of fluorescence instrument. The light is continuous over the desired energy range, it is intense, and it is highly polarized. While the pulse width varies from machine to machine, and many are not as short as desired, this need is not as critical as the others. Many machines also do not have an optimal time structure, but experiments can usually be performed. Also, different electron bunch fill patterns can often be accomodated in the storage ring if a particular repetition rate is required.

PROTEINS, TIME-RESOLVED FLUORESCENCE, AND SYNCHROTRONS

The major advantage that synchrotrons have as the excitation source for time-resolved fluorescence studies on proteins is the ability to excite in the ultraviolet. Consequently, the work on proteins using synchrotron radiation has concentrated on the intrinsic fluorescence of the aromatic amino acids. Even with this advantage over other light sources, only a few studies have been done due to the fact that the intensity decay

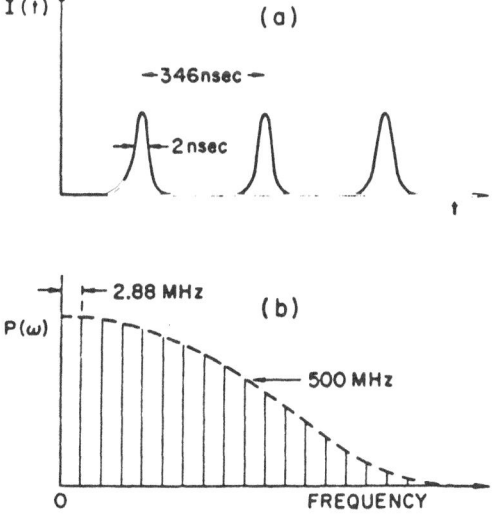

Fig. 5. Panel (a) gives a schematic representation of the light pulses emitted by the ADONE storage ring operating in single bunch mode. Panel (b) gives the power spectrum of the pulse train in panel (a).

characteristics of tryptophan and tyrosine are still not fully understood. The problem is explaining the causes of the multi-exponential decay behaviour of the free amino acids, simple amino acid analogues, and proteins with either a single tryptophan or tyrosine (Beechem and Brand, 1985). Without a complete understanding of the underlying photophysics, it will be difficult to use a change in a decay parameter to explain a property of a protein.

One of the first studies of intrinsic fluorescence of proteins using synchrotron radiation was carried out on hemoglobin and its isolated α and β chains (Alpert and Lopez-Delgado, 1976). The intrinsic fluorescence of these proteins is very weak due to very efficient energy transfer to the heme moiety. The measurements, in the time domain, were facilitated by the high intensity and repetition rate of the ACO storage ring in Orsay, France. A static quenching mechanism was implied by the data, since the decrease observed in the quantum yield between apohemoglobin and hemoglobin was not corroborated by a equivalent decrease in the "average" lifetime.

The excellent deep ultraviolet advantages of synchrotron radiation were utilized in a study of the excitation wavelength dependence of tryptophan's fluorescence lifetime. The pulse instruments at both the ACO storage ring and the SPEAR facility at Standford were used (Alpert et al., 1979; Jameson and Alpert, 1979). The major 3.1 ns lifetime at 20° C and pH 7 was demonstrated to be invariant with excitation between 220 to 320 nm, confirming that fluorescence stems from the lowest energy level of the first excited singlet state independent of the initial excited state.

The SPEAR facility was also used for some pioneering work on protein dynamics using time-resolved fluorescence anisotropy. Several proteins containing a single tryptophan were examined, and it was shown that the tryptophan residue exhibited different degrees of rotational freedom with respect to the rest of the protein (Munro et al., 1979). The single rotational correlation time of 9.9 ns at 20° C for the tryptophan residue in the 20,000 dalton <u>Staphylococcus aureus</u> nuclease B demonstrates that it has essentially no independent mobility. Two rotation times of 0.09 and 1.26 ns were seen for the tryptophan residue in the bovine basic A1 myelin protein (18,000 daltons) suggesting a very unhindered residue. Intermediate of these two examples was the single tryptophan residue of human serum albumin. At 8° C, it showed no independent motion and a single rotational correlation time consistent with rotation of the protein. But at 43° C, a second time of 0.14 ns was seen indicating an increase in mobility of the tryptophan residue with temperature.

Tyrosine to tryptophan resonance energy transfer in bovine serum albumin was studied using the phase/modulation instrument constructed at the ADONE storage ring at Frascati, Italy (Gratton et al., 1984b). Energy transfer was detected by examining the $\tau^P \neq \tau^M$ inequality, particularly the fact that $\tau^P > \tau^M$ with 270 nm excitation which is an indication of an excited-state reaction. Using a fixed modulation frequency, a "transfer excitation spectrum" was recorded by monitoring the phase lifetime as a function of the excitation wavelength. This spectrum was quite similar to the absorption profile of tyrosine, with a maximum near 270 nm (Antonangeli et al., 1983).

The pulse instrument at the SRS facility in Daresbury, England, has been used to examine the time-dependent emission anisotropy of the single tyrosine residue in the peptide hormone angiotensin (Munro et al., 1985). Two rotational correlation times were reported; the shorter time of 0.3 ns when compared to the 3.7 ns component probably indicates some independent motion by the tyrosine residue with respect to the rest of the hormone.

The pulse instrument at ACO has recently been used for two different protein systems. As might be expected, the intensity decay of the seven tryptophans in wheat germ hexokinase LI is complex (Merola and Brochon, 1986). On the other hand, the anisotropy decay appears to be very simple with a single rotational correlation time relating to the entire protein. The two tryptophans in aspartate transcarbamylase from Escherichia coli were also found to have complex decay kinetics (Royer et al., 1987). Two lifetimes, however were found to contribute over 95% of the total fluorescence; by assuming one lifetime for each tryptophan residue, the effects on various portions of the protein were examined upon binding of a substrate analogue and nucleotide effector molecules.

In an attempt to understand the complex fluorescence decay kinetics of tyrosine, a systematic study was conducted using the pulse instrument at the NSLS at Brookhaven National Laboratory in New York (Laws and Sutherland, 1986). Researchers examined phenol, straight-chained phenol analogues, tyrosine, and tyrosine derivatized at the α-amino and α-carboxyl groups as a function of pH below neutrality (Laws et al., 1986). Only in tyrosine and its analogues were multi-exponential kinetics observed that could not be explained by the number of ionic species in solution or the possibility of excited-state proton transfer. The kinetics could be explained in terms of the rotamer model (Gauduchon and Wahl, 1978), where ground-state heterogeneity exists due to the different environments seen by the three configurations about the C^α-C^β bond. The analysis of the data was possible through the use of the linked-function approach (Ross et al., 1986b); ground-state rotamer populations calculated from H-NMR spectra were linked to the amplitudes during iteration. If the rotamer model is the correct interpretation of the kinetic mechanism, then an upper limit of 10^8 s^{-1} must be placed on the rate of interconversion for the rotamers about the C^α-C^β bond. These studies were extended to oxytocin, a small peptide hormone containing a single tyrosine residue (Ross et al., 1986a). The rotamer model was able to explain the complex decay behaviour, including the requirement that one of the tyrosyl rotamers be statically quenched by an interaction with the internal disulfide bond of the hormone. These tyrosine studies were facilitated by synchrotron radiation because it helped overcome the low extinction coefficient and quantum yield of tyrosine as well as the problem of Raman light scattering.

The ADONE phase/modulation instrument has recently been used for studies on two different protein systems. Preliminary lifetime determinations were done on the single tryptophan in the protein biosynthesis elongation factor Tu of prokaryotes (Jameson et al., 1987). The fluorescence decay kinetics were shown to be complex, but the tryptophan residue could be dynamically quenched by added acrylamide. The fluorescence decay characteristics of the single tryptophan of the iron storage protein ferritin is also complex (Rosato et al, 1987). A difference was found in the pattern of the decay parameters between the individual subunit and the assembled apoprotein. Since the tryptophan is near the subunit-subunit interface, this probably indicates an altered environment for the tryptophan residue upon assembly. Furthermore, while the addition of iron to the apoprotein quenched the intrinsic fluorescence, it was a nonlinear process indicating the presence of both static and dynamic quenching of the tryptophan induced by the binding of iron ions.

With the abilities of the technology of genetic engineering, it should be possible to probe specific domains of a protein by introducing a single tryptophan or tyrosine residue and then examining it by time-resolved fluorescence. A similar approach has been taken in the study of lactate dehydrogenase from Bacillus stearothermophilus using the 3RS pulse instrument (Waldman et al., 1987). The native protein contains three tryptophan residues, and its fluorescence decay could be fit by the sum of three exponentials. Site-directed mutagenesis was used to make two mutant proteins, replacing either one or two of the tryptophans with tyrosine. The mutant enzymes are fully active with near normal physical properties and substrate affinities. With one tryptophan removed, the fluorescence decay was double exponential and with two of the three residues replaced the decay was single exponential. Thus in this protein system, it appears that the individual tryptophans do not have complex kinetics but are single exponentials, and that this then allows lifetimes to be assigned to individual residues.

ACKNOWLEDGMENTS

The authors thank Dr. J. B. Alexander Ross for helpful discussions. WRL is supported by NIH grants DK39548 and DK10080. DMJ acknowledges support of NSF grants DMB-8706440 and INT-8408263.

REFERENCES

Alpert, B., and Lopez-Delgado, R., 1976, Fluorescence lifetimes of haem proteins excited into the tryptophan absorption band with synchrotron radiation, Nature, 263:445.
Alpert, B., Jameson, D. M., Lopez-Delgado, R., and Schooley, R., 1979, Tryptophan fluorescence lifetimes as a function of excitation wavelength, Photochem. Photobiol., 30:479.
Antonangeli, F., Bassani, F., Campolungo, A., Finazzi-Agro, A., Grassano, U. M., Gratton, E., Jameson, D. M., Piacentini, M., Rosato, N., Savoia, A., Weber, G., and Zema, N., 1983, A multifrequency cross-correlation phase fluorometer with picosecond resolution using synchrotron radiation, in: "Report LNF-83/68(R) of the Istituto Nazionale di Fisica Nucleare," Laboratori Nazionali di Frascati.
Badea, M. G., and Brand, L., 1979, Time-resolved fluorescence measurements, Methods Enzymol., 61:378.
Beechem, J. M., Knutson, J. R., Ross, J. B. A., Turner, B. W., and Brand, L., 1983, Global resolution of heterogeneous decay by phase/modulation fluorometry: mixtures and proteins, Biochemistry, 22:6054.
Beechem, J. M., and Brand, L., 1985, Time-resolved fluorescence of proteins, Ann. Rev. Biochem., 54:43.
Gauduchon, P., and Wahl, P. H., 1978, Pulse fluorimetry of tyrosyl peptides, Biophys. Chem., 8:87.
Gratton, E., and Lopez-Delgado, R., 1980, Measuring fluorescence decay times by phase-shift and modulation techniques using the high harmonic content of pulsed light sources, Il Nuovo Cimento, 56B:110.
Gratton, E. and Limkeman, M., 1983, A continuously variable frequency cross-correlation phase fluorometer with picosecond resolution, Biophysical J., 22:315.
Gratton, E., Jameson, D. M., and Hall, R. D., 1984a, Multifrequency phase and modulation fluorometry, Ann. Rev. Biophys. Bioeng., 13:105.
Gratton, E., Jameson, D. M., Rosato, N., and Weber, G., 1984b, Multifrequency cross-correlation phase fluorometer using synchrotron radiation, Rev. Sci. Instrum., 55:486.
Jameson, D. M., and Alpert, B., 1979, The use of synchrotron radiation in fluorescence studies on biochemical systems, in: "Synchrotron Radiation Applied to Biophysical and Biochemical Research," A. Castellani and I. F. Quercia, eds., Plenum, New York.
Jameson, D. M., Gratton, E., and Eccleston, J. F., 1987, Intrinsic fluorescence of elongation factor Tu in its complexes with GDP and elongation factor Ts, Biochemistry, 26:3894.
Knutson, J. R., Beechem, J. M., and Brand, L., 1983, Simultaneous analysis of multiple fluorescence decay curves: a global approach, Chem. Phys. Lett., 102:501.

Laws, W. R., and Sutherland, J. C., 1986, The time-resolved photon-counting fluorometer at the National Synchrotron Light Source, Photochem. Photobiol., 44:343.

Laws, W. R., Ross, J. B. A, Wyssbrod, H. R., Beechem, J. M., Brand, L., and Sutherland, J.C., 1986, Time-resolved fluorescence and ^{1}H NMR studies of tyrosine and tyrosine analogues: correlation of NMR-determined rotamer populations and fluorescence kinetics, Biochemistry, 25:599.

Merola, F., and Brochon, J. C., 1986, Polarised pulse fluorimetry study on the conformational properties of wheat germ hexokinase LI, Eur. Biophys. J., 13: 291.

Munro, I., Pecht, I., and Stryer, L., 1979, Subnanosecond motions of tyrptophan residues in proteins, Proc. Natl. Acad. Sci. USA, 76:56.

Munro, I. H., 1983, Synchrotron radiation as a source to study time-dependent phenomena, in: "Time-resolved Fluorescence Spectroscopy in Biochemistry and Biology," R. B. Cundall and R. E. Dale, eds., NATOASI Series A: Life Sciences, Vol. 69, Plenum, New York.

Munro, I. H., and Schwentner, N., 1983, Time-resolved spectroscopy using synchrotron radiation, Nucl. Instrum. Methods, 208:819.

Munro, I. H., Shaw, D., Jones, G. R., and Martin, M. M., 1985, Time resolved fluorescence spectroscopy with synchrotron radiation, Anal. Instrum., 14:465.

Rigler, R., and Ehrenberg, M., 1973, Molecular interactions and structure as analysed by fluorescence relaxation spectroscopy, Quat. Rev. Biophys., 6:139.

Rosato, N., Finazzi-Agro', A., Gratton, E., Stefanini, S., and Chiancone, E., 1987, Time-resolved fluorescence of apoferritin and its subunits, J. Biol. Chem., 262:14487.

Ross, J. B. A., Laws, W. R., Buku, A., Sutherland, J. C., and Wyssbrod, H. R., 1986a, Time-resolved fluorescence and ^{1}H NMR studies of tyrosyl residues in oxytocin and small peptides: correlation of NMR-determined conformations of tyrosyl residues and fluorescence decay kinetics, Biochemistry, 25:607.

Ross, J. B. A., Laws, W. R., Sutherland, J. C., Buku, A., Katsoyannis, P. G., Schwartz, I. L., and Wyssbrod, H. R., 1986b, Linked-function analysis of fluorescence decay curve kinetics: resolution of side-chain rotamer populations of a single aromatic amino acid in small peptides, Photochem. Photobiol., 44:365.

Royer, C.A., Tauc, P., Herve, G., and Brochon, J.-C., 1987, Ligand binding and protein dynamics: a fluorescence depolarization study of aspartate transcarbamylase from Escherichia coli, Biochemistry, 26:6472.

Small, E. W., and Isenberg, I., 1977, Hydrodynamic properties of a rigid molecule: rotational and linear diffusion and fluorescence anisotropy, Biopolymers, 16:1907.

Spencer, R. D., and Weber, G., 1969, Measurements of subnanosecond fluorescence lifetimes with a cross-correlation phase fluorometer, Ann. N. Y. Acad. Sci., 158:361.

Waldman, A. D. B., Clarke, A. R., Wigley, D. B., Hart, K. W., Chia, W. N., Barstow, D., Atkinson, T., Munro, I., and Holbrook, J. J., 1987, The use of site-directed mutagenesis and time-resolved fluorescence spectroscopy to assign the fluorescence contributions of individual tryptophan residues in Bacillus stearothermophilus lactate dehydrogenase, Biochim. Biophys. Acta, 913:66.

Weber, G., 1981, Resolution of the fluorescence lifetimes in a heterogeneous system by phase and modulation measurements, J. Phys. Chem., 85:949.

VACUUM ULTRAVIOLET PHOTOBIOLOGY WITH SYNCHROTRON RADIATION

Takashi Ito

Institute of Physics, College of Arts and Sciences
University of Tokyo, Meguroku, Komaba 3-8-1
Tokyo 153, Japan

INTRODUCTION

The vacuum-ultraviolet (VUV) region of the electromagnetic spectrum is situated between the far-UV and X-ray regions. It covers a vast range of photon energy from 6.20 eV (200 nm) to 41.3 eV (30 nm). In a broader sense, it covers several hundred eV. The spectral regions on both sides of the VUV are relatively easy in biological experimentations since atmospheric conditions pose no obstacles. In contrast, VUV experiments are difficult to perform with living systems because of the necessity of maintaining the target in a vacuum. Besides, photon sources available for VUV experiments have been limited. In the work by Bovie (1916) we see what seems to be the earliest work on biological experiments with VUV radiation. He irradiated microorganisms with VUV radiation through a fluorite crystal plate. Blank and Arnold (1935) reported the killing of Bacillus spores by 110-130 nm photons. Heinmets and Taylor (1952) treated bacterial cells by passing them in aqueous suspension through an electrical arc. More recently, Jagger et al., (1967) studied VUV effects on Streptomyces conidia by successive use of cutoff filters down to 155 nm. These experiments avoided the vacuum problem either by using vacuum-resistant spores or by special innovations for the VUV irradiation. Apart from living materials, biomolecules, like proteins, can be dried in vacuum and thus become suitable for VUV irradiation. Setlow et al., (1959) were probably the first to perform inactivation experiments with enzymes using monochromatic light in the VUV region. Wirths and Jung (1972) and Sontag and Dertinger (1975) irradiated DNA and bacteriophages in vacuum with dispersed VUV radiation using a conventional light source. Although these biological effects, for which a sensitive assay technique is available, could be investigated by conventional VUV light sources, precise product analysis in the chemical sense is almost unthinkable with them due to the lack of intensity after monochromatization. Rare gas discharge lamps such as the one using bromine (Ito and Kobayashi, 1976) or an excimer laser (Green et al., 1987) could provide intense monochromatic VUV radiation but are greatly limited in the number of wavelengths that are available. Synchrotron radiation (SR) is a breakthrough as a VUV radiation source.

Why is the study of the effects of VUV radiation on biological materials important? Miller and Urey (1959) listed electrical discharges as one of the main modes of generation of organic molecules from inorganic materials in the prebiotic period. It is known that VUV radiation is a significant component of the emission from electrical arcs. The synthesis of large molecules by VUV radiation should be attempted in the future. It may also be possible to observe types of damage to DNA and other molecules different from the damage produced by other regions of the spectrum. What would be the evolutionary implication of such damage and how would the presently known repair systems for far-UV and X-rays deal with it? Because of its very high absorption coefficient, VUV may cause localized damage to the cellular structures such as the cell membrane (Ito et al., 1983).

Another aspect, which is more fundamental, has already been treated theoretically by a group of physicists, notably Fano and Platzman (e.g., see Inokuti, 1986). Platzman (1967) urged the study of molecular changes following excited states generated by radiation whose energy is above the ionization potential. He stated that these excited state species may decompose into fragments. It has been commonly observed either by electron loss spectroscopy or by optical measurements that the maximum absorption of electromagnetic radiation by biological molecules occurs for photon energies between 15 and 20 eV (Isaacson et al., 1972; Inagaki et al., 1974). Recent measurements by Iwanami and Oda (1983) showed that the first ionization potential of polypeptides is about 7.3 eV. Therefore, it may be very timely, both from the historical and the academic point of view, to pursue high quality studies on VUV radiation effects with SR.

Development of Instrumentation for Use of VUV Radiation From SR

Use of SR for irradiation requires considerable instrumentation in addition to a storage ring operated at a suitable energy. We installed a modified Wadsworth mount monochromator at the INS-SOR storage ring near Tokyo. The energy of the electrons in this storage ring, 0.38 GeV, is ideally suited for studies that require photons with energies in the VUV range. Our beamline is equipped with a two-stage differential pumping system and two fast closing valves, one upstream from the monochromator and the other upstream from the sample chamber. The present beamline is a modification of the original described some time ago (Ito et al., 1984).

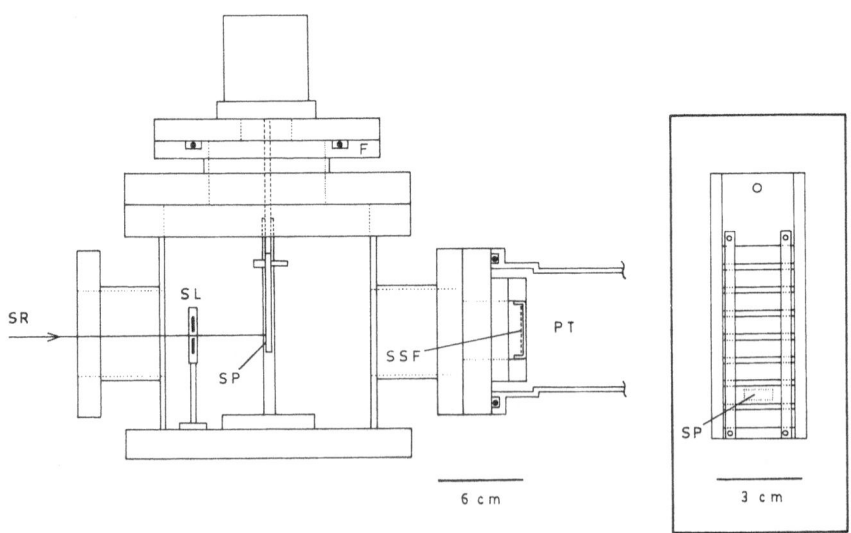

Fig. 1.   Vacuum irradiation chamber for dried materials. SR; synchrotron radiation, SL; slit, SP; sample, SSF; sodium salicylate film, PT; photomultiplier, F; top flange of chamber. Sample holder is shown in the inset.

A premirror and two MgF$_2$ windows were removed and an ion pump was installed for the monochromator instead of the original turbo-molecular pump. A gate valve with an optional MgF$_2$ window was placed between the monochromator and the sample chamber to switch the usable energy regions below and above 9.54 eV (130 nm). Optical surfaces of the monochromator are interchangeable between Al-coated and Au-coated according to the desired energy range (Hieda et al., 1986). A sample chamber has been designed to receive 8 samples vertically in a holder which is controlled from outside to place them into the beam (Fig. 1). The sample chamber is evacuated by a turbo-molecular pump that achieves 10^{-6} Torr in a few minutes. A pressure of 10^{-7} Torr must be achieved in order to connect the vacuum line directly (without the MgF$_2$ window) to the monochromator for irradiation with photons above 9.54 eV.

The sample chamber is, if necessary, equipped with a chopping device, affixed to the top flange (F in Fig. 1), to provide pulsed SR -- normally operated below 100 Hz -- for photoacoustic spectroscopy (Inagaki et al., 1985). The photoacoustic cell is placed at the end of the vacuum line.

To the end of the vacuum line, a plastic chamber that can hold living samples under quasi-physiological conditions (no O$_2$ gas present but supplied with water vapor at saturated pressure) can be attached (Fig. 2) (Ito and Ito, 1981). This wet chamber has been used for the irradiation of yeast cells (Ito et al., 1983) and cultured mammalian cells (Shinohara et al., 1984).

The irradiation system described above provides monochromatic radiation at a reasonably good resolution of about 8% at 22.5 eV and 2% at 8.6 eV. The photon intensity at the sample position was measured as 5.5 x 10^{17}/m^{2}s per 100 mA ring current and 1.3 x 10^{16}/m^{2}s per 100 mA ring current for below and above 10 eV, respectively. Under the routine operating conditions of

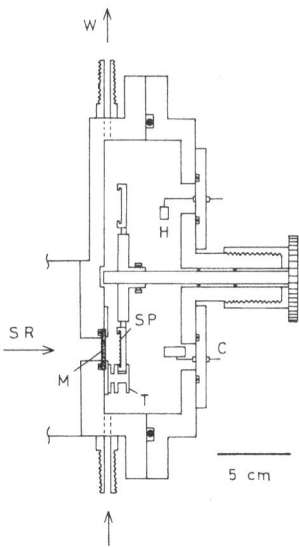

Fig. 2. Moistured irradiation chamber for wet materials. SR; synchrotron radiation, M; MgF$_2$ window, SP; sample, T; adjustment grooves for thickness between SP and M, C; electric connector, H; dew sensor, W; water vapor inlet and outlet (Ito and Ito, 1981).

our storage ring (150 mA ring current), irradiation of the order of 10^{21} photons/m^2 below 10 eV and 10^{20} photons/m^2 above 10 eV can be achieved with the output spectra shown in Fig. 3. This means that the irradiation time required to produce a strand break on DNA ($F_{37} = 10^{-19}$ m^2 for 6.9-8.3 eV) is a matter of minutes or even a matter of seconds for the more sensitive damage of bacteriophages, although it requires a few hours for the production of molecular fragmentation of dinucleotides with a reasonable yield.

Molecular Effects for Photon Energies Between 6.2 and 25 eV

6.2-8.3 eV Region. The effects of photons in this region have been studied using the irradiation system equipped with Al-coated optical elements. Strand breaks of DNA and inactivation of bacteriophages have been the two major systems investigated. Other types of viruses, for example HVJ, were also probed. The inactivation of ATP was studied by following inactivation of its ability to release energy in biochemical reactions and by following molecular decomposition. With E. coli colicin E1 DNA, cross sections of single-strand breaks in dry samples were determined down to 4.88 eV (254 nm) for 8 photon energies. Previously, only two experimental points were available in this range (Wirths and Jung, 1972; Sontag and Dertinger, 1975). This study revealed that the cross section decreases continuously over four decades with an inflection point around 6.9 eV (Hieda et al., 1986). These authors found no peak in the action spectrum around 6.53 eV (190 nm), where a broad absorption band of DNA occurs (see Fig. 9), that is due to absorption by the bases. The action spectrum for phage T1

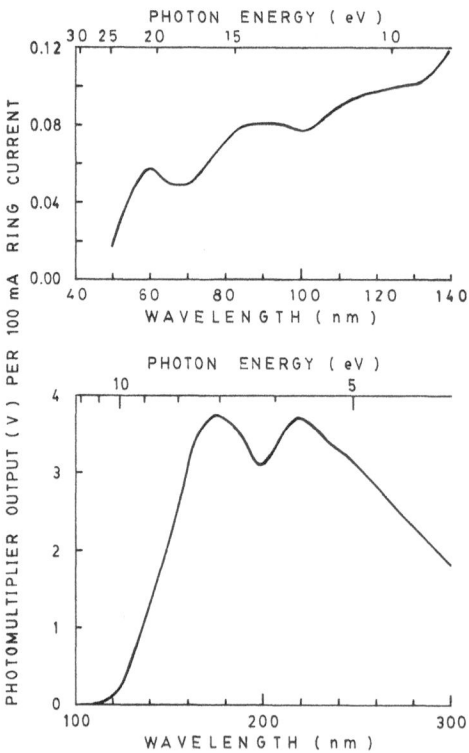

Fig. 3. Output spectra of monochromatized SR in the energy region from 5-10 eV with Al-coated optical system and 10-25 eV with Au-coated optical system. The fluence rate at the photomultiplier output of 1V is 2.7 x 10^{17} photons/m^{2}s.

inactivation assayed on E. coli cells was downward and concave with a minimum around 5.4 eV (230 nm). The cross section ranged in the order of 10^{-19} m^2. Plating irradiated T1 phages on E. coli B_s-1 cells showed that the host-cell reactivatible fraction for VUV effects was smaller than that for far-UV effects. No photoreactivation was observed at 8.3 eV (150 nm) photons in contrast to a photoreactivatible sector of about 0.3 for 254 nm far-UV radiation. The action spectrum was similar to the absorption spectrum of DNA, except around 6.5 eV, where the maximum occurs in the absorption as mentioned above (Maezawa et al., 1984). These results again demonstrate that absorption by the bases is not a major factor in the inactivation. Inactivation of single-stranded DNA virus φ x 174 showed a cross section 5 times larger than T1 at 8.3 eV (150 nm) (Maezawa et al., 1986). The action spectrum for HVJ virus inactivation was similar in shape to T1 inactivation (Megumo, unpublished results).

Product analysis of VUV irradiated simple molecules was first performed with ATP by thin-layer chromatography (Ito et al., 1986). The characteristic feature was the release of adenine. The same was true for VUV-irradiated AMP and even adenosine (see also Dodonova et al., 1982 for related results with VUV from an H_2-lamp). Later it became clear that this was not caused by the breakage of the N-glycosidic bond but rather by the destruction of the sugar moiety (see below). The action spectrum for the inactivation of ATP function as measured by luminescence in the luciferin-luciferase assay showed a monotonic decrease from 8.9 eV to 6.5 eV and no measurable inactivation from 6.5 eV down to 6.2 eV. The quantum yield calculated by the use of the absorption cross section, directly measured by the transmission of thin films of ATP, range from 0.1 (6.9 eV) to 0.2 (8.9 eV). Recent experiments with oligonucleotides will be discussed later in a separate section.

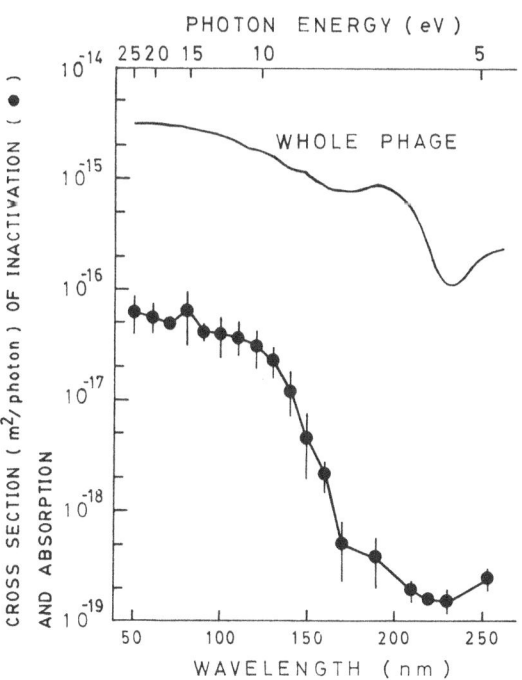

Fig. 4. An action spectrum for the inactivation of bacteriophage T1. Solid line is the calculated absorption of whole phage (Maezawa et al., 1987).

225

8.3-25 eV Region. With the successful improvement of our irradiation
system in 1985, we have extended our studies to photon energies above 10 eV
(124 nm), which is well above the ionization potential of biological molecules. It is not only the boundary of far-UV radiation and X-rays but also
where the majority of the oscillator strength of molecules is located, and
indeed, the energy region where the most exciting phenomena are expected to
be observed.

Experiments on single strand breaks have been extended up to 15.5 eV.
The cross section increased to 1×10^{-16} m^2 for pBR322 DNA (Suzuki and
Hieda, 1986). Experiments on the inactivation of bacteriophages have also
been extended up to 25 eV. The slope of the action spectrum became significantly reduced starting at 10 eV, keeping two decades below the estimated
absorption spectrum of the whole phage as shown in Fig. 4 (Maezawa et al.,
1987). The increased damage to the protein coat in the higher energy range
must be taken into account to explain the action spectrum, since the absorption coefficients are so high in this spectral region that even the thin
layer of protein provides significant shielding for the DNA.

In continuation of ATP experiments mentioned earlier, the photoproduct
analysis was performed extensively with irradiated oligonucleotides, dApdA
(2'-deoxyadenylyl-(3'-5')-2'-deoxyadenosine). A survey was made for the
products over 6.5 to 22.5 eV range. Quite interestingly, the products from
the irradiated dApdA are surprisingly simple; only two products were found
by TLC (Ito et al., 1987). Furthermore, the two observed products, adenine
and 5'-dAMP, occurred approximately in equal quantity. It appears as if the
target molecule fell apart in the same fashion regardless of the photon
energy. The pattern can be explained if the deoxypentose of the adenyl
moiety was destroyed. The hypothetical pathway of this fragmentation is
shown in Fig. 5 as No. 3. More recent results on this type of damage will
be discussed in detail in a later section. Besides these DNA and related
systems, interest has focused on the inactivation of enzymes.

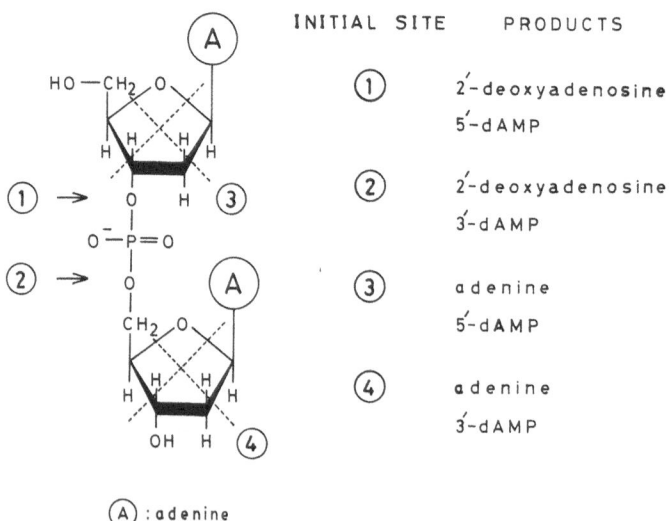

Fig. 5. VUV photodegradation scheme of dApdA. Among the four possible
pathways No. 3 has been proposed (Ito et al., 1987).

Water molecules absorb VUV photons and decompose into hydrogen atoms (H) and ground state hydroxyl radicals (OH). This photochemical reaction is the basis of VUV action in aqueous systems. The peak absorption of liquid water occurs at 7.75 eV and the quantum yield for this dissociation may be assumed roughly as 0.5 (Getoff and Schenck, 1968). The penetration depth (thickness of e^{-1} intensity) at this energy is roughly 0.1 μm. Therefore, the incident photons are absorbed in a thin layer of water, and produce H and OH in equal proportion. Calculations show that under the routine conditions of irradiation, the radicals that originated from different water molecules are not too close to interact with each other in the presence of solute molecules at workable concentrations, although the stirring of the solution is an essential requirement. Below 6.7 eV this mode of absorption becomes negligible, and the direct absorption by solute molecules may become important.

The dissociation of water molecules in the liquid state could be measured by a number of means. The use of the reaction of OH with p-nitrosodimethylaniline (Kraljic and Trumbore, 1965) is a reliable method to quantify VUV-induced OH production in aqueous systems (Ito and Ito, 1980). In order to identify these water radicals we applied a spin-trapping technique using DMPO (5,5-dimethyl-1-pyrroline-N-oxide). Spin adducts, OH-DMPO and H-DMPO, have been identified by ESR after irradiation of water in a thin irradiation chamber with a CaF_2 window (Minegishi et al., 1984; 1985). The addition of OH scavengers, such as sodium formate, reduced the ESR signal of the OH adduct, confirming the production of OH radicals at various photon energies in the VUV. Anomalous OH adduct signals in the presence of O_2 are not explained easily. They appear to be due to radicals produced by side reactions involving O_2 and H, generated in abundance in the VUV photodissociation of water (Minegishi et al., 1986; Minegishi et al., 1987).

The spin-trapping technique has been extensively applied to the identification of the OH adduct structure of pyrimidine and pyrimidine nucleoside with MNP (2-methyl-2-nitrosopropane) as a spin-trap. Well-resolved hyperfine structures of the ESR spectrum have made it possible to identify OH-addition at the C5 position of the 5,6 double bond of the pyrimidine bases. For nucleosides, the ESR spectra are generally broader than those of the bases; therefore, hyperfine structures due to β-proton and γ-proton and γ-nitrogen were sometimes obscured. Nevertheless, OH addition radicals at the C6 of the base moiety as well as C5 position (except 2'-deoxycytidine) were identified (Kuwabara et al., 1986a). No unique radical intermediate has been found in VUV photolysis compared with radiolysis of aqueous solutions. No evidence was obtained by this technique for the presence of radicals in the sugar moiety. An improved analysis combined with gel permeation chromatography of VUV irradiated uridines revealed that four components were separated, each of which gave an ESR spectrum corresponding to C5 and C6 radicals of the base moiety and C4' and C5' radicals of the sugar moiety (Kuwabara et al., 1986b). Since all the signals were affected by the addition of OH scavengers, they must be related to the reaction with OH radicals. However, the observed C4' radical may have originated by the abstraction of hydrogen at C1' of ribose and not directly from C4' (Kuwabara, personal communication) as has been established in radiolysis (e.g., Inanami et al., 1987).

Strand breaks produced in DNA in aqueous systems have also been investigated. Single-strand breaks in E. coli pBR322 and colicin E1 DNA were produced by the attack of OH radicals generated by the VUV photolysis of water in the energy range above 7.29 eV (<170 nm). This was demonstrated by the good agreement between the action spectrum and the absorption spectrum of liquid water. Stern-Volmer kinetics using specific OH scavengers

227

supported the involvement of short-lived OH radicals as the intermediate
active species (Takakura et al., 1987). Below that energy, especially in
the region down to 6.5 eV (190 nm), an anomalous increase of action cross
section was observed. It is apparently not due to direct DNA absorption
alone. Oxygen dissolved in the system has been suggested to play a role
(Takakura et al., 1986). Active species related to the reactions of VUV-
generated H and dissolved O_2, as implied in the anomalous OH production (see
above), could constitute part of the abnormality.

Research on the Cellular Level

Interest has focused on whether or not VUV-induced DNA damage is
repaired by the known repair systems (Ito, 1985). The UV sensitive strain
of yeast (rad 1/rad 1, defective in excision repair for far-UV damage) was
compared with the wild type strain for VUV-induced inactivation. These
yeast strains have attained 85% survival after drying by an improved freeze-
drying technique. This enabled us to use them for irradiations up to 8.0 eV
in a dried state. It was found that the far-UV sensitive cells showed no
enhanced sensitivity to VUV radiation. The induction of gene conversion was
not detected above 7.3 eV, whereas apparent damage in the cell membrane was
observed (Hieda et al., 1984). The action spectra for inactivation indicated that the two strains diverge at about 6.3 eV, at which the cell
showed the minimal sensitivity over the energy range from 5 to 7.3 eV (Hieda
et al., 1986). Apparently, above 6 eV considerably fewer photons would have
reached the nucleus for this size of cell (4 μm in diameter). Consequently,
the usefulness of UV-sensitive strains of yeast is limited. With Bacillus
spores, the size of which is about 1 μm, and which survive in vacuum, action
spectra were obtained up to 20 eV (Munakata, 1986; Munakata et al., 1986).
The spores from three strains, UVS (defective both in excision repair and
spore repair for spore UV-photoproduct), UVP (UVS characters plus defective
in DNA polymerase I), and UVR (wild type), were used for the construction of
action spectra for mutation (His[+] reversion) and inactivation. The action
spectra of the two strains were sufficiently parallel up to 8.3 eV,

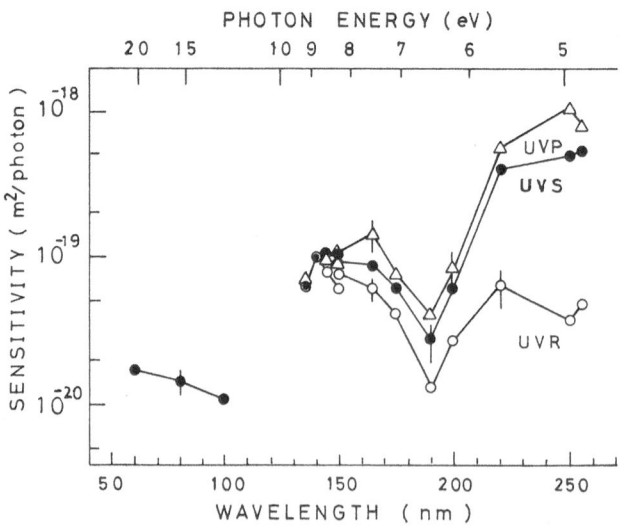

Fig. 6. Action spectra for the inactivation of Bacillus subtilis spores
with different sensitivities to radiations. UVR; wild type strain,
UVS; far-UV sensitive strain, UVP; X-ray sensitive strain
(Munakata, 1986).

to convince us that DNA damage is dominant for the observed effects. As to the differential manifestation of the different strains (Fig. 6), one prominent feature is that a large difference in the far-UV region between UVR and UVS (UVP also) gradually narrowed going toward 6.53 eV (minimum of action spectra), and they seemed to further converge toward 8.3 eV, although UVP still remained more sensitive. It would be interesting to see the transition from far-UV to a seemingly X-ray type response. It is still a major problem as to whether or not the damage in the region above 10 eV is the same as that by ionizing radiation. The interpretation for whole action spectra has not been reached at the present time but the minimum, around 6.5 eV, appears to be due to a significant reduction of photon fluence for spore DNA by shielding due to the intervening substances with conjugated double bonds. The low sensitivity in the higher photon energy (>12 eV) may reflect the general decrease of photon fluence for the DNA.

Vacuum-UV irradiation of wet (living) cells is limited to photon energies of less than 8.3 eV. The fluence at 7.75 eV is reduced by a factor of e^{-1} every 0.1 μm in liquid water (Fig. 7). Similar values may be applied to the living cell since 80% of a cell is water. On top of these unusual situations the cell must be prevented from drying. These difficult requirements for VUV irradiation of living cells were partly overcome by placing the cells into a moisture chamber with a MgF_2 window. The combined transmission of the 1 mm thickness of water vapor at saturated pressure and 2 mm thickness of the MgF_2 plate is estimated to be about 40% in the range from 7.3 to 8.6 eV, indicating VUV irradiation experiments with such a device are feasible. Based on this concept, an irradiation chamber such as the one shown in Fig. 2 was constructed, and used for the irradiation of wet yeast cells and cultured mammalian cells (HeLa), both being placed on a filter. Action spectra have been obtained in the region from 4.9 to 8.9 eV using the wet irradiation chamber for activation and cell membrane damage with wet yeast cells (Fig. 8) (Ito et al., 1983). Both showed maxima at about 7.8 eV which coincides to the calculated maximum in the absorption of water located at a depth of 1 μm from the surface (Ito et al., 1983), indicating the water is probably the common absorbing material responsible for the action. Hydroxyl radicals generated as a result of water photolysis are again suggested as the active species. Above 5.9 eV (<210 nm) the induction of gene

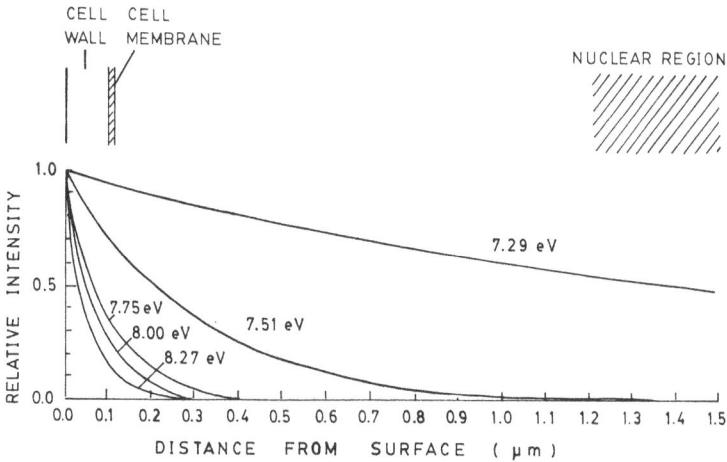

Fig. 7. Attenuation of VUV radiation by liquid water as a function of the distance from the surface. Approximate dimensions are indicated for a diploid yeast cell (Ito et al., 1983).

conversion was not significant. The inactivation cross section below this energy followed the absorption cross section of DNA. Cell membrane damage was not observed, as predicted. For the transient region, namely, 5.9 to 7.3 eV, the action spectra for inactivation may be a composite of actions of different origins in their nature, and the interpretation remains obscure.

In earlier experiments with deuterium lamps as a light source, we measured survival at various thicknesses of water vapor in front of the cell layer on the membrane filter in the wet irradiation chamber. The F_{50} (50% killing fluence) increased exponentially as a function of the thickness of the water vapor. The extrapolation of the slope on a semilogarithmic graph to zero thickness should give F_{50} in the absence of water vapor, which is not possible to measure experimentally with wet cells. Using this fluence, the absolute number of VUV photons that kill the cell (50% killing) was determined as 2.2×10^8 for nondispersed VUV radiation from the deuterium lamp

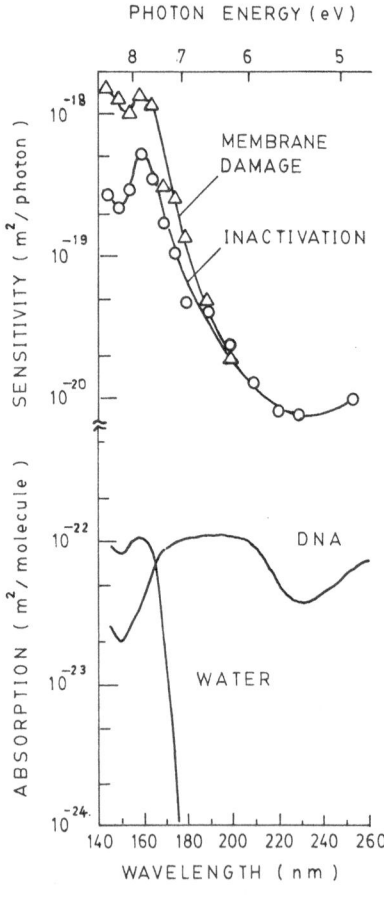

Fig. 8.   Action spectra for inactivation and cell membrane damage of yeast cells. Absorptions of water and DNA have been calculated at the depth of 1 μm in water. The scale for DNA is arbitrary (Ito et al., 1983).

(Ito and Ito, 1981). We later repeated these measurements with monochromatic SR. The F_{50} value ranged from 10^7 to 10^8 photons depending on the incident photon energies (Ito et al., 1983).

Cultured mammalian cells (HeLa) have also been irradiated using the wet irradiation chamber. They survived long enough (1 hour) under the irradiation conditions of the wet chamber, typical irradiation time being several minutes. They even withstood an initial brief evacuation before introducing water vapor. The action spectrum for inactivation showed a minimum at 6.5 eV (Shinohara et al., 1984). During the course of the experiments a number of interesting observations have been made on the behavior of VUV-irradiated cells; cells tend to detach from the membrane filter on which they had grown before irradiation, when irradiated at 7.8 eV but not below 5.9 eV; for a heavily damaged cell group the cells which took up eosin Y increased with the VUV fluence. Both may indicate some sort of damage occurring on the cell membrane. Such observations were never made with the cells irradiated either by far-UV or ionizing radiation (Shinohara et al., 1986). Apparently the damage in the cell surface prevails over the DNA damage in the VUV action.

Spectroscopy for Photoeffect Studies

For the analysis of photoeffects, absorption information is essential. Because of the obvious difficulties in the preparation of materials, however, only a few studies have ever been attempted in the VUV region. The determination of optical constants of DNA (film) was performed by using a conventional light source over 2 to 82 eV range (Inagaki et al., 1974). Transmission measurements with thin films of bases, nucleosides, nucleotides and nucleic acids were reported by Kiseleva et al., (1975) up to 10 eV with a hydrogen lamp as a light source. Sontag and Weibezahn (1975) have determined the absorption of DNA up to 25 eV photon energy by measuring scattered photons with a microwave discharge as a light source. The first use of SR for the direct transmission measurement was reported by Ito and Ito (1986) for the energy region from 6.20 to 8.27 eV with thin films of DNA and related compounds. For example, the absorption cross section of ATP was determined as 7.0×10^{-21} m^2 at 8.27 eV. Recently, transmission spectra were obtained for other biological molecules in the same region (Fig. 9) (Megumi et al., 1987 and unpublished results). There seems to be no good backing material above 10 eV for the transmission measurement. Besides, biological materials are often not obtained as a smooth thin film, which pose a serious problem for transmission measurements.

Photoacoustic measurements avoid most of these difficulties; sample smoothness is not an essential requirement, nor is the thickness. The difficulty with backing materials can be avoided. The photoacoustic measurement is related to the energy absorbed in the object material, since it represents 1-R-T, where R and T are the reflectance and the transmittance, respectively. Furthermore, if the energy is carried away by photoelectron emission, or if energy-consuming photochemical reactions take place, then the photoacoustic signal amplitude would decrease since the energy consumed in these processes does not contribute to the generation of heat. Therefore, information contained in the photoacoustic spectrum is unique for the analysis of photoeffects. From this point of view we have attempted for some time the development of photoacoustic measurements in the VUV region using SR as a light source (Inagaki et al., 1985; 1986). Using this system, spectra of nucleic acid bases (adenine, thymine), DNA, tyrosine, lysozyme, ergosterol, and even whole yeast cells have been measured over the photon energy range up to 8.9 eV. Further extension of the energy range is under way.

MOLECULAR FRAGMENTATION OF DRY OLIGONUCLEOTIDES BY VACUUM-ULTRAVIOLET
RADIATION

Introduction

The indication of a highly selective mode of degradation of deoxydi-
nucleoside monophosphate by VUV photons (see Fig. 5) suggested that further
investigations might provide useful information on the molecular responses
to the excitation induced by VUV photons. There have been no experimental
approaches toward this problem even in the simpler molecules in a dry state.
Obviously, energy-tunable SR is well suited to this investigation. The de-
oxyoligonucleotides used in this study are also similar in structure to DNA.
We thought degradation studies would shed light on the molecular mechanism
of DNA damage by radiation in general. Even the chemical mechanisms leading
to DNA strand breaks, now known to be induced not only by ionizing radiation
but also by VUV radiation, remain unclear. Below is a review of recent progress of our research with oligonucleotides.

Absorption Properties

Absorption measurements are of fundamental importance in any photoeffect study. Several problems have to be solved even for a primitive
system for absorption measurements with SR. The tunability of wavelength
(photon energy) is a great advantage but the experimentation coupled with
the operation of the electron storage ring requires tedious correction procedures for decaying ring current and also confronts the problems associated

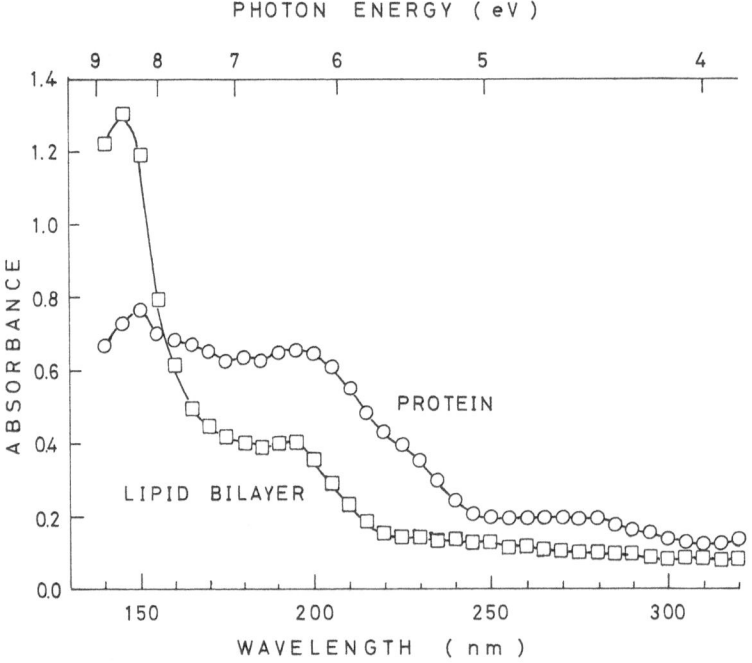

Fig. 9. Absorption spectra of proteins (purified spike proteins) and lipid
bilayers (reconstructed liposomes) from HVJ viruses. The thickness
of these samples is arbitrary (Megumi et al., 1987 and unpublished
results).

with a single beam measurement. Figure 10 is an absorption spectrum recently measured with dApdA placed on a CaF$_2$ plate in the region from 3.88 eV (320 nm) to 8.86 eV (140 nm) with monochromatized SR. The spectrum generally resembles that of DNA in shape but the peak wavelengths are shifted; a peak appearing at about 6.5 eV in DNA is shifted to 5.77 eV in dApdA and a sharp increase occurs at 6.89 eV instead of 8.3 eV with DNA. We also noticed that relative heights of absorption bands depend greatly on the thickness of sample films as also noted by Kiseleva et al. (1975); the absorption due to the π-electron system (4.78 and 5.90 eV) becomes prominent as compared with the absorption beginning at 6.5 eV by ribose phosphates (Kiseleva et al., 1975; Ito and Ito, 1986). Sugar-phosphate groups may be responsible for this behavior. Although the absolute absorption cross section of dApdA has never been determined, we speculate the absorption due to the sugar-phosphate group outweighs that of base moiety by around 7 eV. Because no transparent substrate (backing material) is available, there has been no measurement of absorption spectra above 10 eV by direct transmission. The absorption cross sections or coefficients of DNA and bases above 10 eV may only be obtained by calculations based on the measurements of optical reflectance (Inagaki et al., 1974; Arakawa et al., 1986) as mentioned earlier. These measurements have shown that there is a prominent broad absorption band around 20 eV involving all the valence electrons. Similar types of measurements, however, have not been performed with SR.

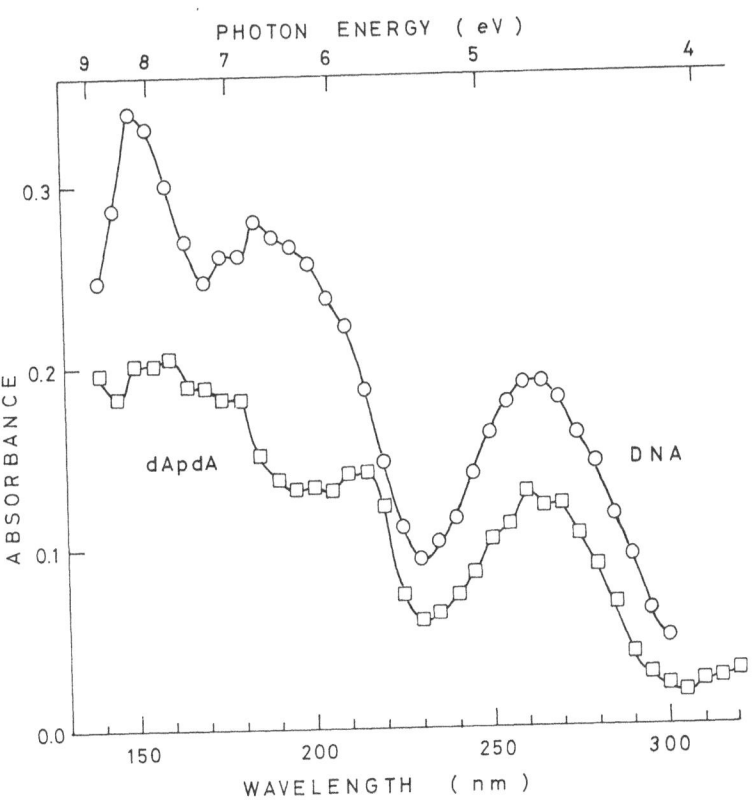

Fig. 10. Absorption spectra of dApdA and calf thymus DNA. The thickness of these samples is arbitrary (Ito et al., unpublished results).

Energy Dependence

The analysis of photofragmentation products were performed with dApdA at several photon energies up to 22.5 eV (Ito et al., 1987). The TLC profiles of all these experiments look the same (Fig. 11), indicating a single fragmentation mode is predominant. From the position of photo-products on the chromatogram, in reference to the authentic substances, the products were identified as adenine and 5'-dAMP (see Fig. 5). The action cross section varied by a factor of 10 going from 8.3 eV to 22.5 eV. Since no precise absorption cross section is known above 10 eV, the quantum ef-ficiency for the fragmentation is not assessed. No fragmentation was detected at the photon energies below 6.52 eV (190 nm), one of the absorp-tion peaks related to the strong $\pi - \pi^*$ transition of the base moiety. A transition in the sugar-phosphate group is assumed to be the absorption responsible for the fragmentation.

Role of the Phosphate Group

In the proposed degradation scheme of dApdA shown as No. 3 in Fig. 5, the deoxypentose of the adenylyl moiety, not another deoxypentose, is indi-cated as the intial site of molecular damage. This selectivity implies that the position of phosphates to the sugar moiety is crucial; the lack of a phosphate group at 5'C somehow could have rendered the molecule vulnerable to destruction. We tested a 5'phospho derivative, pdApdA, and found that this compound is rather resistant to the VUV radiation.

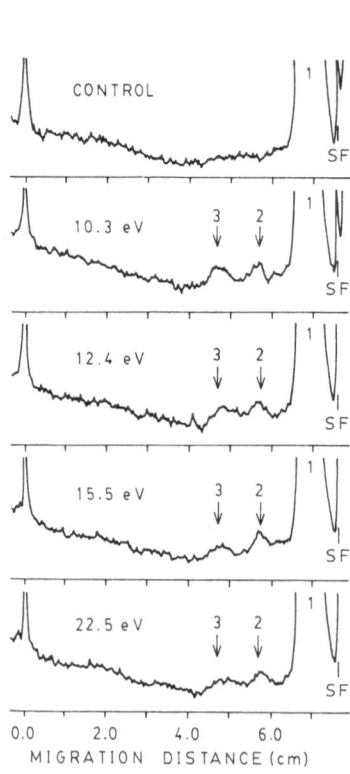

Fig. 11. Chromatogram of irradiated dApdA at indicated photon energies. 1; dApdA, 2; adenine, 3; 5'-dAMP, SF; solvent front (Ito et al., 1987).

The base moiety was found to play a significant role in photofragmentation. When dApdC was used, the release efficiencies of expected photoproducts, adenine and 5'-dCMP, were appreciably affected (by a factor of 2 in terms of action cross section) as compared with dApdA. This factor suggests that the cytosine moiety stabilizes the molecule against photofragmentation (Ito and Saito, unpublished).

The experiments with dCpdA further showed that adenine was released in contrast to the expectation based on the phosphate rule mentioned above. Careful examinations of chromatograms revealed that cytosine was also released to a lesser extent. These results suggest that cytosine affects the sugar moiety attached to it to such an extent that the lack of a phosphate group at the 5' position is overcome, thereby sharing fragmentation at an otherwise stable sugar moiety and releasing the adjacent adenine.

The experiments with dTpdC and dTpdT have indicated that these compounds are very stable, and we have found so far neither thymine nor cytosine released at photon fluence up to 10^{21} photons/m^2. Pyrimidine bases

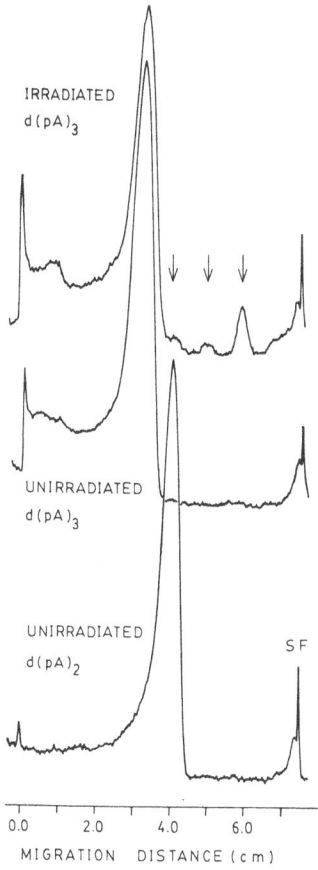

Fig. 12.   Chromatograms of VUV (7.59 eV) irradiated d(pA)$_3$. Arrows indicate the photoproducts.   SF represents solvent front (Ito and Saito, unpublished results).

thus seem to have a stabilizing effect on the molecule. It is to be noted from these results that, if the sugar moiety is the most vulnerable site, the whole molecule is involved in determining the initiation of the fragmentation.

Effects of Chain Length

If the base can be released from either end of the dinucleotide upon VUV radiation, what happens for longer oligonucleotides? This question was only addressed recently. Figure 12 is the first such trial with d(pA)$_3$ irradiated with 7.59 eV photons from a bromine discharge lamp. At least three signals can be identified (arrows); from left to right, d(pA)$_2$, 5'-dAMP and adenine. A simple release of adenine with the equal probability at three possible sites can not be true. The release of adenine from the destruction of the 5'phosphate end seems likely. Further experiments are underway with SR.

Summary on the Characteristics of Fragmentation of Oligonucleotides

End product analysis of oligonucleotides proved to be useful in determining the molecular changes induced by high energy UV photons (>10 eV). Although further systematic investigation will bring about more precise assessment on the photofragmentation of oligonucleotides, even at this stage we can make a few qualitative statements (Fig. 13). (1) The deoxysugar moiety is a key portion that plays a central role in photofragmentation. (2) The presence or absence of a phosphate group may be an important factor in determining the sensitivity of the deoxysugar moiety. (3) Base residues exert their own effects on the stability of the molecule; although further studies are needed for the assessment on the role of the respective bases, pyrimidine bases apparently have a stronger stabilizing effect on the molecule than purine bases. (4) A dependence of chain length is uncertain at present, however, the results hint at some hierarchy as to the order of the initiation of the degradation.

CONCLUSIONS AND PERSPECTIVES

Since 1976 we have focused on investigating VUV effects in cellular and molecular systems with SR by developing a powerful irradiation system. Our objective was to extend the realm of photobiology, which has previously been confined to wavelengths longer than 200 nm (that is, less than 6 eV photon energy), into the deep vacuum-UV region (25 eV). The new dimension is concerned not only with the photon energy itself, but, with a challenge for physical, chemical and biological phenomena not studied before experimentally. Especially important is that this region is responsible for the majority of the oscillator strength of all molecules, hence nearly all the valence electrons will participate in the absorption processes. Moreover, photon absorption in this energy region produces superexcited states, for which virtually no experimental work has been performed to determine the molecular changes that result.

The resulting progress, while initially fragmentary over various biological materials, is gradually taking a coherent form as a branch of photobiology. At the cellular level the short penetration (or strong absorption in the thin surface layer) of VUV photons has already provided a unique tool in the analysis of cell division (Matsumoto et al., 1984). We also expect a chance to discover the role of VUV specific damage on cellular DNA, and its relation to far-UV and X-ray induced damage. At the molecular level the measurement of action spectra was an important achievement. The damage on DNA, either strand breaks or base-type damage, should be intensively studied in the future. The determination of the end structure of VUV-induced DNA strand breaks and the role of base release (or its implications to the strand break), and the conformational factor in the induction of such changes are the immediate problems. A high specificity of VUV-induced strand breaks as compared with those by X-rays was found (Ito and Taniguchi, 1986). While studies using oligonucleotides are still preliminary, they afford an unprecedented opportunity to approach the chemical changes occurring in a highly excited molecule. Applications of other analytical techniques to this problem should be undertaken in the pursuit of end product analysis. At the same time, identification of possible intermediates in the pathways from the highly excited state toward the end product should be attempted.

A unique point in VUV-induced photolysis of aqueous systems as compared with ionizing radiation is that low energy VUV radiation dissociates water molecules only into H and OH. The radiolysis of water, on the other hand, produces secondary electrons, H, OH and other ion species involving these primary species, making the situation more complex than with the VUV radiation. Moreover, the OH produced by radiolysis could be in an excited state, although their reactivity is commonly considered the same as ground state

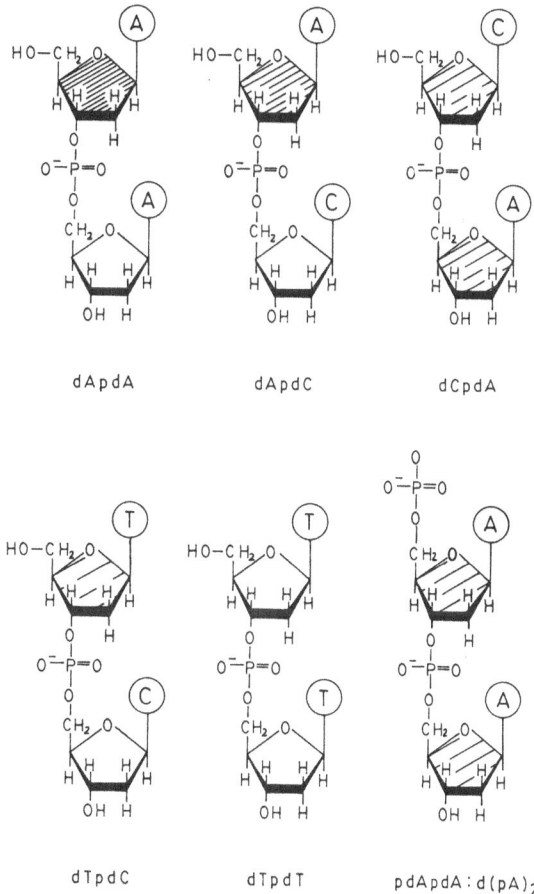

Fig. 13.   Schematic diagrams for VUV sensitivity on the degradation of oligonucleotides. Degrees of shading indicate the vulnerability of deoxysugar moieties.

237

OH.   Spin-trapping studies will continuously find the significance to help
identify the radical intermediate of biological molecules in aqueous
systems.   These studies may also be applied to dehydrated samples.

ACKNOWLEDGMENTS

     I wish to gratefully acknowledge the help of many colleagues, in par-
ticular, K. Hieda, K. Kobayashi, M. Maezawa, and A. Ito.   Recent work with
oligonucleotides was carried out in collaboration with T. Taniguchi and M.
Saito.   In embarking on the study of radiation and photobiology and synchrotron radiation, S. Okada, T. Kada, K. Suzuki, T. Sasali, T. Ishii and H.
Matsudaira have lent constant encouragement and support over the years.   Financial support from the Ministry of Education, Science and Culture over
many years is also acknowledged.

REFERENCES

Arakawa, E. T., Emerson, L. C., Juan, S. I., Ashley, J. C., and Williams, M.
     W., 1986, The optical properties of adenine from 1.8 to 80 eV,
     Photochem. Photobiol., 44:349.
Blank, I. H. and Arnold, W., 1935, The action of radiation in the extreme
     ultraviolet on Bacillus subtilis spores, J. Bacteriol., 30:503.
Bovie, W. T., 1916, The action of Schumann rays on living organisms, Bot.
     Gaz., 61:1.
Dodonova, N. Y., Kiseleva, M., Remisova, L. A., and Tsyganenko, N. M.,
     1982, The vacuum ultraviolet photochemistry of nucleotides,
     Photochem. Photobiol., 35:129.
Getoff, N., and Schenck, G. O., 1968, Primary products of liquid water
     photolysis at 1236, 1470 and 1849 A, Photochem Photobiol., 8:167.
Green, H., Boll, J., Parrish, J. A., Kochevar, I. E., and Oseroff, A. R.,
     1987, Cytotoxicity and mutagensis of low intensity, 248 and 193 nm
     excimer laser, Cancer Res., 47:410.
Heinmets, F., and Taylor Jr., W. W., 1952, Preliminary studies of electrical
     discharge on some chemical compounds and bacteria, Arch. Biochem.
     Biophys., 35:60.
Hieda, K., and Ito, T., 1986, Action spectra for inactivation and membrane
     damage of Saccharomyces cerevisiae cells irradiated in vacuum mono-
     chromatic synchrotron ultraviolet radiation (155-250 nm), Photochem.
     Photobiol., 44:409.
Hieda, K., Kobayashi, K., Ito, A., and Ito, T., 1984, Comparison of the
     effects of vacuum-UV and far-UV synchrotron radiation on dry yeast
     cells of different UV sensitivities, Radiat. Res., 98:74.
Hieda, K., Maezawa, H., Ito, A., Kobayashi, K., Furusawa, Y., and Ito, T.
     1986, Choice of coatings for the optical elements in the irradiation
     system of vacuum-ultraviolet radiation above 50 nm, Photochem.
     Photobiol., 44:417.
Inagaki, T., Hamm, R. N., Arakawa, E. T., and Painter, L. R., 1974, Optical
     and dielectric properties of DNA in the extreme ultraviolet, J. Chem.
     Phys., 61:4246.
Inagaki, T., Ito, A., Motosuga, M., Hieda, K., Kobayashi, K., Maezawa, H.,
     and Ito, T., 1985, Vacuum-ultraviolet photoacoustic spectroscopy of
     biological materials using synchrotron radiation as a light source,
     Photochem. Photobiol., 41:527.
Inagaki, T., Ito, A., Hieda, K., and Ito, T., 1986, Photoacoustic spectra
     of some biological molecules between 300 and 130 nm, Photochem.
     Photobiol., 44:303.
Inanami, O., Kuwabara, M., and Sato, F., 1987, OH-induced free radicals in
     3'-UMP and Poly(U): Spin-trapping and radical chromotography,
     Radiat. Res., 112:36.

Inokuti, M., 1986, VUV absorption and its relation to the effects of ioniz-
    ing corpuscular radiation, Photochem. Photobiol., 44:279.
Isaacson, M., 1972, Interaction of 25-keV electrons with the nucleic acid
    bases, adenine, thymine, and uracil, I. Outer shell excitation,
    J. Chem. Phys., 56:1803.
Ito, A., and Ito, T., 1986, Absorption spectra of deoxyribose, ribosepho-
    phate, ATP and DNA by direct transmission measurements in the
    vacuum-UV (150-190 nm)and far-UV (190-260 nm) regions using synchro-
    tron radiation as a light source, Photochem. Photobiol., 44:355.
Ito, A., Taniguchi, T., and Ito, T., 1986, Wavelength dependence for the
    inactivation of ATP in the vacuum-ultraviolet region above 140 nm,
    Photochem. Photobiol., 44:273.
Ito, T., 1985, Use of synchrotron radiation in photobiology, in:
    "Photobiology 1984," J. W. Longworth, J. Jagger and W. Shropshire,
    Jr., ed., Praeger Publications, New York.
Ito, T., and Ito, A., 1980, Predominance of membrane damage in yeast cells
    in suspension with monochromatic 163 nm vacuum UV light, Radiat.
    Res. 81:416.
Ito, T., and Ito, A., 1981, Effects of 120 to 165-nm vacumm-UV light on wet
    yeast cells, Radiat. Res., 85:161.
Ito, T., Ito, A., Hieda, K., and Kobayashi, K., 1983, Wavelength dependence
    of inactivation and membrane damage to Saccharomyces cerevisiae cells
    by monochromatic synchrotron vacuum-UV radiation (145-190 nm),
    Radiat. Res., 96:532.
Ito, T., Kada, T., Okada, S., Hieda, K., Kobayashi, K., Maezawa, H., and
    Ito, A., 1984, Synchrotron system for monochromatic UV irradiation
    (>140 nm) of biological material, Radiat. Res., 8:65.
Ito, T., and Kobayashi, K., 1976, Induction of lehal and genetic damage by
    vacuum-ultraviolet (163 nm) irradiation of aqueous suspension of
    yeast cells, Radiat. Res., 68:275.
Ito, T., Saito, M., and Taniguchi, T., 1987, A survey of photoproducts of
    an irradiated oligonucleotide by monochromatic photons with the
    energy ranged from 6.5 to 22.5 eV, Photochem. Photobiol. 46:979.
Ito, T., and Taniguchi, T., 1986, Enzymatic quantification of strand
    breaks of DNA incuced by vacuum-UV radiation, FEBS, 206:151.
Iwanami, S., and Oda, N., 1983, Photoabsorption and photoelectron yield
    spectra of polypeptides in the vacuum-UV region, Radiat. Res., 95:24.
Jagger, J., Stafford, R. S., and Mackin, Jr., R. J., 1967, Killing and
    photoreactivation of Streptomyces griseus conidia by vacuum-ultra-
    violet radiation (1500 to 2700A), Radiat. Res., 32:64.
Kiseleva, M. M., Zarochensteva, Ye. P., and Dodonova, N. Ya., 1975,
    Absorption spectra of nucleic acids and related compounds in the
    spectral region 120-280 nm, Biofizika, 28:561.
Kraljic, I., and Trumbore, C. N., 1965, P-Nitrosodimethylaniline as an OH
    radical scavenger in radiation chemistry, J. Am. Chem. Soc., 87:2547.
Kuwabara, M., Minegishi, A., Ito, A., and Ito, T., 1986a, A study of
    aqueous solutions of nucleic acid constituents exposed to mono-
    chromatic 160 nm vacuum-ultraviolet light by spin-trapping method,
    Photochem. Photobiol., 44:265.
Kuwabara, M., Inanami, O., Minegishi, A., Saito, M., Hieda, K., and Ito, T.,
    1986b, VUV-induced free radicals in uridine studied by a method
    combining ESR, spin-trapping and gel permeation chromatography,
    Activity Rep. Synchrotron Rad. Lab., ISSP, Univ. Tokyo, 73.
Maezawa, H., Furusawa, Y., Suzuki, K., and Mori, T., 1986, Killing of dried
    bacteriophages by vacuum-UV radiation above 60 nm, Activity Rep.
    Synchrotron Rad. Lab. ISSP, Univ. Tokyo, 80.
Maezawa, H., Furusawa, Y., Suzuki, K., and Mori, T., 1987, Killing of dry
    bacteriophage T1 by vacuum-UV radiation above 50 nm, Activity Rep.
    ISSP, Univ. Tokyo, 60.
Maezawa, H., Ito, T., Hieda, K., Kobayashi, K., Ito, A., Mori, T., and
    Suzuki, K., 1984, Action spectra for inactivation of dry phage T1

after monochromatic (150-254 nm) synchrotron irradiation in the presence and absence of photoreactivation and dark repair, Radiat. Res., 98:227.

Matsumoto, S., Ito, T., and Ito, A., 1984, Release of P and K from yeast cells irradiated by vacuum UV below 170 nm, Radiat. Environ. Biophys., 23:287.

Megumi, T., Saito, M., and Ito, T., 1987, Absorption spectra of lipid bilayers of Sendai virus in far-UV and vacuum-UV regions, Activity Rep. Synchrotron Rad. Lab. ISSP, Univ. Tokyo, 68.

Miller, S. L., and Urey, H. C., 1959, Organic compound synthesis on the primitive earth, Science, 130:245.

Minegishi, A., Kuwabara, M., Ito, A., and Ito, T., 1984, Spin trapping of OH and H in vacuum-UV irradiated water, Activity Rep. Synchrotron Rad. Lab. ISSP, Univ. Tokyo, 55.

Minegishi, A., Kuwabara, M., Ito, A., and Ito, T., 1985, Spin-trapping study of liquid water irradiated by vacuum-ultraviolet light: Wavelength dependence (150-200 nm) of OH spin adduct formation, Activity Rep. Synchrotron Rad. Lab., ISSP, Univ. Tokyo, 49.

Minegishi, A., Kuwabara, M., Hieda, K. and Ito, T., 1986, VUV-induced OH-DMPO spin adduct in aqueous DMPO solution using 3 mm cell and continuous Ar-flow system, Activity Rep. Synchrotron Rad, Lab., ISSP, Univ. Tokyo, 64.

Minegishi, A., Kuwabara, M., Takakura, K., Saito, M., Azami, A., Hieda, K., and Ito, T., 1987, Spin-trapping of VUV-irradiated liquid water-effect of dissolved air, Activity Rep. Synchrotron Rad. Lab. ISSP, Univ. Tokyo, 58.

Munakata, N., 1986, Vacuum-UV action spectra for inactivation of Bacillus subtilis spores, Activity Rep. Synchrotron Rad. Lab., ISSP, Univ. Tokyo, 76.

Munakata, N., Hieda, K., Kobayashi, K., Ito, A., and Ito, T., 1986, Action spectra in ultraviolet wavelengths (150-250 nm) for inactivation and mutagenesis of Bacillus subtilis spores obtained with synchrotron radiation, Photochem. Photobiol. 44:385.

Platzman, R. L. 1967, Energy spectrum of primary activation in the action of ionizing radiation, in: "Radiation Research," G. Silini, ed., 20, North Holland, Amsterdam.

Setlow, R., Watts, G., and Douglas, C., 1959, "Inactivation of proteins by vacuum ultraviolet radiation," Proc. First Natl. Biophys. Conf., 174, Yale Univ. Press, New Haven.

Shinohara, K., Ito, T., Ito A., Hieda, K., Kobayashi, K., and Okada, S., 1984, An approach to the use of synchrotron radiation for investigating radiation effects on cultured mammalian cells in the range from 160 to 254 nm, Radiat. Res., 97:211.

Shinohara, K., Ito, A., and Ito, T., 1986, Cell surface damage in cultured mammalian cells with synchrotron radiation at 160 nm, Photochem. Photobiol., 44:405.

Sontag, W., and Dertinger, H., 1975, Energy requirements for damaging DNA molecules. III. The mechanisms of inactivation of bacteriophage X 174 DNA by vacuum ultraviolet radiation, Int. J. Radiat. Biol., 27:543.

Sontag, W., and Weibezahn, K. F., 1975, Absorption of DNA in the region of vacuum-UV (3-25 eV), Rad. Environm. Biophys., 12:169.

Suzuki, M., and Hieda, K., 1986, Inactivation and single-strand breaks of pBR322 DNA by vacuum-UV radiation (80 nm), Activity Rep Synchrotron Rad. Lab. ISSP, Univ. Tokyo, 78.

Takakura, K., Ishikawa, M., Hieda, K., Kobayashi K., Ito, A., and Ito, T., 1986, Single-strand breaks in supercoiled DNA induced by vacuum-ultraviolet radiation in aqueous solution, Photochem. Photobiol., 44:397.

Takakura, K., Ishikawa, M., and Ito, T., 1987, Action spectrum for the
   induction of single-strand breaks in DNA in buffered aqueous solution
   in the wavelength range from 150 to 272 nm : dual mechnism, <u>Int. J.
   Radiat. Biol</u>., 52:667.
Wirths, A., and Jung, H., 1972, Single-strand breaks induced in DNA by
   vacuum-ultraviolet radiation, <u>Photochem. Photobiol</u>., 15:325.

DETECTORS FOR HIGH PHOTON RATES

V. Radeka, J. Fischer, J. A. Harder, and G. C. Smith

Instrumentation Division
Brookhaven National Laboratory
Upton, NY 11973

INTRODUCTION

The increasing flux of photons available from synchrotron radiation
sources requires x-ray detectors with greater and greater range in the
counting rate capability. Dynamic studies of biological systems in which
structural changes can be observed on a short time scale are of particular
interest. No detector at present comes close to satisfying all the requirements simultaneously, such as counting rate capability, large dynamic range,
high quantum efficiency over a wide range of energies, high position resolution, fast gating and consecutive time-slicing capability. Each detection
principle developed so far has some good properties and some serious limitations, and the only practical approach for the near future is to find the
best compromise for a particular experiment of interest. For example, some
interesting detectors have a sufficiently large number of position elements
(pixels) and can operate at high counting rates, but have a limited dynamic
range and a long readout time. Thus consecutive time-slicing is not possible, and repetitive excitation and exposures of the sample are necessary.

We concentrate in a brief discussion here on certain electronic detectors which will allow consecutive time-slicing and fast gating and have a
large dynamic range, provide a wide range of sizes and shapes, but have limitations with respect to the number of pixels. We divide this discussion
in two parts. First, some recent developments are described which represent
devices that are operational and available (at least at a National Laboratory). One of these is a multi-element gas proportional detectors for high
counting rates ($\approx 10^8$ sec^{-1}) and the other is a fast position encoding circuit for detectors with delay line position sensing. Next, we discuss the
potential and future outlook of some new developments in gas detectors and
silicon detectors. Both of these types of detector are entirely dependent
on the future development of monolithic low-noise readout circuits.

100 ANODE CHAMBER

This detector is suitable for time-resolved measurements using the high
photon flux available from synchrotron sources. Based on the multiwire proportional counter (MWPC), it operates with a multi-element type of readout
on the anode wires. A side view, schematic diagram of this one-dimensional
detector is shown in Fig. 1. The entrance window is a flat beryllium sheet,
127 mm x 20 mm. Below this is a 4 mm deep absorption and drift region,

followed by the symmetric structure of a MWPC in which the anode cathode spacing is 1 mm and the anode wire pitch is 1.27 mm. The total depth is therefore 6 mm, which gives a good detection efficiency ($\approx$ 70%) for photons up to about 10 keV in energy when the detector is operated with 1 atm. of a suitable xenon/quench gas mixture. The upper cathode is a wire grid with wire spacing 0.635 mm, while the lower cathode is a flat plate of aluminum.

The operating principle is, briefly, as follows. A photon enters the chamber through the beryllium window and interacts with a gas molecule, via the photoelectric effect, in the absorption region. A very small cloud of many electrons is created at the interaction point. These electrons drift toward, and then through, the wire grid, creating an avalanche on the near-

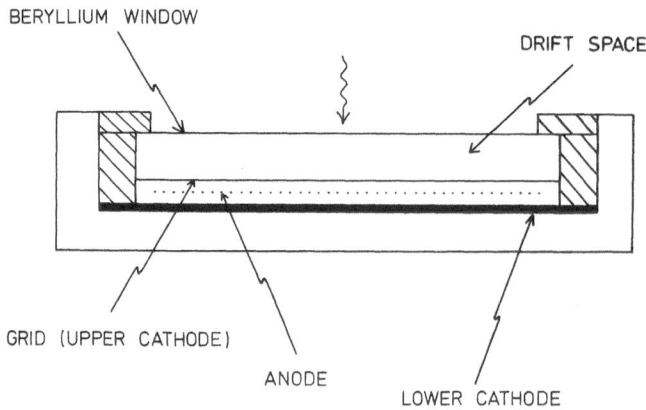

Fig. 1. Schematic diagram of 'readout per anode wire' chamber for very high counting rates (about 10^8 per sec). Drift space is 4 mm deep, anode cathode spacing is 1 mm, anode wire pitch = 1.27 mm (anode wire diameter = 10 microns), upper cathode wire pitch = 0.635 mm (cathode wire diameter = 50 microns). Sensitive area is 127 mm x 20 mm.

est anode wire. Each anode wire is connected to a low noise, fast, charge-sensitive preamplifier, followed by a fast shaping amplifier, discriminator and scaler. The scaler is incremented by one every time an event is recorded by the corresponding anode wire. All 100 anode channels operate independently and after one time slice of the particular experiment has elapsed, the contents of each of the 100 scalers are read into buffer memory so that the next time slice can begin. The read time is only a small fraction of the duration of a time slice, so that a sequence of time slices corresponds to an uninterrupted 'picture' of the dynamics under investigation.

The detector thus operates in the photon counting mode, as opposed to continuous charge integration. The main reason for this is to achieve as high a dynamic range as possible, but photon counting does, nevertheless, have an upper limit beyond which counts are lost for a variety of reasons. This detector achieves about the highest rate that is possible with single photon counting. Some important factors which have made this possible are:

(a) Quite apart from any limitation in the signal shaping electronics, there is a natural limit to count rate on any one spot of an anode wire, due to the phenomenon of space charge saturation. In this the positive ions created in the avalanche reduce the effective field, and hence the gain, at the anode wire surface. The very small inter-electrode spacing used in the detector results in shorter positive ion drift times than normal, helping to alleviate space charge effects. However, to sustain rates of about 10^6 sec^{-1} per wire it is still desirable to operate this detector with avalanche sizes less than a few times 10^5 electrons, which in turn requires very low noise signal amplification. In the signal shaping used here, which has been described in another high rate application (Fischer et al., 1985), each channel has an rms noise less than 2000 electrons. This allows a discriminator setting of about 10^4 electrons, resulting in a high sensitivity for signals of a few times 10^5 electrons.

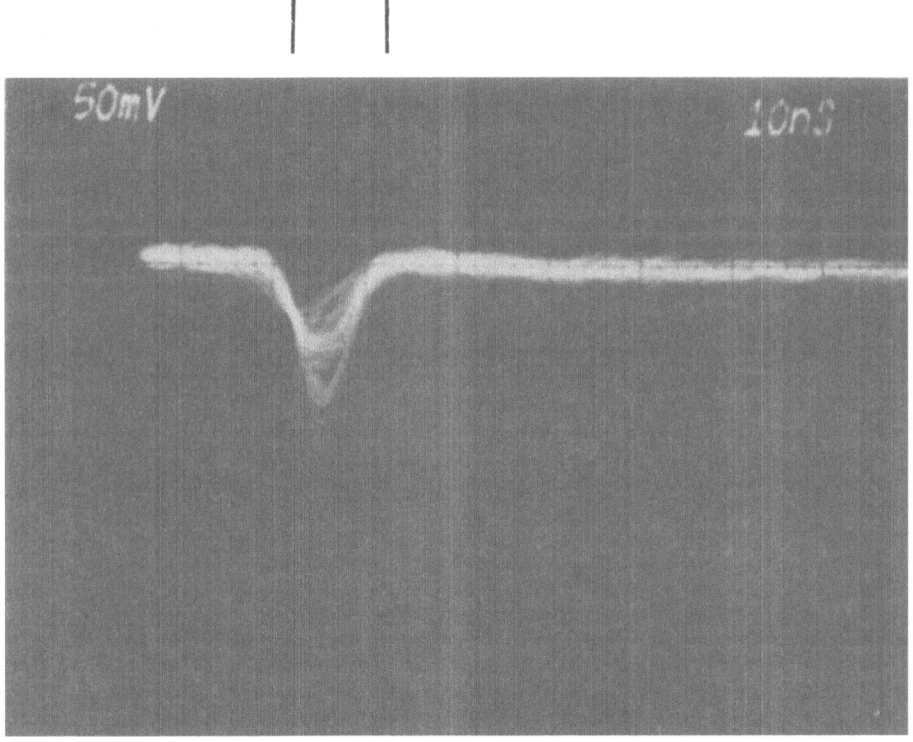

Fig. 2. Waveform from fast shaping amplifier for 5.9 keV x-rays creating avalanches on one anode wire of detector. Gas mixture is $Xe/5\%CO_2/10\%CF_4$. Recovery time of amplifier is about 10 nsec.

(b) The shaping electronics on each anode channel comprises a specially designed common base preamplifier and fast gaussian amplifier (Fischer et al., 1985). The impulse response is a signal with a base width less than 10 nsec. For operation with the gas detector there are pole/zero networks in the shaping amplifier to cancel the long tail of the current pulse due to positive ion motion.

(c) In order to maintain the bandwidth advantage of the electronics it is necessary that, in the detector, the time duration of the electron cloud, passing through the upper cathode into the amplification region, be less than or of the same order as the integration time of the shaping amplifier, about 4 nsec in this case. The factors which determine this time duration are the size of the electron cloud and the electron drift velocity, favorable conditions being a small cloud and high drift velocity. Although electron cloud size is mainly beyond the user's control, an increase in electron drift velocity compared to standard gas mixtures can be obtained by addition of a small percentage of CF_4 to xenon (Christophorou et al., 1980). In the present application we use a gas mixture of $Xe/5\%CO_2/10\%CF_4$, in which the CO_2 acts as an inorganic quencher to avoid the polymerization products associated with organic quenchers. Fig. 2 shows 5.9 keV x-ray signals from one shaping amplifier with this gas mixture. The waveforms return to the baseline in about 10 nsec, thus making possible a count rate of 10^6 sec^{-1} in each channel, with dead time losses of only 1%.

Therefore, with 100 channels operating in identical but independent fashion, the whole detector is capable of accurately recording counting rates in excess of 10^8 sec^{-1} with very small dead time losses (Smith et al., 1985). Statistically significant measurements can be obtained with time slices of order 1 msec or slightly less. Such a system, using small angle x-ray scattering, is a very important tool for dynamic studies of muscle.

TIME-TO-DIGITAL CONVERTER FOR DETECTORS
WITH DELAY LINE POSITION READOUT

Position sensitive radiation detectors using delay line position readout have traditionally used a time-to-amplitude converter (TAC) followed by an analog-to-digital converter (ADC) to transform the timing signals into a digital data word representing particle position. A drawback of this method is that the photon rate at which the system can operate is much lower than the maximum rate of the delay line readout. This is due to the long dead times associated with the TACs and ADCs capable of high resolution and low differential nonlinearity. It is not possible to specify an exact dead time since this depends on actual devices used, but 5-15 μsec is a realistic figure for commercial devices having 10-bit resolution and 1% differential nonlinearity. For most detector systems, this is the dominant dead time component and the primary reason for restricted repetition rate.

Direct time-to-digital converters (TDCs), which operate by counting clock periods between a Start and Stop strobe, can reduce dead time dramatically. However, TDCs which are capable of reasonable resolution require high clock frequencies, and unwanted feedback can cause differential nonlinearities to appear in the position spectrum at multiples of the clock period. Very careful attention must be paid to circuit design and layout to prevent this from happening. We briefly describe here a TDC which operates from a 500 MHz clock (2 nsec time resolution), has less than 0.1% differential nonlinearity and a recovery time of 70 nsec or less.

A detailed account of the TDC has been given by Harder (1988). One of its main features is that all the processing is pipelined, in order for it to be reset, and be ready for the next event, as quickly as possible, after receiving the last detector timing signal. In the first pipeline, when a photon has been absorbed by the detector, the anode signal creates a 'start' strobe which, after synchronization with the TDC clock, is applied to the 'start' input of the TDC, forcing two 500 MHz scalers to begin counting. Also at this time a busy flag is set, causing further start inputs to be ignored. When the 'stop left' and 'stop right' strobes from each end of the

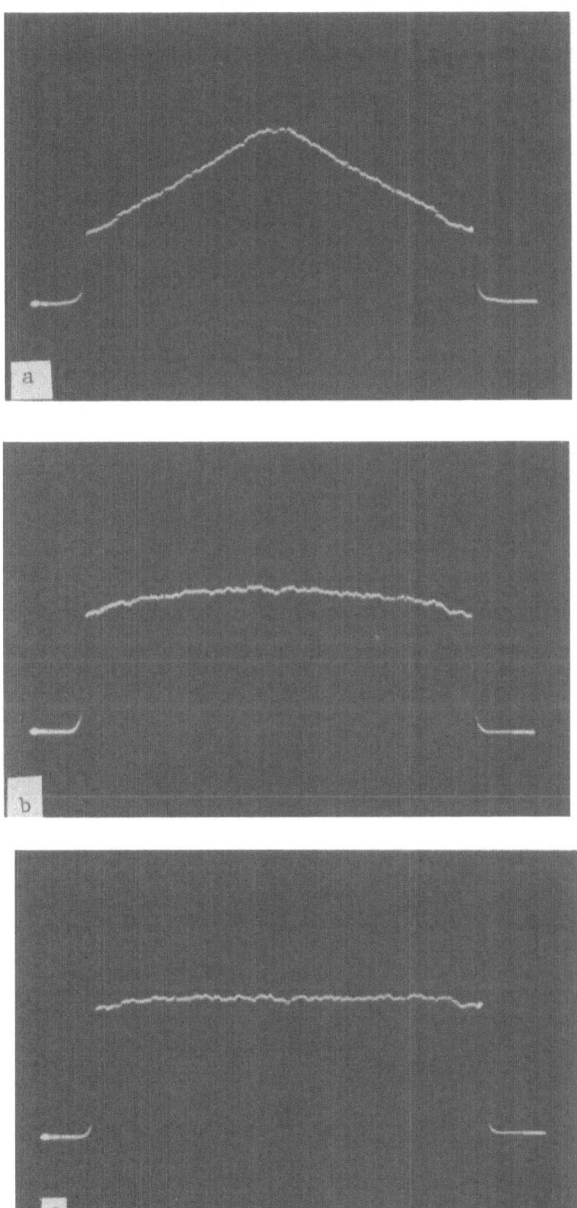

Fig. 3. Position spectra for a uniformly illuminated detector with a 1 μsec delay line. Incident photon rate is 10^6 per sec. (a) Discriminators are not used. (b) Arithmetic sum discriminator only is used. (c) Both arithmetic sum and minimum interarrival time discriminators are used.

detector have caused their respective scalers to halt, both scaler contents are transferred to a register, the scalers and busy flag are reset, and the TDC is ready to process a new event. Further data processing is handled by other pipelines and a minimum of time therefore elapses between consecutive events. In general the maximum dead time of the system is $(\tau + 70)$ nsec, where τ is the temporal length of the delay line, while the average dead time is somewhat shorter.

In the second pipeline slot the stored contents of the left and right scalers are applied to adder and subtractor circuits. The difference represents the position of the absorbed photon along the delay line, while the sum is used as one part of a pile up rejection method. It is easy to show

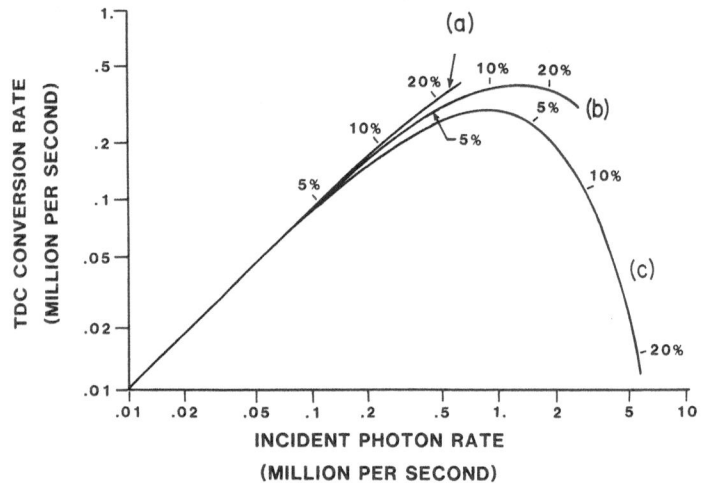

Fig. 4. TDC conversion rate is plotted as a function of incident photon rate for uniformly illuminated detector with a 1 μs delay line. Separate plots are shown for different discriminator settings, with corresponding distortion values indicated at several points. (a) Discriminators not used. (b) Arithmetic sum discriminator only is used. (c) Both arithmetic sum and interarrival time discriminators are used (interarrival time: 500 ns).

that the sum of the propagation times is independent of where along the delay line the absorbed event occurs. Therefore if another photon is detected before the signals from the first one have arrived at the end of the delay line, left and right scalers may be stopped by signals from different photons, leading to an erroneous conversion. A third pipeline slot, the arithmetic sum discriminator, checks for valid conversions using this criterion.

The TDC performance has been measured by connecting it to a detector with a position sensing system incorporating delay line readout with $\tau = 1$ μsec. Details of the detector and position sensing electronics are given by Boie et al. (1982). The detector was uniformly irradiated with photons from a distant source. Figure 3 shows the system response to 10^6 photons sec^{-1}

under three different conditions. In Fig. 3(a) the arithmetic sum discriminator is wide open, and is essentially ineffective, and the humped nature of the spectrum is due to erroneous conversions. When the arithmetic sum discriminator is set to form a narrow window, centered around the transit time of the delay line, the much improved spectrum in Fig. 3(b) is obtained. However, a slight hump is still seen even in this spectrum. This effect is due to the fact that, for a given incident photon, the time required for the resulting signals to reach each end of the delay line is a function of photon position. This time interval has a maximum value τ for photons absorbed at either end of the delay line, and a minimum value of $\tau/2$ for photons at the center of the line. There is thus a higher probability that photons in the center will satisfy the arithmetic sum discriminator than photons at either end. This position sensitive bias may be eliminated with the use of a second discriminator, which determines the minimum interarrival time between anode strobes. If the time interval between consecutive strobes is less than the preselected value, then the conversion cycle associated with the first strobe is aborted and that for the second strobe is never started. In this way detector events occurring too close in time are never processed. The spectra for Fig. 3a and b had no minimum interarrival time setting, while the effect of setting the minimum interarrival time to 700 nsec (i.e. just over half the delay line time) is shown in Fig. 3(c). The uniform irradiation spectrum is now essentially flat.

As might be expected, the price paid for improved spectral quality is a reduction in TDC throughput. The relative trade-offs between speed and performance are described in detail by Harder (1988) and are shown graphically in Fig. 4 for several different discriminator settings. The percentage distortion figure indicated on each plot is defined as: % distortion = 100 $(h-g)/h+g$, where h = maximum height of uniform irradiation response (Fig. 3), and g = minimum height. At photon rates approaching 10^5 sec^{-1}, uniform response can be obtained whether either discriminator is operating or not. Even at event rates as high as several 10^6 sec^{-1}, uniform response can still be obtained by using both arithmetic and interarrival time discriminators, but with progressively decreasing throughput.

FUTURE OUTLOOK

As x-ray sources become more intense, the counting rate capabilities of detectors, whether linear or area, will have to be increased to higher limits. Improvements in detector performance can be achieved by subdivision of the detector into independent sections, or by the use of charge integrating electronics (as opposed to the presently more common form of single event counting) or by a combination of both of these. The development of extremely high rate detectors along these lines will be helped significantly by the solutions being found to similar problems in high energy physics. An important element in this work is the present development of monolithic low-noise readout circuits. These will allow the realistic construction of many more readout channels per unit length than is possible even with the smallest hybrid circuits, and they will also make it easier to perform tasks such as fast, multiple sampling and switching, an essential requirement for dynamic imaging studies.

We will now describe two different detector developments which follow the ideas mentioned above, whose motivation stems from the need to observe high particle multiplicities in relativistic heavy ion physics. Very similar devices, however, would meet the requirements of extremely high rate, position sensitive x-ray detectors.

Figure 5 shows a view of a two-dimensional multiwire chamber for heavy ion physics (Debbe, personal communication). Situated midway between two cathodes is a wire plane consisting alternately of an anode wire, then a field wire. The position sensing element is the lower cathode, which consists of a row of conducting 'pads' directly underneath each anode wire, plus a thin guard strip directly below each field wire. The pads are coupled together resistively to allow charge division, every tenth pad or so forming a node or readout channel. The upper cathode is formed by a thin flat window.

Each anode wire forms a long, narrow cell, which is read out, via the cathode pads, independently of all the other anode wires. With a conventional centroid finding electronic analysis fed by each node along a row of pads, any one wire would not be capable of multi-event capability, but the design of a fast centroiding electronic system, capable of dealing with two or more simultaneous events on the same wire is feasible, provided any two events do not appear at a common node. In the present prototype detector, anode cathode spacing is 2 mm, and anode field wire spacing is 2 mm, giving an anode cell width of 4 mm. A practical area x-ray detector would need a cell size of order 1 to 2 mm or less to yield useful position resolution; this would require a proportionate reduction in the anode cathode spacing, and place quite severe but manageable tolerance requirements on wire and electrode spacing. Position accuracy along the anode wire for 8 keV x-rays would be of the order 100-200 μm, while across the anode wires, the cell size determines the resolution. It is feasible to consider centroid finding electronics which will operate at approximately 10^6 sec^{-1} per cell. A 100 anode wire detector would therefore offer a two-dimensional chamber with a rate capability of 10^8 sec^{-1}.

The second development is concerned with silicon pixel detectors. Fig. 6 illustrates two silicon 'pad' detectors being developed for multi-particle detection in heavy ion physics (Kraner, personal communication). Both have an array of 512, independently readable pads, with sides of order 2-3 mm. The detectors, about 7 cm in diameter, are fabricated using principles which generally follow accepted planar technology (Kemmer, 1984; Beuttenmuller et al., 1986) and consist of 300 μm fully depleted silicon. These detectors

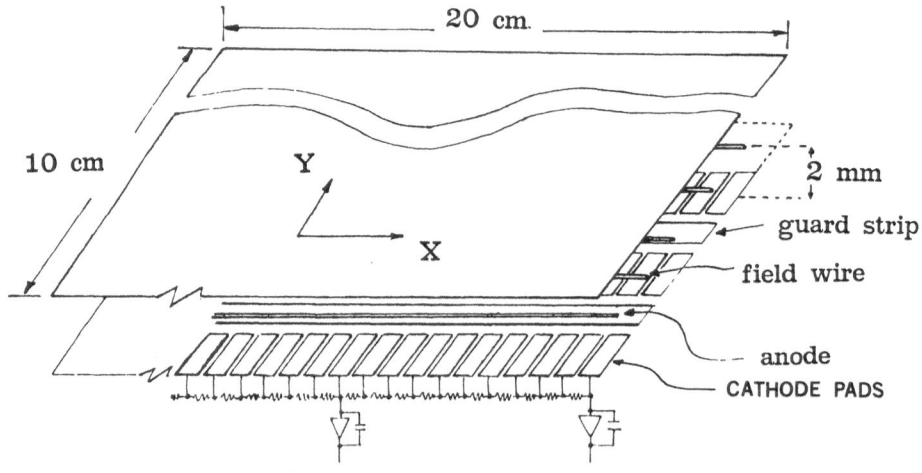

Fig. 5. Two-dimensional, multi-wire chamber, with cathode 'pad' readout, currently under development for multi-particle detection in heavy ion physics. A detector suitable for x-ray scattering or diffraction experiments would retain the same approximate overall size, but wire spacing and pad size would be reduced to the order of 1 to 2 mm (see text for details).

are fabricated on high resistivity silicon (≈ 5 kohm cm) in contrast to transistors and integrated circuits which use very low resistivity silicon (≈ 1 ohm cm). The pad pattern, which is applied to the detector using lithography, is generated using computer graphics. An important factor here is that the pad pattern is, to a large extent, under the complete control of

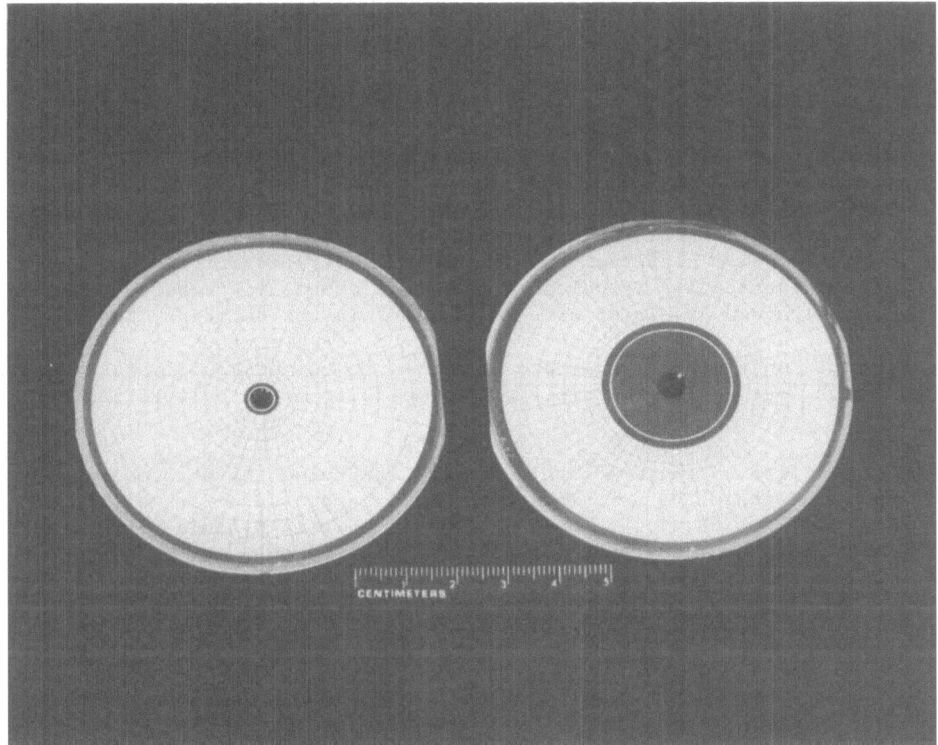

Fig. 6. Two variants of a silicon pad detector for multiparticle detection in heavy ion physics. The central holes allow the direct particle beam to pass through. Detectors of this type, but with various electrode patterns, are suitable for dynamic studies. X-ray detectors utilizing a smaller pad size and many thousands of readout channels would require the use of monolithic circuit technology.

the user; for example, if it is known that a particular experiment will produce a high density of diffraction peaks in one area, and relatively few in another, the pad sizes can be tailored accordingly to make efficient use of the number of readout channels available. Electrical contact is made to each pad by ultrasonically bonding a thin aluminum wire to it, and the other end feeds a preamplifier via a connection board.

An extremely useful application for a device such as this would be as a very high rate position sensitive detector in dynamic studies. Each pad/preamp would act as a charge integrator. At the end of the first time slice charge would be switched to a capacitor C_{M1} (Fig. 7) by closing S_0 and S_1. This continues for as many time slices as there are capacitors, at which time the charge on C_{M}'s are multiplexed through a buffer amplifier to an output storage register. This electronic chain is duplicated for all pads. The typical count rates and time slices attainable are simply calculated as follows. If each charge preamp has a feedback capacitor of, say, 10 pF, and we allow a maximum of 1 volt to develop across it, then a charge of 10 pC can be stored. A 10 keV photon will deposit about 0.5 fC in silicon, so that the maximum number of photons detectable per pad, per time slice is $10^{-11}/5.10^{-16} = 2.10^4$. Thus, with 1 μsec time slices, an incident flux of $\approx 10^{10}$ sec^{-1} pad^{-1} could be sustained, and with 1 nsec time slices, an incident flux of $\approx 10^{13}$ sec^{-1} pad^{-1} could be sustained. Correctly fabricated detectors will have leakage currents of $\approx$ 1 nA per pad, contributing 1 fC in 1 μsec. Noise is therefore very low and a high dynamic range in count rate is achieved.

A realistic detector for x-ray scattering and diffraction would require more pads of smaller size than the detector illustrated in Fig. 6. Even with present fabrication techniques, one could probably envisage 0.5 - 1 mm size pads, although the major problem lies not in the detector, but the electronics. A detector with several thousand, or even tens of thousands, of pads could only sensibly be read out with the use of monolithic circuits. With such devices, multiple sampling and switching of capacitors can be realized.

When integrated and monolithic circuit technology have become suitably advanced within the next few years, silicon pad detectors will offer unique features which will be most useful for certain dynamic x-ray studies. The developments of monolithic circuits for this purpose may allow detectors with 10^3 elements in the next 2-3 years, and with larger numbers of elements on a 5-year time scale. Some of the important features include the user's control of the pixel pattern, the 300 μm depletion region which results in very high detection efficiency for 5-15 keV x-rays (10% efficiency even at 30 keV) and both a high dynamic range and high count rate capability.

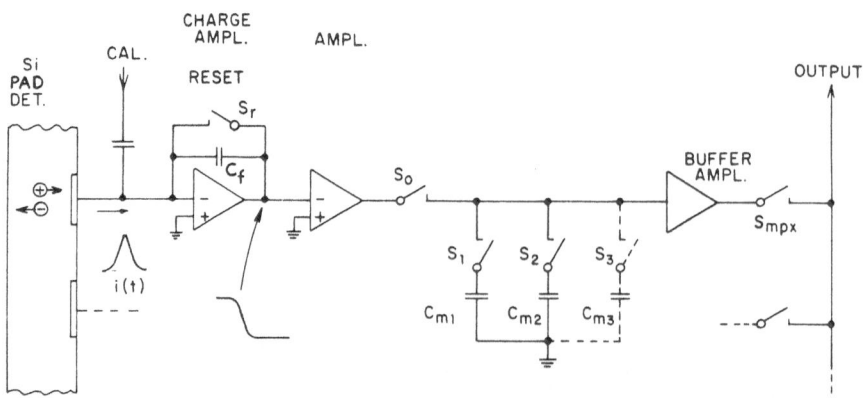

Fig. 7. Functional circuit diagram for multiple sampling of stored charge on a silicon pad detector, and subsequent switching of charge onto capacitors, for use in high rate, dynamic x-ray studies.

ACKNOWLEDGMENTS

The working devices mentioned in the first part of the paper result from a detector development program in the Instrumentation Division, BNL. Dmitri Stephani and Lee Rogers have developed much of the readout electronics and Gene Von Achen and Fred Merritt constructed the detectors. In the last part of the paper the new gas detector is due to a collaboration of BNL and the University of Pittsburgh, while the silicon pad detectors are being developed by Hobie Kraner and Zheng Li of the Instrumentation Division, BNL. We greatly appreciate the encouragement of Ben Schoenborn, Head of Structural Biology, BNL, who has provided considerable motivation for the x-ray detector development.

REFERENCES

Beuttenmuller, R., Kraner, H. W., Ludlam, T. W., Polychronakos, V. A., van Dijk, J., Radeka, V., Chesi, E., Piuz, F., Tschulik, A., and M. J. Esten, 1986, The development and testing of a silicon trigger detector with 400 elements for the HELIOS (NA34) experiment, IEEE Trans. Nucl. Sci., NS-33, No. 2, 1045.

Boie, R. A., Fischer, J., Inagaki, Y., Merritt, F. C., Radeka, V., Rogers, L. C., and Xi, D. M., 1982, High resolution x-ray gas proportional detectors with delay line position sensing for high counting rates, Nucl. Instr. and Meth., 201:93.

Christophorou, L. G., Maxey, D. V., McCorkle, D. L., and Carter, J. G., 1980, Xe-containing fast gas mixtures for gas-filled detectors, Nucl. Instr. and Meth., 171:491.

Debbe, R. (personal communication).

Fischer, J., Hrisoho, A., Radeka, V., and Rehak, P., 1985, Proportional chambers for very high counting rates based on gas mixtures of CF_4 with hydrocarbons, Nucl. Instr. and Meth., A238:249.

Harder, J., 1988, A fast time-to-digital converter for position-sensitive radiation detectors with delay line readouts, Nucl. Instr. and Meth., A265:500.

Kemmer, J., 1984, Improvement of detector fabrication by the planar process, Nucl. Instr. and Meth., 226:89.

Kraner, H. W., (personal communication).

Smith, G. C., Fischer, J., and Radeka, V., 1985, Developments in gas detectors for synchrotron x-ray radiation, Trans. Am. Cryst. Assoc., 21:41.

X-RAY OPTICS FOR SYNCHROTRON RADIATION

Andreas K. Freund

European Synchrotron Radiation Facility
B.P. 220
F-38043 Grenoble Cedex, France

INTRODUCTION

There has been dramatic recent progress in development of dedicated synchrotron radiation sources that have low emittance and high brightness. This calls for equivalent progress in development of beamline instrumentation, in particular, of optical elements. Trying to adapt conventional X-ray optics to the severe conditions required by present-day synchrotron sources solely by improving the quality of the optical devices is a first step, which is not sufficient. Totally new designs of beam-defining devices are necessary, which take into account the specific properties of the radiation: the small size of the source and beam divergence, linear or circular polarization, the wide spectral range, and the high power of beams emitted by the various source-defining devices of modern storage rings.

Several reviews on this topic show the transition from conventional to completely new schemes (Matsushita and Hashizume, 1983; Bonse, 1986; Caciuffo et al., 1987; Freund, 1987). They are all necessarily incomplete, because the field is very wide and often there are several different approaches to the solving of the various problems occurring in the design and construction of beamlines. This paper will try to give a brief introduction to the present situation in general terms and will illustrate more recent achievements by a few examples. The emphasis will be on the medium and high X-ray energy range provided by high energy storage rings, while VUV and soft X-ray optics will not be described here.

ROLE OF X-RAY OPTICS

X-ray optical elements are inserted between the source and the sample and sometimes also between the sample and the detector. These elements transform the beam emitted by the source to meet the experimental requirements at the sample as closely as possible, and eventually to achieve the best possible matching between sample and detector. The transformation is performed in terms of an optimization among various parameters characterizing an X-ray beam, such as intensity, and resolution in energy, angle, and space. This process corresponds to tailoring the shape of the phase space volume occupied by the photons according to the desired experimental conditions. Reducing this volume and decreasing the number of particles it contains cannot be avoided but must be minimized. The transformation is

---

Dedicated to Professor Ulrich Bonse on the occasion of his 60th birthday.

governed by Liouville's theorem, which states that the particle density in phase space can never be increased if only conservative forces are applied to the ensemble. Thus there is an upper bound to the performance of the optical system while, on one hand, the lower bound is determined by the performance of the individual elements, and on the other hand, by the degree of matching between the elements along the beamline.

In an ideal optical system, there should be no decrease of phase space density. However, real optics have limited efficiencies that are responsible for loss of useful photons on their way from the source to the sample. Therefore, the best optics are no optics. Because this is usually not adequate for an experiment so at least the number of optical elements should be minimized. This minimization begins at the source because there is some flexibility in its design (Elleaume, 1988), which permits optimization of the properties of the source for the proposed experiment. Therefore, the first step consists in choosing the most suitable parameters of the insertion device (wiggler, undulator, wavelength shifter) available at present-day storage rings. For instance, quasi-monochromatic radiation can be obtained from undulators.

Detector properties also must be taken into account when defining the optics. For instance, the spatial resolution of a position-sensitive device must match the energy-space correlation in energy-dispersive absorption spectroscopy. Because the various parameters (such as energy range and resolution, angular resolution, and intensity) are not independent of each other, the whole optimization of the source-optics-detector is a complex task and the best solution for a beamline will be determined iteratively.

Figure 1a shows schematically the effect of the performance of the optics on emittance, which is another word for the volume in phase space. The ideal optics would always provide the same emittance at the sample as

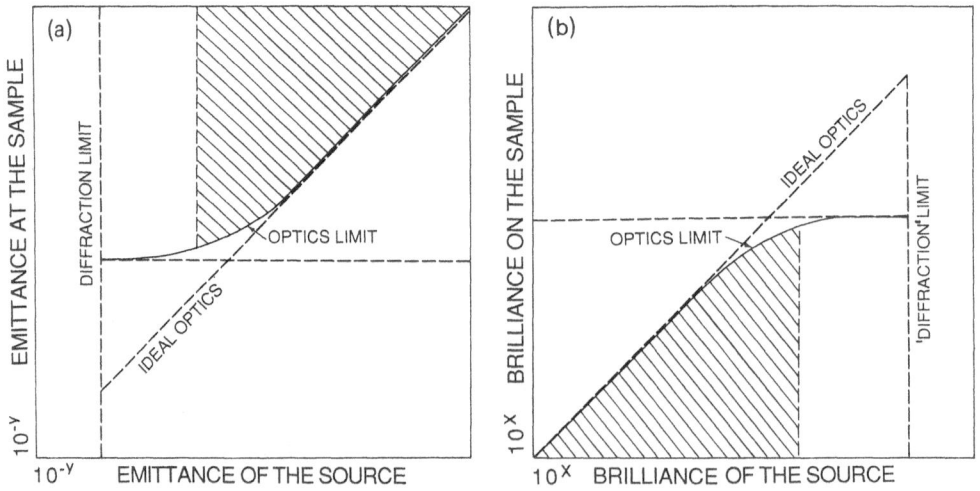

Fig. 1. (a) Schematic picture of the limitation of the emittance obtained at the sample due to the limited performance of the optics.
(b) Limitation of the transferable brilliance by the optics. The scale factors y and x are arbitrary.

that given by the source. Below a certain value of the emittance, the limits of real optics do not allow the further improvement of the experimental conditions that would otherwise be possible by decreasing the emittance of the source further.

Limits of the optics and the source can be fundamental or technological. The fundamental limit of source emittance is the so-called diffraction limit, which is attained when the emittance equals the photon wavelength. Similarly, the limit of very simple imaging optics based on diaphragms is reached when the diameter of the diaphragm is of the order of the wavelength of the radiation used. Some numerical examples for fundamental and technological limits are given later. The question of whether an experiment is limited by the source or by the optics depends on the type of experiment.

The increase of emittance or phase space volume caused by the non-ideal imaging processes of optical elements decreases resolution, resulting in a dilution of the photons in phase space. In addition, there are effects due to photon dissipation by absorption or incoherent and inelastic scattering that decrease further the phase-space density (also called brilliance), of X-ray beams. These effects are illustrated in Fig. 1b in a similar way as for emittance in Fig. 1a. At present, the main limitation of the maximum brilliance transferable from the source to the sample is not determined by these dissipation effects, but rather by the power associated with highly intense synchrotron radiation. The power density of wiggler beams today exceeds that existing on the sun's surface! The thermal load on the first optical element (and also on the second, if the first is a mirror) degrades the optical quality and thus affects both emittance and brilliance. Many materials that do not withstand the radiation damage produced by powerful X-ray beams and beam-defining devices must be reprocessed or replaced after some time.

Associated with the losses of photons due to absorption and unwanted scattering is a simultaneous increase in background intensity that decreases the quality of experimental data. This aspect is of particular importance for experiments that are signal-limited, such as trace element analysis or nuclear resonance experiments. Therefore, those materials for beam-defining devices that present absorption edges in the energy range of interest should always be avoided. Equally important for a clean beam is the absence of higher-order wavelengths reflected by the monochromator (contamination by harmonics). Other undesirable effects are produced by so-called parasitic scattering, which is the diffraction of X-rays, in the direction of the sample, by lattice planes other than those of the main reflection of a monochromator crystal. These effects give the well-known glitches which represent an important experimental problem in EXAFS techniques. Therefore, the quality of optical elements is not only characterized by high reflectivity or transmission inside the spectral, angular, or spatial window of interest, but also by a low reflectivity or transmission outside that window.

To optimize the design of a beamline and of its optics, one has to know both the detailed properties of the phase space and the real performances of present beam-defining devices or ones that can be developed with a reasonable effort and within a reasonable time. These properties are then incorporated into graphic schemes which permit matching the devices amongst themselves and predicting the overall performance of the beamline optics. Such schemes are, for instance, DuMond diagrams or phase space diagrams, which are described in the above reviews and, more recently by Suortti and Freund, (in press). The results of these graphic methods can be refined further by ray tracing codes, which approach veritable computer simulations of X-ray scattering experiments (Lai et al., 1988). While phase space descriptions, including reciprocal space diagrams, give a more idealized picture of a beamline scenario, the ray tracing methods take into account the various effects limiting the real efficiencies of optical devices, such as aberrations, roughness, and slope errors of mirrors, and thus yield a more realistic prediction of beamline performance.

A final important point concerns stability. With decreasing emittance and increasing brilliance, the mechanical problems related to precision angular and translation positioning, which must be both stable and reproducible, become a major issue in beamline construction. Optical elements require anti-vibration supports and must be shielded from acoustic noise. The temperature of the beam-defining devices and their environment has to be stabilized. Short-and long-term variations of position, direction, and flux of the X-ray beams must be detected, and efficient feedback devices must be designed to compensate for these changes. The concept of active optics has to be developed. Because it is more and more difficult to define an adequately stable network of reference points when emittance becomes very small, such a concept will rather be based on the internal consistency of the various optimization parameters of such an active system. At all crucial points of a beamline, pick-up devices are installed to monitor beam properties and to provide the feedback signals for the alignment system. All these signals are continuously surveyed and processed by the experimental computer, which reacts by adjusting the optics and thus maintaining the beamline in optimal operating conditions or permitting us to correct the data if the offsets observed cannot be corrected in real time.

The criteria for good x-ray optics materials and devices can be summarized as follows:

    i)   Ability to reflect photons of a desired wavelength or energy at a
         suitable angle.
   ii)   Angular and spectral resolution must be matched to the
         experimental needs.
  iii)   High transmission (reflectivity) inside the ranges defined in (i)
         and (ii) and low transmission outside.
   iv)   Energy tunability within a specified range.
    v)   Focusing possibilities are almost always desirable.
   vi)   Good performance under severe radiation and heat load.
  vii)   Availability in suitable quality and size or, at least, reasonable
         prospectives for development.
 viii)   Ease in preparation and mounting.
   ix)   Supports of high mechanical stability, precision and
         reproducibility (feedback controls, encoders).

After having described these more general problems I will now discuss several relevant properties of individual devices for beam definition.

USE AND PERFORMANCES OF OPTICAL ELEMENTS

Table 1 lists the various devices presently used for X-ray beam conditioning and the energy ranges for which they are most efficient. These ranges are determined by fundamental and technological limitations: For instance, a single crystal monochromator cannot reflect X-rays of wavelengths longer than twice the d-spacing of its lattice planes. On the other hand, the improvement of multilayer preparation techniques could allow their application to higher X-ray energies. The most ordinary device is a slit, which may be of quite simple construction when used to eliminate stray radiation, but which becomes a huge beam shaper consisting of up to one-meter-long watercooled copper jaws used in grazing incidence to limit the size of powerful wiggler beams. At the other extreme, it is not easy to fabricate very narrow pinholes of submicron diameter, for example, for microprobe experiments.

258

Table 1. X-ray optical devices (presently existing or under development), the energy and wavelength ranges where they are used (R-reflection, T-transmission geometry, xxx-frequently, xx-moderately, x-occasionally) and the beam parameters which they define (α-divergence, Δx-spatial resolution, E-energy, ΔE-energy spread, p-polarization).

	Devices for Beam Definition	Soft X-Rays 0.3-3 keV (40-4 Å)		X-Rays 3-30 keV (4-0.4 Å)		Hard X-Rays 30-300 keV (0.4-0.04 Å)		Beam Parameters
		R	T	R	T	R	T	
Beam Shapers	Pinholes, Diaphragms	xxx		xxx		xxx		α,Δx
	Soller Slits	x		xxx		xxx		α
Total Reflection	Mirrors	xxx		xxx				α,Δx,(E,ΔE)
	Guide Tubes		x		x			α,(Δx,E,ΔE)
Linear and Planar Microstructures	Gratings	xxx	xx	x				α,Δx,E,ΔE
	Fresnel Plates	xx	xx	x				α,Δx,E,ΔE
Bragg Optics	Multilayers	xxx		xx	x			α,Δx,E,ΔE
	Single Crystals	x		xxx	xx	x	xxx	α,Δx,E,ΔE,p
Combined Systems	Multilayer Gratings	x		x(?)			x(?)	α,Δx,E,ΔE
	Bragg-Fresnel-Optics	x(?)		x(?)			x(?)	α,Δx,E,ΔE,p

Here the discussion will focus on the following devices: mirrors, monochromators and multilayers. The latter, also called layered synthetic microstructures, are sometimes referred to as "multilayer mirrors", sometimes as "multilayer monochromators", and are thus situated between single-layer mirrors and the three-dimensional lattices of single crystal monochromators. Because gratings are not very efficient at medium and higher X-ray energies they will not be described here. The same holds for Fresnel lenses. However, at the end of this paper some potential applications of Bragg-Fresnel optics will be outlined.

Fig. 2 shows the energy and wavelength dependence on the glancing angle and the typical ranges covered by the three major kinds of beam-defining devices. These DuMond diagrams are based on Snell's law for mirrors and on Bragg's law (including refraction) for single crystals and multilayers. At a given glancing angle, mirrors reflect X-rays of all energies smaller than the critical energy characteristic of a given material, while single crystals and multilayers select sets of narrower or wider energy bands corresponding to Bragg reflections and their harmonics.

According to the experimental requirements, one or more of these devices can be used in flat or focusing geometries. Some examples of phase space and reciprocal space diagrams will be given to illustrate the properties of flat or curved monochromators and their effect on beam transformation. Special problems arising from heat load on optical elements and present approaches to solve these problems also will be mentioned.

Mirrors

Mirrors are widely applied to focus X-ray beams and to eliminate higher-order harmonics of Bragg reflections from crystal monochromators. They have the advantage that the heat and radiation load is dispersed over a much bigger surface than in the case of crystal monochromators because the reflection angle is very small, of the order of some mrad. Curved mirrors also are used to match the divergence of the white beam to the acceptance of perfect crystal monochromators.

The interaction of X-rays with a medium can be described in terms of a complex index of refraction, given by $n = 1 - \delta - i\beta$. The wavelength-dependent quantities δ and β account for the dispersive properties of the material and for absorption, respectively:

$$\delta = \frac{r_o}{2\pi} N\lambda^2 \tag{1}$$

$$\beta = \frac{\mu}{4\pi} \lambda \tag{2}$$

where $r_o = 2.818 \times 10^{-15}$ m is the classical electron radius, λ is the X-ray wavelength, μ is the linear absorption coefficient and N is the number of electrons per unit volume of the reflecting material. Close to the absorp-

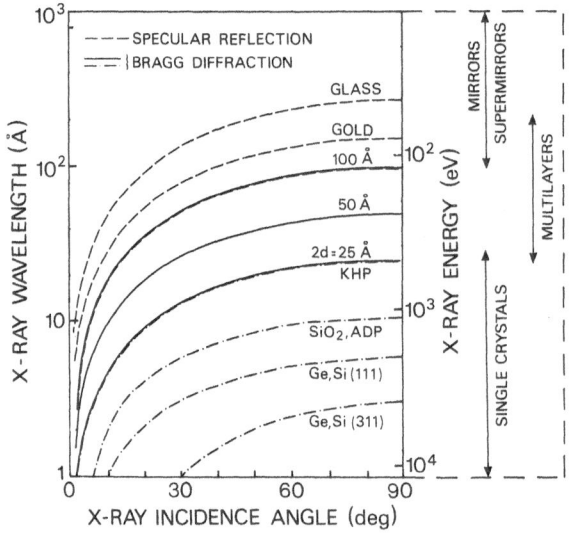

Fig. 2.   The relation between the angle of reflection and X-ray wavelength and energy for three major kinds of beam-defining devices.

tion edges, where δ and β vary strongly with wavelength, N must be corrected for anomalous scattering. In fact, N is replaced by the real part of the scattering factor, F_o', for forward scattering, divided by the volume v_o of the unit cell in the case of crystalline substances or per unit volume in the case of amorphous materials. For monoatomic crystals the complex structure factor $F_H = F_H' + iF_H''$ can be written as:

$$F_H = F_G \cdot (f_o + f' + if'') \tag{3}$$

where F_G is the geometrical structure factor which includes a correction for thermal motion (Debye-Waller factor) and stands for the number of electrons participating in the elastic scattering process. It is dependent on the scattering vector through the Debye-Waller factor, but is independent of the X-ray wavelength as is also the atomic form factor, f_o, in the energy range concerned here. By contrast, the real and imaginary parts of the anomalous scattering correction, f' and f'', respectively, depend on wavelength but are to first order independent of the scattering vector. If the material contains several kinds of atoms, F_H can be represented by a sum over the structure factors contributed by each species.

In the absence of absorption ($\beta = 0$) the refractive index is real and smaller than unity. As a consequence, for a given wavelength λ_c, total external reflection occurs for angles of incidence θ smaller than a critical value, θ_c. By using Snell's law one obtains $\cos \theta_c = (1 - \delta)$ and since δ is typically of the order of 10^{-5}:

$$\theta_c = \sqrt{2\delta} = 2.34 \cdot 10^{-3} \sqrt{\frac{Z\rho}{A}} \; \lambda_c \qquad (4)$$

where Z, ρ and A are the atomic number, the density expressed in CGS units, and the atomic mass of the reflecting material, respectively. The (ideal) reflectivity is equal to 1 for $\theta \leq \theta_c$ and 0.5% for $\theta = 2\theta_c$.

However, due to absorption the sharp cutoff is broadened and the complex reflection coefficients as a function of the glancing angle are given by the Fresnel equations (Compton and Allison, 1963; Born and Wolf, 1964) which also involve polarization. At small glancing angles the polarization dependence can be neglected and a single expression for the reflectivity of X-rays is obtained (Parratt, 1954):

$$r = \frac{h - (\theta/\theta_c) \; \sqrt{2(h - 1)}}{h + (\theta/\theta_c) \; \sqrt{2(h - 1)}} \qquad (5)$$

where,

$$h = (\theta/\theta_c)^2 + \sqrt{[(\theta/\theta_c)^2 - 1]^2 + (\beta/\delta)^2} \qquad (6)$$

Eq. 5 is valid for a perfectly smooth and uniform mirror surface and can be used for wavelengths smaller than about 10 Å. For bigger wavelengths the polarization dependence should be taken into account (Henke, 1972).

From the above it is obvious that with increasing Z and λ the integrated reflectivity of a mirror increases while the efficiency as a high energy bandpass filter decreases, because the enhanced absorption leads to a less steep cutoff and to a smaller reflectivity near θ_c. As pointed out by Witz (1969), the total flux reflected by the mirror can be estimated by the product $\theta_c \cdot r(\theta_c)$ for a cylindrical mirror and by $\theta_c^2 \cdot r(\theta_c)$ for a toroidal mirror.

For a quartz mirror $\theta_c = 5.1$ mrad and $r = 0.5 (\lambda = 1.54$ Å) while for the gold-coated surface $\theta_c = 10$ mrad and $r = 0.4$. Therefore, an increase of the reflected intensity by a factor of between 1.6 and 3.1 can theoretically be obtained in focusing geometries by using heavy element coating. On the other hand, L-edges of these elements may affect the reflection properties of the mirrors so that elements other than Au, namely Pt, Rh and Ir also have been used. To smooth the irregularities at absorption edges, thin additional coatings were proposed by Fukamachi et al. (1986).

Fig. 3 shows theoretical and experimental reflectivity curves r(θ) at
λ = 1.54 Å both for fused quartz, pyrex glass, and float glass in 3a, and
for Au-coated and uncoated pyrex glass mirrors in 3b (Matsushita et al.,
1984). When comparing the experimental results obtained for the gold-coated
pyrex glass mirror with the theoretical curve in Fig. 3b, one observes a
dramatic disagreement due to surface imperfections. The disagreement in
Fig. 3a is much less. Common to all curves is that the experimental values
lie systematically below the theoretical ones assuming a perfect surface.
Fig. 4 shows angle-resolved scattering curves obtained from the same samples
which were recorded with a high resolution triple axis X-ray diffractometer.

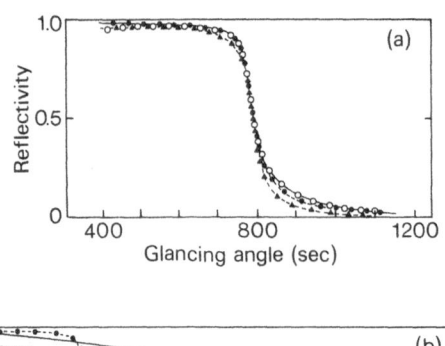

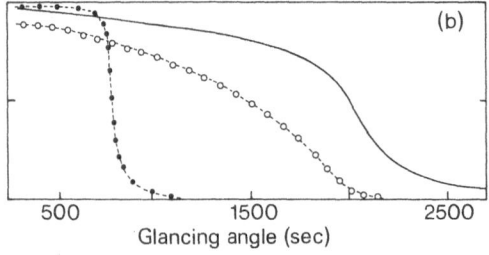

Fig. 3.   (a) Observed reflection curves for mirrors of fused quartz (O),
          pyrex glass (●) and float glass (▲). The solid line is the cal-
          culated curve for a fused-quartz mirror. (b) Calculated (solid
          line) and observed (O) reflection curves from the gold-coated
          mirror are shown together with the observed reflection curve(●) of
          the substrate pyrex glass. The broken lines in (a) and (b) are
          guides to the eye. λ = 1.54 Å (After Matsushita, et al., 1984).

Part of the photons lacking in the center of the peak are found in the tails
while the curve corresponding to the float-glass mirror is substantially
broadened. Both effects are explained by microroughness down to the Å range
and by a waviness of the surface on a much bigger scale, respectively. The
scattered curves are a Fourier transform of the scattering medium. The
efficiency of the gold-coated mirror is much more affected by diffuse scattering than that of quartz or pyrex because both the absorption and scattering cross sections are higher. From the measured reflectivities it
can be estimated that for the above case the real increase in integrated

intensity due to Au coating is by factors of 1.2 and 1.8 only for cylindrical and toroidal mirrors, respectively.

The real surface can be modeled by a faceted structure where the distribution of the size and misorientation of the facets depend on the fabrication process. The type of the distribution functions, their amplitude and spatial frequency must be known to calculate the observed X-ray reflection properties. While the scattering by imperfect surfaces is easy to understand qualitatively, quantitative evaluations and surface metrology are very complex (Nevot and Croce, 1980; Takacs, 1986, 1988; Church, 1988; Christensen et al., 1988; Sinha et al., 1988).

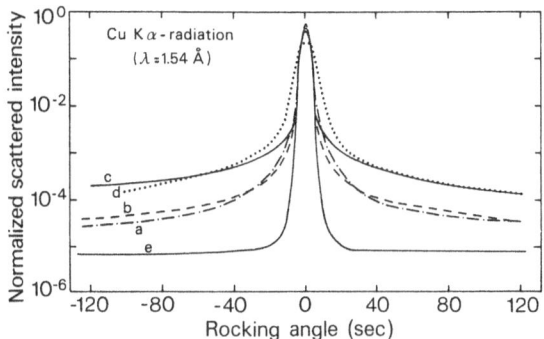

Fig. 4. Angle-resolved scattering curves. Curve a is for the fused quartz mirror; b for the pyrex-glass mirror; c for the gold-coated pyrex-glass mirror; d for the float-glass mirror and e is the instrumental curve. The glancing angles of the X-ray beam to the mirror surface are 1200" for the gold-coated pyrex-glass mirror and 640" for the other mirrors (After Matsushita et al., 1984).

The most important application of mirrors is to focus X-ray beams emitted by the source onto the sample. The efficiency of focusing is determined by fundamental effects such as aberration and coma, and by mirror imperfections such as microroughness and slope errors. The best conditions are obtained with an elliptical shape of the mirror surface (Fig. 5). The relations between the various parameters of the focusing geometry are:

$$a^2 = (d + \sqrt{d^2 - (f_1-f_2)^2(f_1+f_2)^2 / 4}) / 2, \qquad (7)$$

$$b^2 = a^2 - (f_1 + f_2)^2 / 4, \qquad (8)$$

$$\text{and } d = (f_1 + f_2)^2 / 2 + (2f_1f_2\theta)^2 / (f_1+f_2)^2. \qquad (9)$$

The source S of size s is imaged on I of size i so that s/i = p/q. Liouville's theorem is confirmed by the fact that the ratio of the angles Ω_s and Ω_i is given by q/p and thus $s\Omega_s = i\Omega_i$. For a $\to \infty$ the shape becomes a parabola and a parallel beam ($f_1 \to \infty$) is imaged on a point.

For different shapes of the mirror surface the image size is increased by spherical aberrations. Replacing the ellipse by a circle of radius R_m gives rise to a broadening B (Rosenbaum and Holmes, 1980):

$$B = 3 q L^2 \sin 2 \theta \, |1/q^2 - 1/p^2|/8 \qquad (10)$$

where L is the length of the reflecting segment. For a magnification M = q/p = 1 the circle is an excellent approximation. If p and q are not too different, then for small θ and with L = $p\alpha_o/\theta$ (neglecting source size) B is proportional to $qp^2\alpha_o^2/\theta$ where α_o is the divergence of the incident beam. This favours short distances p and q and big reflecting angles θ. For L = 0.5 m, p = 30 m, q = 10 m and θ = 10 mrad, the broadening is 0.3 mm. The corresponding radius of curvature is given by:

$$R_m = 2 pq/(p + q) \sin \theta \qquad (11)$$

which for the above example is 1.5 km. Spherical aberrations must occur in order to conserve phase space density because for a circle the angles Ω_s and Ω_i are equal (see Single Crystal Monochromators).

Focusing can be achieved not only in the scattering plane (meridional or in-plane focusing) but also perpendicular to that plane (sagittal or out-of-plane focusing). The corresponding radius of curvature is:

$$R_s = 2 pq \sin \theta/(p + q) = R_m \sin^2 \theta \qquad (12)$$

which, using the above numbers, gives 15 cm. For double focusing, toroidal mirrors are used which can be manufactured by grinding a cylinder of radius R_s in a substrate which is subsequently bent to a radius R_m. For our example this bending corresponds to a deflection of only 20 µm at the center which shows the high precision needed for mirror preparation techniques and supports. High figure accuracy and small surface roughness are difficult to achieve at the same time. The best results recently reported are slope errors of about 5 µrad rms and a microroughness of a few Å rms for super-polished surfaces (Takacs, 1988; Tirsell et al., 1988). This means that for q = 10 m the focus will be broadened by 0.1 mm which is about equal

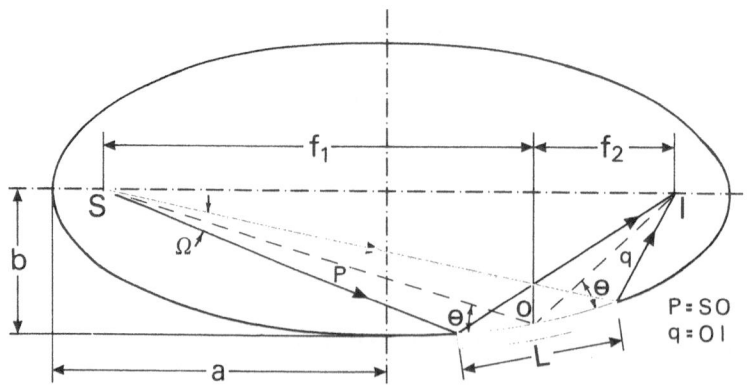

Fig. 5.   Focusing geometry of an elliptically shaped mirror. (See text for details).

Table 2. Materials used for X-ray monochromatization. Only a few of them are suitable for synchrotron radiation (Si, Ge, Be, Graphite ...). P, PP and M mean availability as nearly perfect, perfect (dislocation-free) and mosaic crystals, respectively.

Crystals (hkl)	Crystal system	2d spacing (Å)	Quality
α-quartz SiO_2 (1010)	Hexagonal	8.52	PP-P
ADP $NH_4H_2PO_4$ (200)	Tetragonal	7.50	P
β-alumina Al_2O_3 (0330)	Hexagonal	2.748	P-M
Beryllium (002)	Hexagonal	3.58	P-M
EDDT $C_6H_{14}N_2O_6$ (020)	Monoclinic	8.808	
Fluorite CaF_2 (111)	Cubic	6.306	P-M
Germanium Ge (111)	Cubic	6.533	PP-M
Graphite C (0002)	Hexagonal	6.708	M
Indium antimonide InSb (111)	Cubic	7.481	P
Lead stearate $(CH_3(CH_2)_{12}COO)_2$ Pb	Smectic soap multilayer	100.4	M
Lithium fluoride LiF (200)	Cubic	4.027	P-M
Mica $K_2O.3Al_2O_3.6SiO_2$ (002)	Monoclinic	9.84	P-M
KHP $KOOC.C_6H_4.COOH$ (1010)	Hexagonal	26.632	P
Rock salt NaCl (200)	Cubic	5.641	P-M
Silicon Si (111)	Cubic	6.271	PP-M
Beryl $3BeO.Al_2O_3.6SiO_2$	Hexagonal	15.954	PP-M

to the source size of modern storage ring facilities such as that of the ESRF in Grenoble.

When the mirror is used as the first optical element in the beamline, problems related to radiation damage and thermal gradients are added to manufacturing difficulties. High-power wigglers yield a total power of several KW and power densities up to a few 100 W/mm^2 perpendicular to the beam. Gradients parallel and perpendicular to the mirror surface create strains which deform the surface and degrade the material, even if the mirrors are cooled. A substantial part of the radiation is absorbed by the mirror and gives rise to radiation damage. From this and other points of view the best material at present is SiC (Takacs et al., 1984; Mourikis et al., in press; Sato et al., 1988).

Single Crystal Monochromators

Bragg diffraction from single crystals is perfectly suited for X-ray monochromatization because a variety of crystals provides a range of d-spacings, permitting a selection of wavelengths in the whole range of interest and at angles that are convenient for most applications. Table 2 summarizes the materials that are commonly used in X-ray instrumentation. However, not all of them can resist the radiation and heat load of synchrotron X-ray beams, and among those that do only Ge and Si are available as big and highly perfect crystals. GaAs, SiO_2, ADP, CaF_2, InSb, LiF, KHP, NaCl and Cu can be considered as nearly perfect crystals, whereas graphite cannot be manufactured with a mosaic spread smaller than about 5 mrad. Recent progress in crystal growth yields Be single crystals with a mosaic spread of about 1 mrad.

The wavelength is selected according to the familiar Bragg law:

$$m\lambda = 2d_m \sin \theta \tag{13}$$

where d_m is the spacing of a set of planes in m-th order and θ is the Bragg angle. The resolution achieved by the monochromator is determined by an angular contribution and by a variation of the length of the reciprocal lattice vector giving rise to reflection. This is obtained as:

$$\Delta\lambda/\lambda = \Delta E/E = [(\Delta\theta \cot\theta)^2 + (\Delta d/d)^2]^{1/2} \qquad (14)$$

The quantity, $\Delta\theta$, corresponds to the angular distribution of the beam in the scattering plane, to be convoluted with the distribution of the mosaic structure, i.e., of the lattice plane misorientations if the crystal is not perfect. The quantity $(\Delta d/d)$ represents the diffraction pattern of a perfect crystal, which has to be convoluted with the strains or variations in d-spacing in the case of an imperfect crystal. Usually the effects of particle or mosaic block size are neglected. To first approximation, all distribution functions are assumed to be Gaussians, and $\Delta\theta$, $\Delta\lambda$ and $(\Delta d/d)$ are their full widths at half height. In addition to the angular and wavelength spreads provided by a monochromator, two other quantities are important: the peak reflectivity, $r_p = r(0)$, and the integrated reflectivity, R_H, which is the area below the reflection curve $r(\Delta)$ recorded when the crystal is rocked through the Bragg position in a parallel and monochromatic beam.

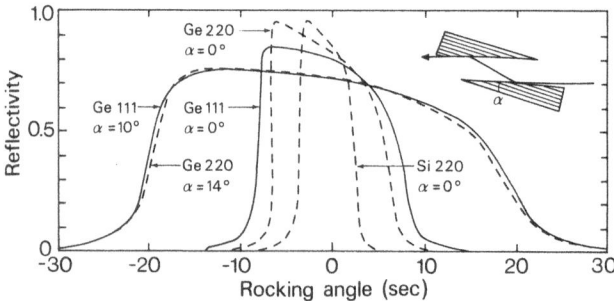

Fig. 6.   Calculated double-reflection curves of grooved monochromators for symmetric and asymmetric reflections of silicon and germanium and for X-rays of 1.54 Å wavelength (After Kohra et al., 1978).

The reflection properties of perfect crystals can be exactly calculated. The original diffraction theories by von Laue, Ewald and Darwin, also called "dynamical" theories, date back to the beginning of this century and are described in several textbooks and review articles, e.g., by Zachariasen (1945), Batterman and Cole (1964), and James (1963). We mention some results of these theories that are of importance for basic monochromator properties. Because mostly reflection geometry is used (the reflectivity in reflection geometry is always higher than in transmission, focusing is easier to achieve), the following results are given for the Bragg case. The best available Si crystals yield diffraction patterns which are very close (10^{-8}) to the calculated reflection curves.

First, some characteristic parameters will be defined. The extinction length l_{ext} is the distance of penetration of an X-ray beam into a perfect crystal set at a Bragg position. Because of multiple diffraction effects

266

inside the crystal, the intensity is decreased by 1/e at this distance which is given by:

$$l_{ext} = v_0 \,/\, (|F_x|\lambda C) \qquad (15)$$

where v_0 is the unit cell volume and $F_x = r_0 \, F_H$; r_0 is the classical electron radius and F_H the (complex) structure factor (see Eq. 3). The parameter C is the polarization factor which for synchrotron radiation is usually unity in the vertical scattering plane and equals $\cos 2\theta$ if diffraction takes place in the horizontal plane. This is due to the linear polarization of synchrotron radiation in the orbit plane, which is the usual case. The corresponding extinction thickness t_{ext} of a perfect crystal can be written as:

$$t_{ext} = v_0 \,/\, (2d \, |F_x| \, C) \qquad (16)$$

and is thus independent of λ for normal polarization except if anomalous scattering contributions to F_x cannot be neglected. Therefore, a crystal is said to be thick if its thickness $t_0 \gg t_{ext}$. Similarly, a crystal is said to be non-absorbing if the absorption thickness defined by:

$$t_{abs} = (\sin \theta) \,/\, (2\mu) \qquad (17)$$

again for the symmetric Bragg case, is big compared to the extinction thickness (μ is the linear absorption coefficient).

Due to refraction, the center of the peak is shifted from θ to an angle θ_B by an amount:

$$\theta_B - \theta = \delta \, (1 + 1/b) \,/\, \sin 2\theta \qquad (18)$$

where θ is the geometrical Bragg angle and δ is defined in Eq. 1. The parameter b stands for the asymmetry factor defined by:

$$b = \sin (\theta - \alpha) \,/\, \sin (\theta + \alpha) \qquad (19)$$

where α is the angle between the Bragg planes and the crystal surface as shown in Fig. 6. (For the Laue case, $\theta_B = \theta$). Because δ depends on wavelength, fundamental and harmonic reflections occur at slightly different Bragg angles. This gives a very useful possibility to suppress higher-order wavelengths by slightly detuning the second crystal in a double monochromator (Hart and Rodrigues, 1978; Materlik and Kostroun, 1980).

The reflectivity distribution for the symmetric case, a "thick" crystal and very small absorption ($t_{abs} \gg t_{ext}$, "Darwin solution") is:

$$r(y) = 1 \qquad\qquad\qquad r \; |y| \le 1$$
$$r(y) = [|y| - (y^2 - 1)^{1/2}]^2 \qquad \text{for } |y| \ge 1 \qquad (20)$$

where y is a variable along the diffraction vector $\underline{\tau}$. The full width at half height of this curve recorded on an angular scale is:

$$\omega_D \approx 2y = (2 \, |F_x|\lambda^2 C) \,/\, (\pi \sin 2\theta) \qquad (21)$$

and is commonly called the Darwin width. For the asymmetric case $\omega_{D,a} = \omega_D |b|^{1/2}$. The diffraction patterns of perfect crystals have far ranging tails which can be strongly decreased by many successive reflections inside a grooved monolithic double-crystal monochromator (Bonse and Hart, 1965; Deutsch, 1980).

Calculated rocking curves of such a channel-cut monochromator based on only two reflections are shown in Fig. 6 for symmetric ($\alpha = 0$) and asymmetric ($\alpha \neq 0$) geometries. They are the product of two single reflection curves and may be interpreted as the angular acceptance of the monochromator for a monochromatic beam. Details on the transformation of divergent and polychromatic beams by symmetric and asymmetric reflections follow later. Several features can be seen in Fig. 6. First, the centers of the rocking curves are displaced with respect to the geometrical Bragg angle corresponding to zero rocking angle. This is due to refraction (see above). Second, the diffraction curves are asymmetric and their peak reflectivity is not unity. This is due to absorption. Third, the width of the rocking curves can be varied for a given reflection, due to the miscut angle, α, between lattice planes and crystal surface. In the case presented here the width is increased. The opposite is possible as well by inverting the geometry.

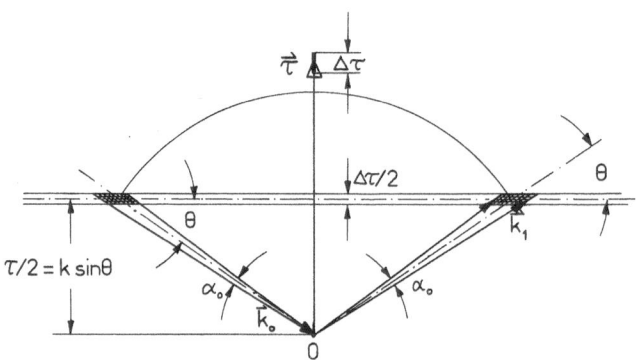

Fig. 7. Diffraction properties of a perfect crystal monochromator in reciprocal space. $\underline{\tau}$ is the reciprocal lattice vector, $\Delta\tau$ is the "true Darwin width" and $\underline{k}_o$ and $\underline{k}_1$ stand for the wavevectors contained in the incident and reflected beams, respectively (After Freund, 1983).

The integrated reflectivity for very weakly absorbing crystals is:

$$R = (8 \, |F_x| \, \lambda^2 \, C) \, / \, (3\pi \, |b|^{1/2} \, v_o \, \sin 2\theta). \qquad (22)$$

Equation 21 can be written as:

$$\omega_D = (\Delta d/d)_D \, \tan \theta. \qquad (23)$$

The functional form of Eq. 23 suggests that the Darwin width is not an angular property of the crystal (which is physically true for perfect crystals!) but can be compared to a one-dimensional broadening of particle size or to an uncertainty of the d-spacing. This effect can be graphically represented by a smearing of the reciprocal lattice point parallel to the diffraction vector, $\underline{\tau}$, which expresses the limited number of lattice planes $N_p = t_{ext}/d$ participating in the interference process. The quantity $(\Delta d/d)_D$ can thus be compared to the resolution achieved with N_p lines of an optical grating and is just the relative wavelength resolution obtained with a perfect crystal from a parallel "white" beam. In the symmetric Bragg case:

$$(\Delta d/d)_D = \Delta\lambda/\lambda = \Delta E/E = (4d^2 \, F_x \, C) \, / \, (\pi v_0) = 0.64 \, / \, N_p. \qquad (24)$$

Instead of $4/\pi = 1.27$, the value of $3/\pi \sqrt{2} = 1.35$ sometimes is used, depending on the theory chosen.

Fig. 7 is a reciprocal space diagram showing the selection of a wavelength band by a flat perfect crystal from a "white" incident beam of divergence α_0. The beam divergence is conserved after reflection. The big axis of the resolution "ellipsoid" is parallel to the reflected wavevector $\underline{k}_1$ if $\Delta\tau/\tau = (\Delta d/d)_D > \alpha_0$ ctg θ and perpendicular to $\underline{\tau}$ if $(\Delta d/d)_D < \alpha_0$ ctg θ. This means that the Darwin width and the beam divergence both give rise to a wavelength spread but that in both cases the angle-wavelength correlation is different, which is important for the optimization of the beamline optics. Monochromator and source contributions to the total band pass $\Delta\lambda/\lambda$ transmitted by a flat perfect crystal (source size neglected) are:

$$\Delta\lambda/\lambda = [\alpha_0^{\,2} \, \mathrm{ctg}^2\theta + (\Delta d/d)_D^{\,2}]^{1/2}. \qquad (25)$$

The contributions are often said to be matched if they are equal, which means that the section through the resolution volume in the diffraction plane has a shape which is close to a circle. However, as shown later, this is not always the best solution for a given experiment. The reflected intensity is proportional to the hatched areas in Fig. 7.

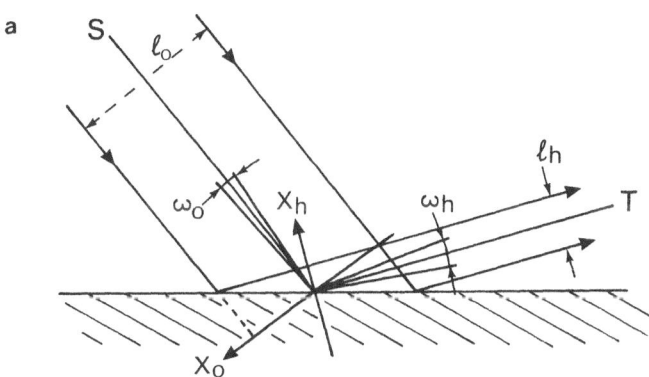

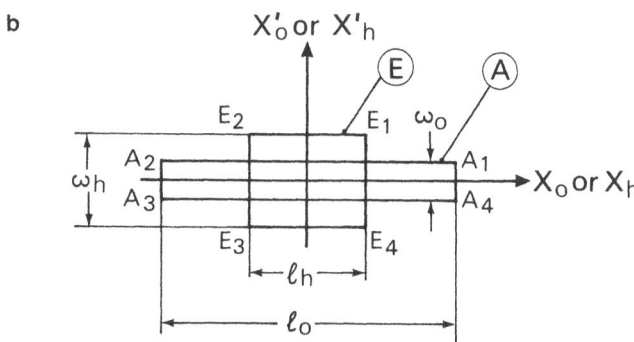

Fig. 8. Asymmetric reflection by a perfect crystal (a) in real space, (b) in position-angle space. A and E are the acceptance and emittance windows of the crystal (After Matsushita and Kaminaga, 1980).

While reciprocal space diagrams are useful and adequate to illustrate symmetric reflections and energy-angle correlations, the asymmetric diffraction properties of perfect crystals and their effects on beam transformation are better viewed in phase-space diagrams where a coordinate in momentum space is plotted against a coordinate in real space. This real-space coordinate may be an angular against a positional coordinate, for instance x' against x, or z' against z. The method has been developed by accelerator physicists and later adapted to X-ray optics for synchrotron radiation (Hastings, 1977). Furthermore, by adding a wavelength coordinate (Matsushita and Kaminaga, 1980) even three-dimensional representations in position-angle-wavelength space can be constructed which are quite complete but may then lack simplicity, which is one of the virtues of graphic approaches. Matsushita and Hashizume (1983) gave an exhaustive discussion of this method so that here the description will be restricted to two relatively simple examples, namely asymmetric reflection from a flat crystal and focusing Johansson geometry. Parallel representations in phase space and reciprocal space, including crystal thickness and penetration depth effects, also were described by Suortti and Freund (in press).

The first example is shown in Fig. 8a and b illustrating asymmetric reflection in real and in phase space. SO and OT define the directions of exact incidence and reflection of the central ray. Off-axis displacements are denoted as x_o for the incoming and as x_h for the reflected beam, respectively. The angles between the rays and SO and OT are x_o' and x_h', respectively. All rays contained in the angular range defined by $|x_o'| \leq \omega_o/2$ are accepted and reflected in the angular range given by $|x_h'| \leq \omega_h/2$. The widths ω_o and ω_h are $\omega_D / \sqrt{b}$ and $\omega_D \sqrt{b}$, respectively, where ω_D is the Darwin width for symmetric reflection and b is the asymmetry factor mentioned above (Eq. 19). From Fig. 8 it is clear that the lateral width decreases from l_o before to l_h after reflection while from dynamical theory follows that the angular width or acceptance/emittance increases by the same factor b. This is shown by the phase space diagram in Fig. 8b where x' is plotted against x. The rectangles A and E contain the directional and positional coordinates of the photons in the beams before and after reflection. The fact that A and E have the same area expresses Liouville's theorem.

The final beam transformation produced by asymmetric reflections depends on the properties not only of the crystal but also of the incident beam: white or monochromatic, parallel or divergent/convergent. For

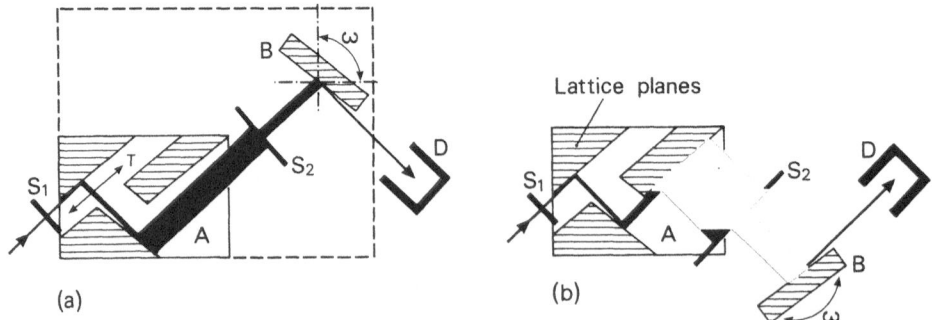

(a) (b)

Fig. 9. Production of highly parallel and monochromatic beams by successive asymmetric reflections. By translating the collimating monolithic structure A along T either single, double (a) or triple reflection (b) can be achieved. S_1 and S_2 are slits, B is the sample (angle of rotation ω) and D is the detector. A and B can be parts of the same crystal block (broken lines in (a)) or independent crystals. In the first case the axis of rotation is best defined by a weak link.

instance, a white and perfectly parallel beam remains parallel after reflection and selects a wavelength band $\Delta\lambda/\lambda = \omega_0 \text{ ctg } \theta$. On the other hand, a strictly monochromatic beam of divergence α_0 that is bigger than ω_0 is transformed into a beam of divergence ω_h. Very narrow and monochromatic beams can be produced by successive asymmetric reflections. This effect is shown in Fig. 9a and b. In the above case symmetric reflection from a perfect crystal does not change the beam divergence while asymmetric reflections can increase or decrease the angular width. Other examples for multiple reflection systems are given later.

For calculating the focusing conditions of curved single crystals the same equations can be used as for mirrors, adding the angle α for asymmetric reflections which are not possible for mirrors. In the case of circular cylinders of radii R_m and R_s the following focal lengths are obtained for meridional and sagittal focusing, respectively:

$$f_m = (R_m/2) \sin (\theta \pm \alpha) \tag{26}$$

$$f_s = R_s/(\sin (\theta \pm \alpha) \tag{27}$$

While sagittal focusing has practically no effect on the wavelength nor on energy resolution, meridional focusing can be both dispersive (chromatic) and nondispersive (monochromatic, achromatic). The latter is achieved by the so-called Guinier condition:

$$p/q = \sin (\theta \pm \alpha) / \sin (\theta \mp \alpha) \tag{28}$$

where the signs depend on whether the magnification of the system is smaller (upper signs) or greater (lower signs) than unity. The radius of curvature is given by:

$$R_m = p/\sin (\theta \pm \alpha) = q/\sin (\theta \mp \alpha). \tag{29}$$

The second example of phase-space description is shown in Fig. 10 for a focusing geometry. Again parallel representations of asymmetric reflection in real and angle-position space are given, but now the crystal is shaped and bent to fulfill the Rowland condition: a single crystal plate is ground to a radius of curvature $2R_m$ and then bent so that the Bragg planes form a circular cylinder of radius R_m. The optical axes SO and OF correspond to the central ray of a beam emitted by the source S and reflected to the focus F, both situated on the Rowland circle of radius $R_m/2$. This so-called Johansson monochromator is free from geometrical aberrations if penetration depth effects are neglected. In order to be reflected, a ray incident on the crystal at a position x_0 must have the direction $\zeta_0' = x_0 / [R_m\sin(\theta + \alpha)]$ within a range given by $|x_0' - \zeta_0'| \leq \omega_0^c/2$, where α is the angle between the Bragg plane and the crystal surface. The position-angle correlation of the reflected X-ray is $\zeta_h' = x_h / [R \sin(\theta - \alpha)]$ within a range $|x_h' - \zeta_h'| \leq \omega_h^c/2$. The widths ω_0^c and ω_h^c are the asymmetric reflection widths for a bent crystal which may be bigger than the Darwin widths ω_0 and ω_h corresponding to a flat crystal. The phase space diagram in Fig. 10b shows that Ω is conserved for the beam directly before (A) and after the crystal (E) and at the focus (F). Furthermore, at F rays of all directions within Ω converge to a spot whose width is determined by ω_0^c and ω_h^c. Also, in this case, Liouville's theorem is verified. However, the various factors affecting the width of the beam at F must be considered more carefully. In fact, there are several contributions to the size of the focus: source size, (i.e., both lateral width and depth), natural beam collimation, Darwin width, penetration depth into the crystal, degree of monochromaticity of the beam emitted by the source. For each specific case the individual contributions must be calculated separately and their interdependence must be taken into account. For example, in the case of a white point source and symmetric reflections, the Darwin width does not increase the beam divergence and hence, the focal size. However, if the source is monochromatic and has a finite size, the Darwin width has an effect on the size of the focus.

If the wavelength and also the radius of curvature must be changed frequently, the Johann geometry is more flexible than the Johannson geometry. Here an initially flat lamella is used which is bent to a radius R_m. Their geometrical aberrations must be taken into account.

Figure 11 shows the dispersive case of the beam transformation by curved crystals in real and reciprocal space (symmetric reflection). An incident white beam of width W and angular divergence α_0 is focused onto a point F at a distance FM = q from the center of the curved crystal.

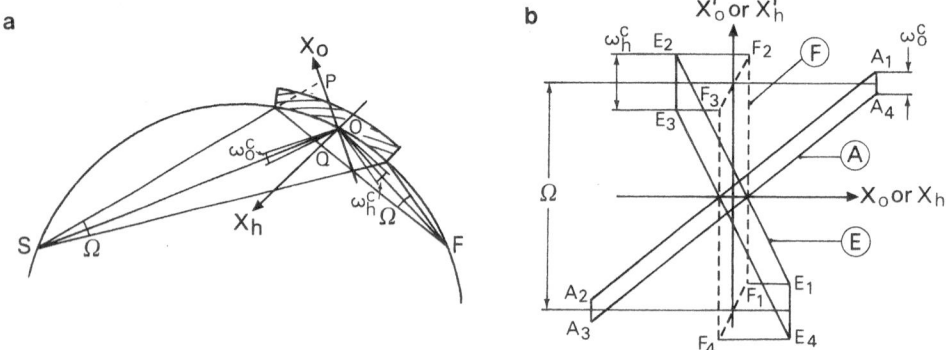

Fig. 10. Focusing properties of a curved crystal in Johansson geometry: (a) in direct space and (b) in position-angle space. The crystal accepts an X-ray beam described by A, changes its properties to E and focuses it to F. (After Matsushita and Kaminaga, 1980).

Neglecting the Darwin width, the finite width, w, at the focus is due to the divergence α_0 and/or the source size and to the penetration depth, T, of the radiation into the crystal. The angle ε is the change in lattice plane orientation across the diffracting segment. If the bending radius is big in the sense that the change of lattice plane orientation over one extinction length is smaller than the Darwin width, the radiation penetrates by about one extinction length into the crystal. At the other extreme, the penetration depth for strong curvatures is given by t_{abs} which is usually small, e.g. < 0.1 mm for Si(111) and λ > 1 Å. Both t_{ext} and t_{abs} may become important for shorter wavelengths. However, for quantitative evaluation of the penetration depth, the results of dynamical theory applied to distorted crystals should be used (Takagi, 1962, 1969; Taupin, 1967; Gronkowski and Malgrange, 1984). Simplified model calculations are often sufficient (Caciuffo et al., 1987; Suortti and Thomlinson, 1988).

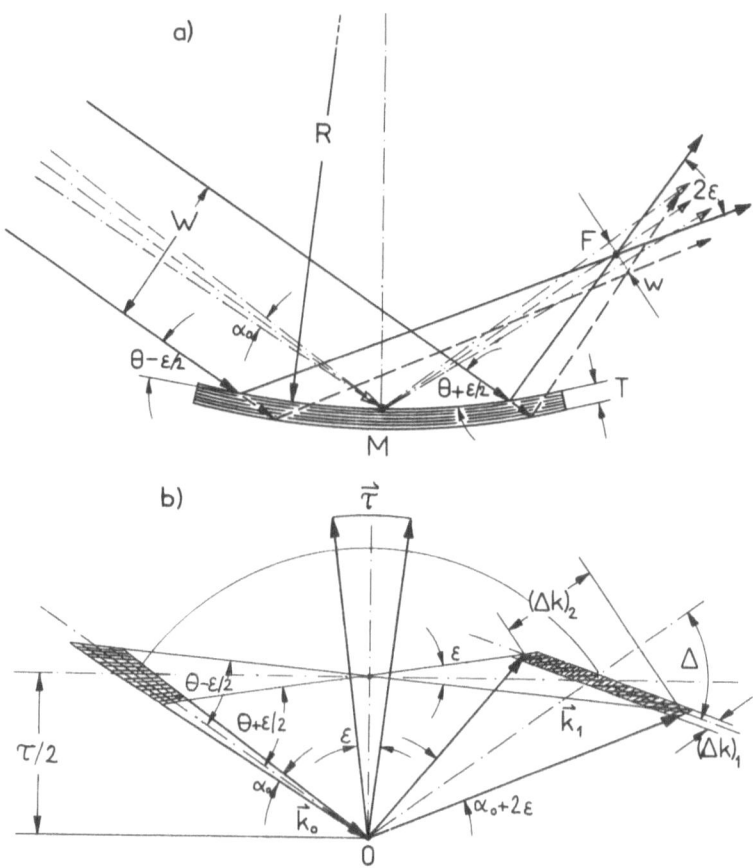

Fig. 11. Dispersive focusing geometry in direct space (a) and its effect on
beam transformation, in particular on angle-wavelength correlation
in reciprocal space (b). The symbols are explained in the text.
The reciprocal space diagram is simply obtained as the sum of many
flat crystal diagrams (see Fig. 7) distributed over an angle ε.
Therefore, the same diagram can be used to describe diffraction by
both a bent and a mosaic crystal (After Freund, 1983).

Chromatic or energy-dispersive focusing can be used to observe energy-dependent phenomena instantaneously instead of scanning the energy of a monochromatic beam. This technique is presently applied to EXAFS experiments (Tolentino et al., 1988) and is best illustrated in reciprocal space. The diagram in Fig. 11b shows how the reciprocal space volume on the left hand side selected by the curved crystal is transformed into an ellipsoid whose major axis is inclined by an angle Δ to the reflected wavevector $\underline{k}_1$. This angle defines the wavelength-angle correlation and depends on the ratio of beam divergence before reflection, α_o, and ε. If $\alpha_o/\varepsilon < 1$, which is the usual situation, the angle Δ is given by:

$$\mathrm{tg}\Delta = \mathrm{tg}\theta \ / \ [1-q/(R \sin \theta)] \tag{30}$$

where R is the radius of curvature and q is the distance between crystal and focus. For $\alpha_o/\varepsilon > 1$, $\Delta = \theta$. It can also be seen that the total wavelength spread $\Delta\lambda/\lambda = \Delta k/k$ is composed of a contribution $(\Delta k)_1/k$ due to beam divergence (or also to a source size effect) and of another term $(\Delta k)_2/k$ due to geometrical effects (convolution of Gaussian distributions is again assumed):

$$(\Delta k)_1/k = [(T \ (\mathrm{ctg}^2 \ \theta - \mu_p)/R_m)^2 + (\alpha_o \ \mathrm{ctg} \ \theta)^2]^{1/2} \tag{31}$$

where T is the penetration thickness and μ_p is Poisson's ratio. The Darwin width is neglected. The quantity:

$$(\Delta k)_2 \ / \ k = \varepsilon \ \mathrm{ctg} \ \theta \tag{32}$$

becomes zero for special geometries. In this case:

$$q = R_m \sin \theta \tag{33}$$

which means that the image of the source lies on the Rowland circle defined above and the focusing is strictly monochromatic ($\Delta = 90°$).

Taking into account the angular properties of synchrotron radiation from bending magnets and wigglers, sagittal focusing leads usually to bigger gains in flux at the sample than meridional (vertical) focusing and energy resolution is affected only to second order. Cylindrically curved crystals at a magnification, M, near one-third satisfy the Bragg condition of the central (vertical) rays for a horizontal radiation fan of arbitrary divergence angle (Sparks et al., 1980). Crystals curved to approximate a conical surface can be used to intercept large horizontal divergences with a satisfactory match to the Bragg condition for M from 1/3 to 2. These crystals cause even less focal aberrations at M $\approx$ 1 than is obtained with cylindrical crystals at M = 1/3. In general, the focusing efficiency improves at lower Bragg angles and smaller horizontal divergence. The conditions for simultaneous meridional and sagittal focusing by doubly-curved crystals are (Sparks et al., 1982):

$$R_m = [(p + q)^2 - 4 \ pq \ \sin^2 \ \theta]^{1/2} \ / \ \sin 2\theta \tag{34}$$

$$R_s = 2 \ pq \ \sin \ \theta \ / \ (p + q). \tag{35}$$

The magnification again is given by M = q/p. Asymmetric geometries are generally not considered for sagittal focusing.

The resolution ellipsoid in reciprocal space (see the hatched areas in Figs. 7 and 11) can be of isotropic or anisotropic shape. Matching the Darwin width of a flat perfect crystal to the effective angular divergence of the synchrotron radiation beam involving natural collimation and source size, results in an isotropic intensity distribution in reciprocal space.

274

This may not always be possible or, as mentioned above, not desirable for the intensity-resolution optimization. In any case, the art of assembling optical elements for a given experiment consists in tailoring the shape and orientation of the resolution ellipsoids so that they all match along the beamline. This is done by choosing flat and/or focusing elements and designing their properties accordingly to achieve what is called focusing in reciprocal space, and to combine it with real space focusing (or defocusing!) whenever this is possible.

The simplest example for reciprocal space focusing is the so-called nondispersive arrangement of a two-crystal diffraction experiment where the first crystal is stationary while the second (sample) crystal is scanned through the Bragg position. If the first crystal receiving the white beam is flat and perfect, it can easily be seen that, independent of the beam divergence, a maximum of both reflectivity and resolution is obtained if the Bragg angles of the two crystals are equal: $\theta_1 = \theta_2$. This is important, in particular, if the beam divergence is much bigger than the angular acceptance of the first crystal. If the first crystal is curved or a mosaic crystal the optimum arrangement is achieved for tg θ_2 = tg Δ_1 (see Eq. 30). The second crystal is assumed to be perfect or nearly perfect. From these considerations, it is clear that the nondispersive setting of two flat crystals is a special case of a much more general geometrical arrangement of curved crystals. Freund (1983) gave an example demonstrating how focusing both in real and reciprocal space can be obtained simultaneously.

The nondispersive or (+n, -n) setting is a special case of a general geometrical arrangement of many crystals: ($\pm$n, $\pm$m, $\pm$l, ...). The signs denote the direction of reflection (to the right or to the left) while n, m, l ... stand for the type and order of reflection and are integers. The beam divergence accepted, $\Delta\theta$, the divergence at the exit, $\Delta\psi$, and the bandpass, $\Delta\lambda/\lambda$, for the dispersive (+n,+m) arrangement of two crystals are (Matsushita and Hashizume, 1983):

$$\Delta\theta = T_1\omega_2/\sqrt{b_2} + |\omega_1/\sqrt{b_1} - T_1\omega_1\sqrt{b_1}| \qquad (36)$$

$$\Delta\psi = T_2\omega_1\sqrt{b_1} + |\omega_2/\sqrt{b_2} - T_2\omega_2/\sqrt{b_2}| \qquad (37)$$

$$\Delta\lambda/\lambda = (\omega_1\sqrt{b_1} + \omega_2/\sqrt{b_2})/T_s \qquad (38)$$

where T_s = tgθ_1 + tgθ_2; T_1 = (tgθ_1) / T_s; T_2 = (tgθ_2) / T_s. The subscripts 1 and 2 refer to the first and second crystal, respectively, and ω_1 and ω_2 are the Darwin widths for symmetric reflection. For the (+n, +n) arrangement the above equations give $\Delta\theta = \Delta\psi = \omega_1 = \omega_2 = \omega$ and $\Delta\lambda/\lambda = \omega$ ctgθ. The nondispersive-dispersive-nondispersive setting of four crystals (+1,-1,-1, +1) was used to produce a highly parallel and monochromatic beam for the study of nuclear reflections by Siddons et al. (1987). As shown in Fig. 12 the X-ray beam is pre-monochromatized by a Si(111) channel-cut crystal which takes most of the heat off the white beam. The high degree of monochromatization is achieved by two channel-cut Si(10,6,4) monochromators set in the dispersive mode. By rotating both channel-cuts it is possible to change the wavelength and to keep the exit beam at a fixed vertical position. The energy resolution obtained was 0.005 eV at 14.413 keV ($\Delta E/E = 3.5 \times 10^{-7}$) and the angular divergence was 2 μrad (0.4 arcsec).

At Bragg angles close to 90° the angular term in the energy spread becomes very small and the resolution is given by the Darwin width only (Eq. 14). An also quite high energy resolution (but without high angular resolution) was obtained by the back reflection technique using curved and grooved Si crystals (/// reflection, E = 13.8 keV) in order to measure phonons: $\Delta E/E = 8 \times 10^{-7}$ (Dorner et al., 1986; Burkel et al., 1987). The beam geometry is shown in Fig. 13. Due to the focusing technique and to the relaxed angular

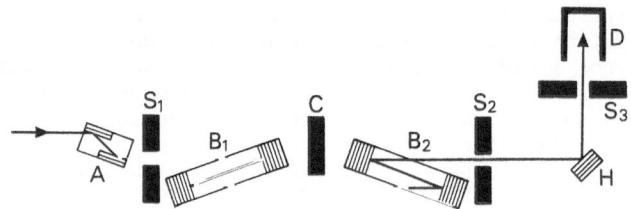

Fig. 12. Very high energy and angular resolution achieved with dispersive
geometry. A - Si(111) promonochromator; B1, B2 - Si(10, 6, 4)
channel-cut crystals; C - beam stop; H - hematite crystal; D -
detector; S1, S2, S3 - slits (After Siddons et al., 1987).

resolution about two orders of magnitude in flux can be gained compared to
flat crystal geometry without loss in energy resolution. In this case
energy tuning is very limited and is performed by changing the temperature
of the crystal. Very careful crystal preparation and stable and accurate
bending are needed.

Both examples show how high energy resolution can be achieved using
different techniques, where the first gives simultaneously very high angular
resolution, while the second allows us to obtain several orders of magnitude
higher flux at the expense of angular resolution. Schulke (1986) gave other
geometries for inelastic scattering experiments matching different requirements. Further examples for multiple reflection configurations and their
effect on beam properties were proposed by Hart et al. (1984). Beam polarizers based on monolithic multiple reflection systems were described by Hart
(1978), Hart and Rodriques (1979), Materlik and Kostroun (1980), and Mills
(1988).

Table 3 gives some crystal data for several materials selected according to their suitability as monochromators for synchrotron radiation. These
data also served for calculating the diffraction properties of mosaic crystals presented later. The reflection widths of perfect crystals like Si

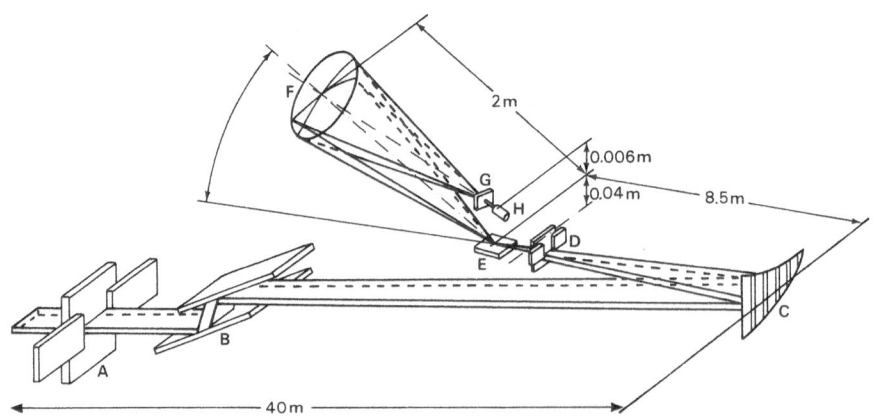

Fig. 13. High energy resolution experiment using back reflection geometry.
A - white beam-defining slits; B - Ge(111) premonochromator
(asymmetric); C - curved Si crystal; D - crossed slits; E -
sample; F - focusing Si analyzer crystal; G - pinhole; H -
detector (After Burkel et al., 1987).

Table 3. Relevant quantities for some reflections of materials used as
X-ray monochromators and for CuKα radiation (λ = 1.542 Å) (Freund, 1988).

Crystal	d (Å)	$F_x r_o$ (10^{-12} cm)	l_{ext} (μm)	t_{ext} (μm)	t_{abs} (μm)	Q (cm^{-1})	$\Delta\lambda/\lambda$ (10^{-5})	ω_D (sec)
Be(002)	1.7916	0.906	23.1	5.0	1200	0.0148	2.29	2.25
Be(004)	0.8958	0.544	38.6	16.6	2390	0.00471	0.34	1.20
Be(110)	1.1428	0.692	30.4	10.3	1874	0.00671	0.71	1.34
PG(002)	3.3540	4.87	9.43	1.1	110	0.157	19.82	9.65
PG(004)	1.6770	2.69	17.0	3.9	222	0.0262	2.74	2.92
PG(006)	1.1180	1.85	24.8	8.5	331	0.0101	0.84	1.64
Cu(111)	2.0870	21.87	2.81	0.52	3.91	1.146	25.7	21.0
Cu(200)	1.8074	20.23	2.90	0.62	4.51	0.872	17.8	17.3
Cu(220)	1.2780	15.23	4.01	1.21	6.38	0.396	6.71	10.5
LiF(200)	2.0132	8.17	11.5	2.21	59.3	0.0812	6.46	5.52
LiF(400)	1.0066	3.38	25.1	9.60	119	0.00988	0.67	1.64
LiF(220)	1.4235	5.72	14.8	4.00	83.8	0.0310	2.26	3.00
Si(111)	3.1355	16.70	12.5	1.53	8.7	0.0827	13.1	6.83
Si(220)	1.9201	18.32	11.3	2.27	14.2	0.0631	5.37	4.86
Ge(111)	3.2664	40.97	5.76	0.68	2.94	0.409	30.7	5.4
Ge(220)	2.0002	48.84	4.83	0.93	4.79	0.375	13.7	11.8

and Ge match more or less the divergence of synchrotron X-ray beams and therefore they are most appropriate as monochromators. The matching can be improved by asymmetric geometries but these allow wavelength variation only in a limited range.

Figure 14 shows a quantitative comparison between the Darwin widths of perfect Si and Ge crystals (FWHM) and the expected beam divergences produced by several sources at the ESRF in Grenoble. Good matching is achieved for undulators at medium energies and for the other sources at low energies only, while at high energies there is a mismatch by an order of magnitude. Here Ge crystals give twice more flux than Si crystals, and even higher Z materials would be better, but when attention has to be paid to the presence of absorption edges and to problems with crystal quality. In the medium- and low-energy ranges paraboidal mirrors can be used either to increase or to decrease the beam divergence and so improve the situation (Hastings et al., 1983; Trela et al., 1988). Following the same argument it should be noted that divergence α_o, source size, s, and distance, p, between source and monochromator should also fulfill a matching condition which is α_o = s/p. For an undulator at 10 keV Fig. 14 indicates α_o = 20 μrad so that an optimum length p = 10 m is obtained with s = 0.2 mm. For wigglers at 20 keV α_o = 200 μrad which means that undulators should be located in low β sections and wigglers in high β sections, where β is the source size divided by the electron beam divergence. However, because the above matching conditions do not apply for all experiments, flexibility in source design is desirable (Elleaume, 1988).

In the high energy range, mirrors are not a practicable way to improve the matching. Here flux can be gained without loss of energy resolution by increasing the reflection width of the monochromators, by using either mosaic crystals or crystals with a variable lattice spacing. Both possibilities were proposed in the past. Gradient crystals have the advantage that they do not increase the beam divergence (similar to perfect crystals in symmetric reflection) while the angular dispersion of a mosaic crystal decreases spatial resolution and thus the flux on the sample. However,

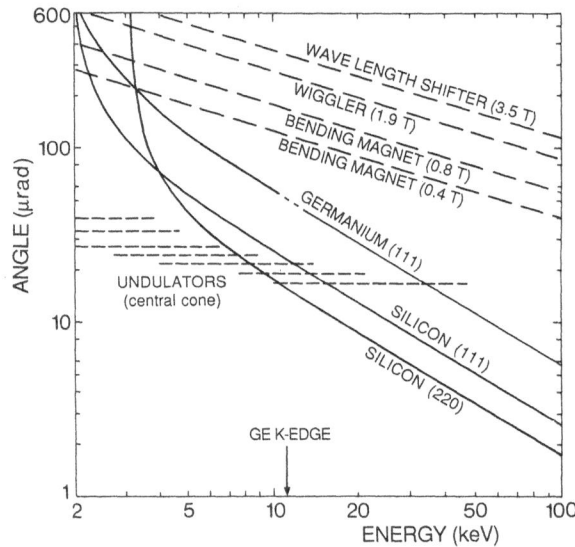

Fig. 14. Vertical opening angles of beams (FWHM) generated by several insertion devices at the ESRF and Darwin widths of Si and Ge perfect crystals as a function of X-ray energy.

mosaic crystals are easier to obtain and the increase in beam divergence due to the mosaic spread can be minimized by using two crystals in n, -n geometry (Hohlwein et al., 1988). Multilayers could also be considered if they were available with small d-spacings. Common to both is that they are not perfect. The first and the second element in the double monochromator arrangement must have the same, uniform defect structure so that their diffraction patterns match. Otherwise there will be losses in the transmitted intensity. The integrated reflectivities of imperfect crystals are absorption-limited rather than extinction-limited. Therefore, mosaic crystals become efficient at high energies where they permit a gain in flux and thus can compensate for the flux decrease of the source.

Because the gain in reflectivity is limited by absorption, low Z materials like graphite and beryllium are most suitable. Freund (1988) calculated the performance of several available materials. The diffraction pattern for the symmetric Bragg case is given by:

$$r\ (\Delta) = a/\ [1\ +\ a\ +\ \sqrt{1\ +\ 2a}\ \coth(A\ \sqrt{1\ +\ 2a})] \qquad (39)$$

and the maximum peak reflectivity, r_{pm}, is obtained for $\Delta = 0$ and $t \to \infty$ where t is the crystal thickness:

$$r_{pm} = a_o/(1\ +\ a_o +\ \sqrt{1\ +\ 2a_o}) \qquad (40)$$

with $a = G(\Delta)\ Q/\mu$; $a_o = a(\Delta = 0)$; $A = \mu t/\sin\ \theta$; $Q = (F_x/v_o)^2\ \lambda^3/\sin\ 2\theta$. $G(\Delta)$ is a normalized Gaussian distribution function that describes the angular variation of the mosaic blocks replacing the real defect structure in Darwin's mosaic model. For unpolarized radiation Q has to be multiplied by $(1 + \cos^2 2\theta)/2$. Equations 39 and 40 should be good approximations if $t_{abs} / t_{ext} \gg 1$ and if primary extinction is small.

The maximum peak reflectivity is plotted as a function of the mosaic spread γ in Fig.15a for CuKα radiation ($\lambda = 1.542$ Å) and for several materials using reflections with d-spacings close to 2 Å. Fig. 15b gives the results for d-spacings around 1 Å. They show that quite high reflectivities can be obtained at small mosaic spreads and that Be and highly oriented pyrolytic graphite (HOPG) are the most efficient materials.

However, these curves must be used with care because real crystals often exhibit nonuniform mosaic distributions and also primary extinction

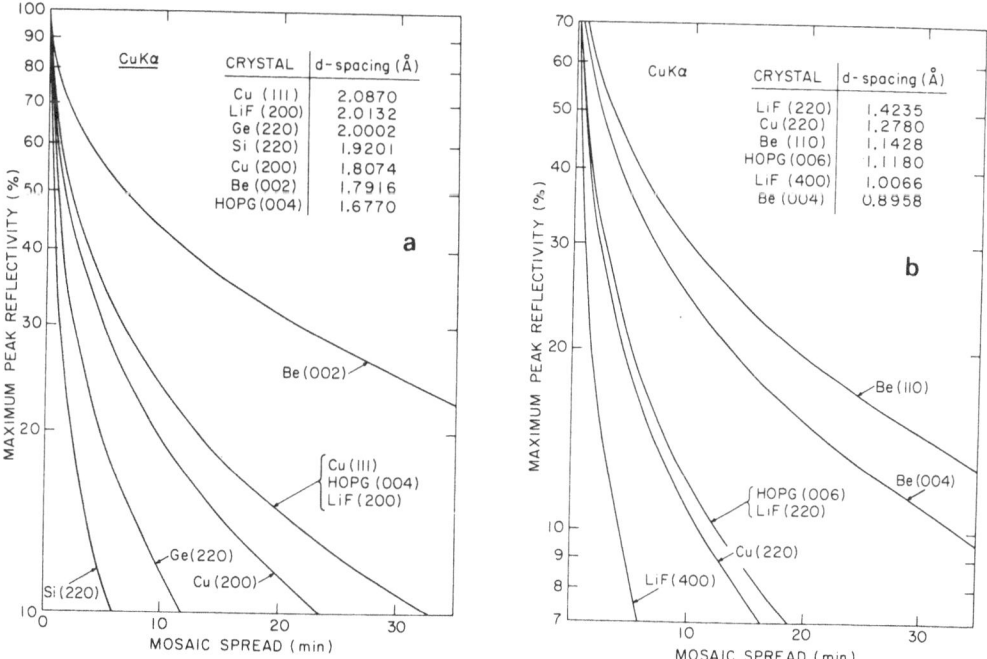

Fig. 15. Maximum peak reflectivity that theoretically can be obtained from mosaic crystals as a function of the intrinsic mosaic spread for various crystals with bigger (a) or smaller (b) d-spacings. (After Freund, 1988).

may occur, in particular for small values of γ. Only experimental tests will reveal the importance of these effects. While Be can be obtained in principle with any mosaic spread starting at about 0.5 mrad, HOPG is available with γ > 5 mrad only, which limits its range of applications. Recently Schneider (1988) reported mosaic spreads of about 10 μrad obtained by oxygen diffusion into initially perfect Si crystals. These crystals were successfully used at short wavelengths where the Darwin width is much smaller than the divergence of synchrotron beams (Schneider et al., in press).

For short wavelengths, the penetration depth necessary to get high integrated intensities from the crystals becomes quite big. For instance, to obtain 90% of r_{pm} from a Be(110) crystal 2.6 mm and 5.5 mm thick specimens are needed for CuKα and MoKα radiation, respectively, and for a mosaic spread of about 4 mrad. On the other hand, the gain in integrated reflectivity with respect to a perfect Si(440) monochromator (similar d-spacing) would be by factors of about 80 and 200, respectively. Therefore, whenever an increase of divergence, energy resolution and beam cross section is tolerable, substantial gains in count rate can be achieved by replacing perfect crystals by mosaic crystals. In real imperfect crystals, strains or variations in d-spacing are always small compared to lattice tilts produced by dislocations, which is the reason why Darwin's model works in general. Therefore, in reciprocal space the same diffraction diagram is obtained as for diffraction by a bent crystal (Fig. 11b) replacing ε by the full width at half maximum of r(Δ) (Eq. 39).

A further application of mosaic crystals is as analyzers inserted between the sample and detector. For instance, for diffuse or Compton scattering experiments these crystals permit background reduction, or the elimination of fluorescence radiation from the sample with an energy resolution better than that of Si(Li) diodes. For these applications the increase of beam cross section due to the increased beam divergence can be accepted because distances can be kept small.

As discussed above, there are very stringent requirements for high positional and angular stability which are common to all optical devices. Precision mounts and bending devices avoiding anticlastic curvature, very accurate translation and rotation units, feedback systems for keeping the energy fixed and antivibrational supports with automatic levelling and height control, and ambient temperature stability are mandatory for beamlines designed for low emittance - high brilliance synchrotron radiation sources (ESRF, PEP, APS). I will not describe these problems and the various solutions here, but refer the reader to the review articles cited earlier and to the proceedings of the yearly conferences on synchrotron radiation instrumentation.

Multilayers

Layered synthetic microstructures are the youngest of all optical devices for X-ray instrumentation. Originally, technologies for their fabrication were developed for other purposes, mainly for thin film composite structures for electronic devices. However, during the past two decades specific research and development for VUV and X-ray optical applications increased very rapidly. Present-day multilayers are frequently used in the VUV and soft X-ray range (Barbee, 1986; Dhez, 1989) and upgrade substantially the instrumentation in various fields of experimental research: astronomy, VUV, X-ray and neutron spectroscopy and scattering in physics, chemistry and biology. Furthermore, a multilayered structure is an interesting object by itself: surface and interface studies can be carried out and standing wave experiments become possible on materials that normally would not be accessible to this technique because high quality single crystals are unavailable. A convincing demonstration of present and potential applications to X-ray and neutron scattering and of the recent progress in fabrication processes was given in a recent conference organized by SPIE in San Diego (SPIE proceedings Nos 982, 983 and 984, 1988).

High reflectivity is obtained for high scattering contrast between usually two types of alternating layers. For neutrons there is no systematic relation between the real and the imaginary part of the refractive index, while for X-rays both scattering power and absorption increase with increasing Z. Therefore, multilayers are more efficient for neutrons where even spin polarization of the beam can be obtained by using magnetic materials. For X-rays, layers of high and low Z should alternate, but this is not the only condition for obtaining good and stable multilayer monochromators. There should be no interdiffusion between the layers, the two materials must be chemically compatible and withstand the high power of synchrotron X-ray beams. My aim here is not to go into detail and also not to describe the advantages and disadvantages of the three major techniques used for multilayer fabrication, which are evaporation, sputtering and molecular beam epitaxy. For the present state-of-the-art, the Symposium Proceedings, Vol. 103, of the Materials Research Society (Barbee et al., 1988) can be consulted. The main problems encountered are interface roughness and variations of layer thickness, which both affect reflectivity and become very important when the d-spacing is decreased, which is necessary for monochromators for X-rays of higher energies. At present, d-spacings of bilayers are in the order of 15 Å, and 10 Å or less are envisaged.

Bragg's law also holds for multilayers and the lattice spacing d_m has to be replaced by the spacing d_1 of a bilayer which is $t_A + t_B$ where t_A is the thickness of layer A and t_B the thickness of layer B, respectively. The refraction correction is slightly different from that for single crystals. The corrected Bragg equation can be written as:

$$m\lambda = 2d_1\sin\theta \; [1 - (2\bar{\delta} - \bar{\delta}^2) / \sin^2\theta]^{1/2} \tag{41}$$

where $\bar{\delta}$ (cf. Eq. 1) is now a mean value:

$$\bar{\delta} = (t_A\delta_A + t_B\delta_B) / d_1 \tag{42}$$

Fig. 16 shows two examples of experimental reflection curves obtained at fixed angles as a function of energy. The first example is a stack of 260 W-C bilayers spaced by 15 Å ($t_W = 6.7$ Å; $t_C = 8.3$ Å). Here the glancing angle is 51.7 mrad to obtain 8 keV (Fig. 16a). Both the peak reflectivity (19%) and the bandwidth (0.6%) are smaller than in the next case (Bilderback et al., 1983). The second multilayer consists of 30 bilayers of Mo and C spaced by 56 Å ($t_{Mo} = 26$ Å; $t_C = 30$ Å). At a glancing angle of 15 mrad, a peak reflectivity of 70% is achieved for 8.1 keV radiation (Fig. 16b) and the bandwidth is 5.4%. Although these results are already several years old, they are still representative for the expected hard X-ray efficiencies. These can be very close to the theoretical values for not too many thick layers (Underwood et al., 1988) and by more than a factor of two smaller when a bigger number of thinner layers is required.

The calculation of the reflectivity and transmission of multilayers is very similar to Darwin's or Ewald's treatments of diffraction by perfect crystals (Spiller, 1988). In both cases, a set of differential equations has to be solved which can be done either by a matrix method or by recursion. The first one is exact and corresponds to the full dynamical theory, while the second one is an approximation similar to the kinematical theory. In the kinematical case the multilayer is treated like a planar diffraction grating where all the planes contribute with equal weight to the diffracted beam. Then the reflected amplitude at each boundary is approximated by:

$$r = \pm \, [(\delta_1 - \delta_2) + i(\beta_1 - \beta_2)] \, e^{-z/z_{max}} / \, (2 \sin^2\theta) \qquad (43)$$

where $\delta_{1,2}$ and $\beta_{1,2}$ refer to both layer materials and were defined earlier (Eqs. 1 and 2); z is the depth of the interface below the surface and z_{max}

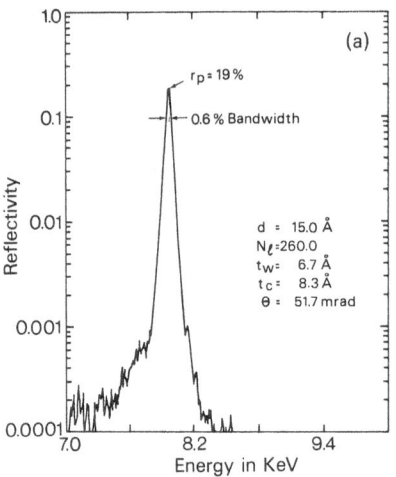

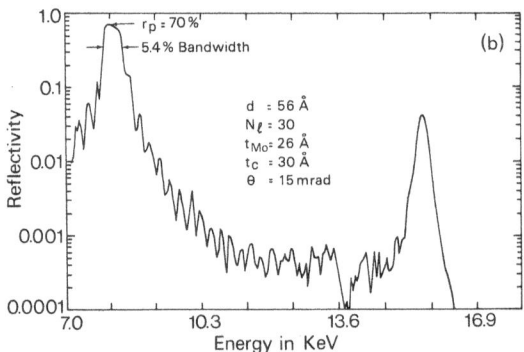

Fig. 16. Typical reflectivity profiles obtained with multilayers (after Bilderback et al., 1983).

is the total penetration depth. The alternating sign in Eq. 43 stands for the phase shift of 180° between adjacent boundaries. The value of z_{max} is estimated by:

$$z_{max} = 1.58 \ \lambda/[2\pi \ (\beta_1 + \beta_2) \sin \theta] \tag{44}$$

which becomes, replacing β by the absorption coefficient μ,:

$$z_{max} = \pi \ / \ [(\mu_1 + \mu_2) \sin \theta]. \tag{45}$$

The total reflectivity as a function of $q = (4\pi/\lambda)m\sin\theta$ is then obtained by summing over all layers N_ℓ.

$$r(q) = \sum_{j=0}^{N_\ell} r_j e^{iqz_j} \tag{46}$$

The shortcoming of this theory is that z_{max} depends only on absorption and not on scattering. Therefore, it should be a good approximation for weak reflectivities only. This can be seen in Fig. 17, where the results of exact calculations are compared with those of the kinematical approximation. Similar to strong reflectivities from mosaic crystals, the low-order peaks of multilayers are affected by extinction which results in an effective value of z_{max} below the value given by absorption. It is certainly possible to introduce a correction similar to the secondary extinction correction for diffraction by imperfect single crystals, and then the approximation should give good estimates for multilayer efficiencies.

The spectral resolution of an ideal multilayer structure is approximately $1/N_1 = d_1/z'_{max}$ where z'_{max} is the corrected penetration depth. From the measured data presented above it can be concluded that either the multilayers were not ideal, or $z'_{max} < d_1 N_1$, or both. Three types of imperfections can be defined schematically by analogy to the properties of imperfect crystals. Far-ranging variations of layer spacings lead to a broadening and maybe also to a shift of the reflection profiles in the direction of the diffraction vector $\underline{\tau}$, while far-ranging lattice tilts or curvature of the layers produce a similar effect in direction perpendicular to $\underline{\tau}$. The third kind of imperfection is short-ranged disorder, e.g., microroughness, which, similar to point defects in crystals, gives rise to diffuse scattering affecting the tails of the reflection curves. This can be described mathematically by a static Debye-Waller factor. However, the simple kinematical theory is not able to give satisfactory agreement with experimental results of real multilayer reflectivity and their dependence on microroughness (Rosen et al., 1988; Vidal and Vincent, 1984; Christensen et al., 1988). If we assume that the multilayer is perfectly flat and that there are only fluctuations in the thickness and roughness of the layer, the same reciprocal space diagram as that of perfect crystals is obtained replacing $(\Delta d/d)_D = (\Delta \tau/\tau)$ by $1/N_1$ or by d_1/z'_{max}, depending on what is the limiting factor. However, there can also be large-scale fluctuations in layer orientation as observed by Pianetta et al. (1985) and by Hornstrup et al. (1988). Much attention has to be given to substrate quality, and careful characterization of the substrate and the multilayer by visible light interferometry, TEM and X-ray scattering is needed. Uniformity is very important, in particular for double monochromators.

The most attractive feature of multilayers is a great flexibility at their fabrication. They can be grown on curved surfaces to produce focusing elements, their thickness can be graded in depth and/or laterally, and the constituting materials can be varied over a wide range to achieve optimum performance for a given application. Thus the range from a "supermirror" where the angle of specular reflection is increased by layer coatings of increasing thickness, until Bragg reflectors with adjustable energy

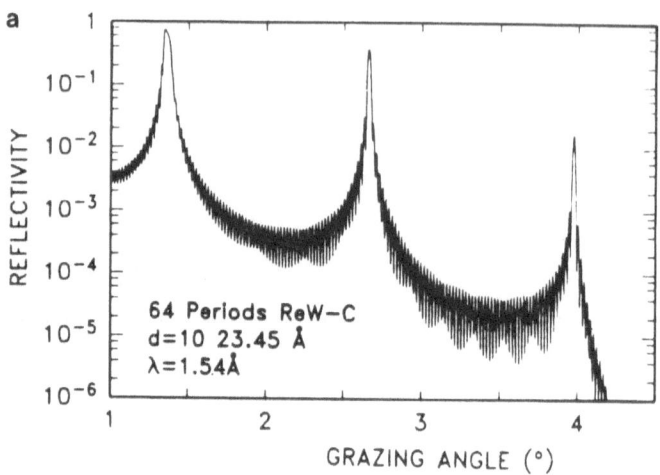

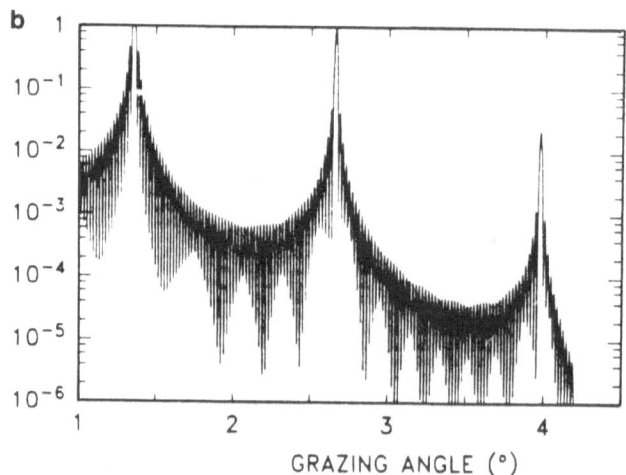

Fig. 17. Calculated diffraction patterns of (perfect) multilayers using (a) "dynamical" and (b) "kinematical" theory (after Spiller, 1988).

resolution is covered, and even higher-order contamination can be avoided by a periodical change of the layer spacing. The same focusing conditions as for mirrors apply. The reflection properties of curved multilayers were recently described by Marshall (1986). Their performance was experimentally demonstrated by Underwood et al. (1988). Spherical multilayers were arranged in a Kirkpatrick-Baez geometry to build an X-ray microprobe. A spatial resolution of 5 μm was achieved as compared to about 2 μm for a perfect optical system. The blur at the focus was assigned to both spherical aberration and multilayer quality.

Bragg-Fresnel Optics

This review would be incomplete without mentioning the various new possibilities provided by combining Bragg diffraction with Fresnel optics. This field was opened very recently by the development of technologies for microelectronics structures and is at the very beginning of its development.

According to the name, Bragg-Fresnel optics is based on a superposition of Bragg diffraction and dispersion by a Fresnel structure, either linear or planar, which is grooved into the surface of the Bragg diffracting element. Thus a Bragg-Fresnel element is a multifunctional optical device that permits us to monochromatize, disperse or focus X-ray beams. Its main advantage, however, is that the various possibilities of beam transformation arising from the specific properties of Fresnel structures now become accessible also to the medium and higher energy X-ray range. Because specular reflection is replaced by Bragg diffraction, the glancing angles are much bigger and efficiency is substantially increased (Jark, 1986).

The limiting case of a Fresnel structure is a linear grating. Recently, Barbee (1988) produced simple laminar amplitude gratings and diffracted X-rays of up to 1.5 keV from the multilayer-covered tops of the grating bars. Various experiments performed on these multilayer gratings demonstrated that they act as X-ray prisms and are high efficiency dispersion elements. By combining the Bragg equation and the grating relations it can be shown that optical constants of the materials constituting the multilayer can be determined by simply measuring the angle between the zero and first order interference lines of the dispersed beam.

Similar diffraction gratings were fabricated earlier by Aristov et al. (1986a) who studied X-ray diffraction from a periodic surface structure grooved into a perfect Si single crystal. The same group showed that it was possible to achieve a spatial resolution of 5 μm (M = 0.1) using a multilayer-based Bragg-Fresnel element for 8 keV X-rays (Aristov et al., 1986b). Later, a resolution of 7.5 μm was obtained with single-crystal based Bragg-Fresnel elements for 8 keV X-rays (Aristov et al., 1987). Aristov et al. (1988) gave a review of the very wide range from high spatial resolution to microinterferometry and high energy resolution, showing how spherical and chromatic aberrations can be avoided by in-depth profiling of multilayer Bragg-Fresnel lenses.

The limiting spatial resolution that can be obtained in Bragg-Fresnel optics is given by the width of the outermost zone of the zone structure. Present technologies yield fractions of a micron. With aspect ratios of up to 10 the height would be of the order of X-ray extinction lengths in single crystals. Thus submicron resolution should become possible. Not only will technological problems have to be solved, but also three-dimensional diffraction theory will need further development to calculate the detailed optical properties of single-crystal Bragg-Fresnel elements. Nevertheless, it can be anticipated that this novel type of truly three-dimensional and multifunctional X-ray optics will very much help to make better use of the low-emittance X-ray sources provided by modern storage rings, and thus overcome limitations of more conventional optical elements (cf. Fig. 1a).

THERMAL PROBLEMS

Probably the most severe problem to be solved for the efficient use of future high brilliance storage rings is the thermal load produced by very intense photon beams. Already, at existing sources, in particular wigglers, there are problems with cooling the first optical element(s) (beam shapers, crystals, mirrors, windows). Total powers of several kW have to be removed, but the most important difficulty is to reduce the thermal gradients across the surface illuminated by the X-ray beam. These gradients cause internal stresses which deteriorate optical surfaces and produce deformations degrading the spatial and spectral resolution of optical devices. At beamline VI at Stanford the power density is about 20 kW/cm^2 perpendicular to the beam, which means that at the exit of an existing wiggler the power density is about 5 times that of the sun's surface (E_e = 3 GeV, i = 200 mA, B = 1.75 T). It will be still higher for ESRF and APS wigglers (Elleaume, 1988).

The total power, P_t, is given by:

$$P_t = 0.633 B_o^2 E_e^2 I L \qquad (47)$$

while the power density is:

$$P_s = 10.8 B_o E_e^4 I N_p G(K)/d^2 \qquad (48)$$

where:

$$G(K) = [(K^7 + 24K^5/7 + 4K^3 + 16K/7) / (1+K^2)]^{7/2} \qquad (49)$$

P_t is obtained in kW and P_s in W/mm^2 if E_e is the electron energy in GeV, B_o the magnetic field in Tesla, I the stored electron current in A, L the length of the insertion device in m, and d the distance from the center of the insertion device to the point of observation. N_p is the number of periods and K the deflection parameter.

Several methods are used or envisaged to solve the heat load problem: passive cooling by radiation and conduction, active cooling by water or gas flow or by heat pipes, filters based either on reflection or on transmission (absorption), premonochromators (e.g., grazing incidence multilayers and maybe gratings, the latter having a quite low efficiency for hard X-rays), and special designs of monochromators. Their optimum application depends on the type of experiment and on the energy range of interest, and also on the source. To assess the thermal and mechanical stress distributions produced by the X-ray beam, finite element analysis (Edwards et al., 1985; Youngman, 1988) is very useful and helps to design the optimum cooling system. Feedback controls are needed to compensate for the variation of the X-ray flux as a function of time (short-time and long-time stability).

A recent mirror cooling design was reported by Di Gennaro et al. (1988) for beamline VI at SSRL for a total absorbed power of 2.4 kW and a peak absorbed power density of 520 W/cm^2. Direct cooling by convection was achieved using internal water channels in a brazed, dispersion-strengthened copper and OFHC copper substrate with a polished electroless-nickel surface. No damage was observed on aluminium-alloy mirrors covered with electroless nickel for power densities up to 0.5 W/mm^2 above which only SiC was able to withstand the heat load (Saile, 1988). The various virtues of SiC as mirror substrate material were described by Takacs et al. (1984). The cooling efficiency can be increased by using liquid gallium instead of water (Smither et al., 1988) but the risks of gallium contamination and subsequent

286

corrosion must be weighed against the cooling advantage. Special problems occur when the optical devices have to be compatible with ultra-high vacuum. The temperature distribution on the surfaces of mirrors and monochromators should be monitored so that the beam can be shut off in the case that the temperature exceeds a limiting value due to cooling failure. This equipment can also be used to check the results of finite element analysis. Significant engineering efforts are needed to ensure safety, reliability, stability and optical system precision and flexibility at the same time.

Three types of filters for reducing heat load should be mentioned. The first is made of a set of graphite blades whose thickness increases in direction of the beam. So the first foil is the thinnest one, of the order of 1 μm thick. Those following are progressively thicker because the power is gradually reduced. Graphite can be heated to temperatures in excess of 1000°C without problems, and is cooled only by radiative losses if it has to be in UHV. The second filter is also based on specific absorption properties but for somewhat higher useful energies and consists of a box containing krypton gas (K-edge at 14.3 keV) of controlled pressure (Suortti, 1987). The heat is eliminated by convection towards the cooled walls of the box. The third type works according to the principle of reflection and transmission by thin foils (Lairson and Bilderback, 1982). Set at a given, tunable angle, the foil reflects most of the power of the X-ray beam. When used as a beam splitter, it also acts as a higher-order filter for the reflected beam. The latter two filters still need further development.

More and more, cooled premonochromators have become the standard equipment in multiple-crystal fixed-exit monochromators (Heald, 1988). The crystals are attached, either from the side or from below, to cooled copper blocks with good thermal contact (indium foil, liquid gallium layer). Another possibility is to drill holes in silicon or to grind channels (Bilderback et al., in press; Smither et al., in press). The problem here is to make tight, reliable and solid connections to the cooling system without deforming the crystals. Cooling by jets appears to be most efficient. Asymmetric reflection geometries permit a decrease of the power density received by the monochromator crystal (Lemaire et al., 1988). Premonochromators consisting of multilayered materials also have been proposed, but their performance with respect to radiation load and increased temperatures is not yet established (Ziegler et al., in press).

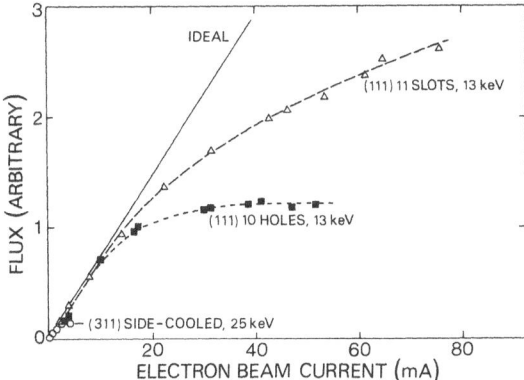

Fig. 18. Limitation of flux transmitted through a double crystal, gallium-cooled Si monochromator as a function of the incident flux generated by an undulator (After Bilderback et al., in press).

Special design of monochromators also can reduce the absorbed power. The idea is to reduce the thickness of the reflecting part of the monochromator to about twice or three times the extinction thickness. Because the absorption thickness is always much bigger than the extinction thickness, most of the radiation is transmitted through the crystal except the soft radiation which has to be filtered away upstream. In the Bragg case t_{ext} is independent of λ, so that the thickness is optimized for a whole wavelength range. Promising results were recently obtained using a boron-stop etching technique to produce less than 20 µm thick silicon blades (Graeff, 1987). These crystals are not easy to mount and to cool while maintaining stable alignment and crystalline perfection.

The most elegant and efficient solution to these problems would be to cool silicon and germanium to low temperatures because they show zero thermal expansion at 125 K and 50 K, respectively. Thus any temperature variation in a range close to these temperatures would not create lattice deformations. Simultaneously, the thermal conductivity of silicon increases by one order of magnitude when cooling to 125 K. With a suitable design, cryogenic Si monochromators might withstand perhaps 100 to 1000 times more heat loading than an equivalent optics system at room temperature (Bilderback, 1986). However, recent calculations show that the limitations arising from the so-called critical heat flux of the possible coolants makes the total power to be evacuated incompatible with the requirement for low temperature (Lemaire et al., 1988).

At present, various approaches are actively studied by different groups to solve the heat problem. At existing synchrotron storage rings losses up to a factor 3 are observed that are due to thermal deformations of monochromators, even if these are cooled with liquid gallium. This is shown in Fig. 18 where the flux after the second crystal of a double-crystal monochromator is plotted against the electron beam current of the CESR storage ring at CHESS. At 76 mA the power density of an undulator beam after 1.4 mm of beryllium is about 30 W/mm^2 with a total power of 360 W at 18.5 m from the source. Starting at 10 mA corresponding to 3 W/mm^2 the deformation of the first, gallium-cooled crystal becomes so big that the second crystal does not accept anymore all of the radiation reflected by the first crystal. This results in the deviation from the straight curve for the ideal optics. An improved cooling geometry (11 slots instead of 10 holes) still gives very unsatisfactory results at higher currents although the flux increases with increasing current. This shows clearly that the maximum transferable brilliance is limited by the optics rather than by the source (cf. Fig. 1b). It is hoped that technological developments are still possible that will achieve the performance of the optics needed for the efficient conditioning of very bright synchrotron x-ray beams.

REFERENCES

Aristov, V. V., Nikulin, A. Y., Snigirev, A. A., and Zaumseil, P., 1986, Experimental investigation of X-ray Bragg diffraction on the periodic surface relief of a perfect crystal, Phys. Stat. Sol., (a) 95:81.
Aristov, V. V., Gaponov, S. V., Salashchenko, N. N., and Erko, A. I., 1986, Profiled multilayer mirrors for X-ray imaging and spectroscopy, Optics News, (b) 12:128.
Aristov, V. V., Basov, Y. A., Redkin, S. V., Snigirev, A. A., and Yunkin, V. A., 1987, Bragg zone plates for hard X-ray focusing, Nucl. Instr. and Meth., A261:72.
Aristov, V. V., Erko, A. I., and Martinov, V. V., 1988, Principles of Bragg-Fresnel multilayer optics, Rev. Phys. Appl., 23:1623.
Barbee, T. W., Jr., 1986, Multilayers for X-ray optics, Opt. Eng., 25:989.

Barbee, T. W., Jr., 1988, Combined microstructure X-ray optics; multilayer diffraction gratings, in: "Multilayers: Synthesis, Properties and Non-Electronic Applications," Materials Research Society Symposium Proceedings, T. W. Barbee, Jr., F. Spaepen and L. Greer, eds., Materials Research Society Symposium Proceedings, Pittsburgh.

Barbee, T. W., Jr., Spaepen, F., and Greer, L., 1988, eds., "Multilayers: Synthesis, Properties and Non-Electronics Applications," Materials Research Society Symposium Proceedings 103, Pittsburgh.

Batterman, B. W., and Cole, H., 1964, Dynamical diffraction of X-rays by perfect crystals, Rev. Mod. Phys., 36:681.

Bilderback, D. H., 1986, The potential of cryogenic silicon and germanium X-ray monochromators for the use with large synchrotron heat load, Nucl.Instr. and Meth., A246:434.

Bilderback, D. H., Lairson, B. M., Barbee, T. W., Jr., Ice, G. E, and Sparks, C. J., 1983, Design of doubly focusing, tunable (5-30 keV),wide bandpass optics made from layered synthetic microstructures, Nucl. Instr. and Meth., 208:251.

Bilderback, D. H., Henderson, C., White, J., Smither, R. K., and Forster, G. A., 1989, Undulator heat loading studies on X-ray monochromators cooled with liquid gallium,Proc. Synchr. Rad. Conf. 1988, Tsukuba, Rev. Sci. Instr., (in press).

Bonse, U., 1986, X-ray optics, in: "Proc. Int. School of Crystallography," Section D1, Erice.

Bonse, U., and Hart, M., 1965, Tailless X-ray single-crystal reflection curves obtained by multiple reflection, Appl. Phys. Lett., 7:238.

Born, M., and Wolf, E., 1964, "Principles of Optics," Pergamon Press, London.

Burkel, E., Peisl, J., and Dorner, B., 1987, Observation of inelastic X-ray scattering from phonons, Europhys. Lett., 3:957.

Caciuffo, R., Melone, S., Rustichelli, F., and Boeuf, A., 1987, Monochromators for X-ray synchrotron radiation, Phys. Rep., 152:1.

Christensen, F. E., Hornstrup, A., and Schnopper, H. W., 1988, Surface correlation function analysis of high resolution scattering data from mirrored surfaces obtained using a triple-axis X-ray diffractometer, Appl. Optics, 27:1548.

Church, E. L., 1988, Fractal surface finish, Appl. Optics, 27:1518.

Compton, A. H., and Allison, S. K., 1963, "X-Rays in Theory and Experiment," Van Nostrand, Princeton.

Deutsch, M., 1980, The asymmetrically cut Bonse-Hart camera, J. Appl. Cryst., 13:252.

Dhez, P., 1989, Multilayered mirrors for high brightness X-ray sources, Proc. SR 88 Meeting, Novosibirsk (1988), Nucl. Instr. and Meth., (in press).

Di Gennaro, R., Gee, B., Guigli, J., Hogrefe, H., Howells, M., and Rarback, H., 1988, A water-cooled mirror system for synchrotron radiation, Nucl. Instr. and Meth., A266:498.

Dorner, B., Burkel, E., and Peisl, J., 1986, An X-ray backscattering instrument with very high energy resolution, Nucl. Instr. and Meth., A246:450.

Edwards, W. R., Hoyer, E. H., and Thompson, A. C., 1985, Finite element analysis of the distortion of a crystal monochromator from synchrotron radiation thermal loading, SPIE Proceedings, 582:281.

Elleaume, P., 1988, Design considerations for the insertion devices and beamline frontends of the ESRF, Nucl. Instr. and Meth., A266:125.

Freund, A. K., 1983, On the use of focusing for small-angle scattering experiments, Nucl. Instr. and Meth. 216:269.

Freund, A. K., 1987, "X-Ray Optics," part 1, ESRF Report, Grenoble.

Freund, A. K., 1988, Mosaic crystal monochromators for synchrotron radiation instrumentation, Nucl. Instr. and Meth., A266:461.

Fukamachi, T., Nakano, Y., and Kawamura, T., 1986, Energy dependence of X-ray reflectivity from multilayer mirrors, in: "X-Ray Instrumentation for the Photon Factory," S. Hosoya, Y. Iitaka and H. Hashizume, eds., KTK Scientific Publishers, Tokyo.

Graeff, W., 1987, Communication at the Workshop "X-Ray Optics for the ESRF," Grenoble, unpublished.

Gronkowski, J., and Malgrange, C., 1984, Propagation of X-ray beams in distorted crystals (Bragg case), Acta Cryst., A40:507.

Hart, M., 1978, X-ray polarization phenomena, Phil. Mag., B38:41.

Hart, M., and Rodrigues, A. R. D., 1978, Harmonic-free single-crystal monochromators for neutrons and X-rays, J. Appl. Cryst., 11:248.

Hart, M., and Rodrigues, A. R. D., 1979,, Tuneable polarizers for X-rays and neutrons, Phil. Mag., B40:149.

Hart, M., Rodrigues, A. R. D., and Siddons, D. P., 1984, Adjustable resolution Bragg reflection systems, Acta Cryst., A40:502.

Hastings, J. B., 1977, X-ray optics and monochromators for synchrotron radiation, J. Appl. Phys., 48:1576.

Hastings, J. B., Suortti, P., Thomlinson, W., Kvick, A., and Koetzle, T. F. 1983, Optical design of the NSLS crystallography beamline, Nucl. Instr. and Meth. 208:55.

Heald, S., 1988, A versatile two/four crystal monochromator for X-ray absorption spectroscopy, Nucl. Instr. and Meth., A266:457.

Henke, B. L., 1972, Ultrasoft X-ray reflection, refraction and production of photoelectrons (100-1000 eV region), Phys. Rev., A6:94.

Hohlwein, D., Siddons, D. P., and Hastings, J. B., 1988, A graphite double crystal monochromator for X-ray synchrotron radiation, J. Appl. Cryst., (in press).

Hornstrup, A., Christensen, F. E., Wood, J. L., Bending, M., and Schnopper, H. W., 1988, Measurements of X-ray surface scattering and the diffraction properties of selected multilayers, SPIE Proceedings, 984:174.

James, R. W., 1963, The dynamical theory of X-ray diffraction, in: "Solid State Physics," Vol. 15. Academic Press, New York.

Jark, W., 1986, Enhancement of diffraction grating efficiencies in the soft X-ray region by a multilayer coating, Opt. Comm., 60:201.

Kohra, K., Ando, M., Matsushita, T., and Hashizume, H., 1978, Design of high resolution X-ray optical system using dynamical diffraction for synchrotron radiation, Nucl. Instr. and Meth., 152:161.

Lai, B., Chapman, K., and Cerrina, F., 1988, Shadow: new developments, Nucl Instr. and Meth., A266:544.

Lairson, B. M., and Bilderback, D. H., 1982, Transmission X-ray mirror - a new optical element, Nucl. Instr. and Meth., 195:79.

Lemaire, A. D., Wijsman, A., and van Zuylen, P., 1988, Cooling of silicon crystals for X-ray monochromators, Report 832.020 TNO Institute of Applied Physics, Delft.

Marshall, G. F., 1986, Monochromatization by multilayered optics on a cylindrical reflector and on an ellipsoidal focusing ring, Opt. Eng., 25:922.

Materlik, G., and Kostroun, V. O., 1980, Monolithic crystal monochromators for synchrotron radiation with order sorting and polarizing properties, Rev. Sci. Instr., 51:86.

Matsushita, T., and Hashizume, H., 1983, X-ray monochromators, in: "Handbook on Synchrotron Radiation," E. E. Koch, ed., Vol. 1, E. E., North-Holland, Amsterdam.

Matsushita, T., and Kaminaga, U., 1980, A systematic method of estimating the performance of X-ray optical systems for synchrotron radiation I and II, J. Appl. Cryst., 13:465.

Matsushita, T., Ishikawa, T., and Kohra, K., 1984, High resolution measurement of angle-resolved X-ray scattering from optically flat mirrors, J. Appl. Cryst., 17:257.

Mills, D. M., 1988, Phase-plate performance for the production of circularly polarized X-rays, Nucl. Instr. and Meth., A266:531.

Mourikis, S., Koch, E. E., and Saile, V., 1989, Surface temperature and distortion of optical elements exposed to high power synchrotron radiation beams, Proc. SRI 1988 Conf., Tsukuba, Rev. Sci. Instr., (in press).

Nevot, L., and Croce, P., 1980, Caracterisation des surfaces par reflexion rasante de rayons X. Application a l'etude du polissage de quelques verres silicates, Rev. Phys. Appl., 15:761.

Parratt, L. G., 1954, Surface studies of solids by total reflection of X-rays, Phys. Rev., 95:359.

Pianetta, P., Redaelli, R., and Barbee, T. W., Jr., 1985, Performance of layered synthetic microstructures in monochromator applications in the soft X-ray region, SPIE Proceedings, 563:393.

Rosen, D. L., Brown, D., Gilfrich, J., and Burkhalter, P., 1988, Multilayer roughness evaluated by X-ray reflectivity, J. Appl. Cryst. 21:136.

Rosenbaum, G., and Holmes, K. C., 1980, Small-angle diffraction of X-rays and the study of biological structures, in: "Synchrotron Radiation Research," H. Winick and S. Doniach, eds., Plenum, New York.

Saile, V., 1988, "Communication at the Workshop on Thermal Problems of Intense Synchrotron Radiation Beams," ESRF, Grenoble, unpublished.

Sato, S., Yanagihara, M., Iijima, A., Takeda, S., Koide, T., and Maezawa, H., 1989, SiC mirror development at the photon factory, Proc. SRI 1988 Conference, Tsukuba, Rev. Sci. Instr., (in press).

Schneider, J. R., 1988, The applications of γ-ray diffractometry to the study of condensed matter and perspectives for the use of short wavelength synchrotron radiation, in: "Proceedings of the Workshop on Applications of High Energy X-Ray Scattering," A. Freund, ed., ESRF, Grenoble.

Schneider, J. R., Nagasawa, H., Berman, L. E., Hastings, J. B., Siddons, D. P., and Zulehner, W., 1989, Test of annealed Czochralski grown silicon crystals as X-ray diffraction elements with 145 deV synchrotron radiation, Nucl. Instr. and Meth., (in press).

Schulke, W., 1986, Inelastic X-ray scattering with synchrotron radiation: the scientific case, current experiments and projects, Nucl. Instr. and Meth., A246:491.

Siddons, D. P., Hastings, J. B., and Faigel, G., 1987, A new apparatus for the study of nuclear Bragg scattering, Nucl. Instr. and Meth., A266:329.

Sinha, S. K., Sirota, E. B., Garoff, S., and Stanley, H. B., 1988, X-ray and neutron scattering from rough surfaces, Phys. Rev., B38:2297.

Smither, R. K., Forster, G. A., Kot, C. A., and Kuzay, T. M., 1988, Liquid gallium metal cooling for optical elements with high heat loads, Nucl. Instr. and Meth. A266:517.

Smither, R. K., Forster, G. A., Bilderback, D., Bedzyk, M., Finkelstein, F., Henderson, C., White, J., Berman, L., Stefan, P., and Oversluizen, T., 1989, Liquid gallium cooling of silicon crystals in high intensity photon beams, Proc. SRI 1988 Conf., Tsukuba, Rev. Sci. Instr., (in press).

Sparks, C. J., Jr., Borie, B. S., and Hastings, J. B., 1980, X-ray monochromator geometry for focusing synchrotron radiation above 10 keV, Nucl. Instr. and Meth., 172:237.

Sparks, C. J., Jr., Ice, G. E., Wong, J., and Batterman, B. W., 1982, Sagittal focusing of synchrotron radiation with curved crystals, Nucl. Instr. and Meth., 194:73.

Spiller, E., 1988, Characterization of multilayered coatings by X-ray reflection, Rev. Phys. Appl., 23:1687.

Suortti, P., 1987, Communication at the Workshop, "X-Ray Optics for the ESRF," Grenoble, unpublished.

Suortti, P., and Freund, A.K., 1989, On the phase space description of synchrotron X-ray beams, Proc. SRI 1988 Conf., Tsukuba, Rev. Sci. Instr., (in press).

Suortti, P., and Thomlinson, W., 1988, A bent Laue crystal monochromator for angiography at the NSLS, Nucl. Instr. and Meth., A269:639.

Takacs, P. Z., 1986, Metrology of reflection optics for synchrotron radiation, Nucl. Instr. and Meth., A246:227.

Takacs, P. Z., 1988, these proceedings.

Takacs, P. Z., Hursman, T. L., and Williams, J. T., 1984, Application of silicon carbide to synchrotron radiation mirrors, Nucl. Instr. and Meth., 222:133.

Takagi, S., 1962, Dynamical theory of diffraction applicable to crystals with any kind of small distortion, Acta Cryst., 15:1311.

Takagi, S., 1969, A dynamical theory of diffraction of a distorted crystal, J. Phys. Soc. Jpn., 16:1239.

Taupin, D., 1967, Prevision de quelques images de dislocations par transmission des rayons X (cas de Laue symetrique), Acta Cryst., 23:25.

Tirsell, K., Berglin, E. J., Fuchs, B. A., Holdener, F. R., Humpal, H. H., Karpenko, V. P., and Kulkarni, S., 1988, Highly polished, grazing incidence mirrors developed for synchrotron radiation beamlines at Stanford Synchrotron Radiation Laboratory, Opt. Engin., 27:985.

Tolentino, H., Dartyge, E., Fontaine, A., and Tourillon, G., 1988, X-ray absorption spectroscopy in the dispersive mode with synchrotron radiation: optical considerations, J. Appl. Cryst., 21:15.

Trela, W. J., Bartlett, R. J., Michaud, F. D., and Alkire, R., 1988, An X-ray beamline for the energy range 5-20 keV, Nucl. Instr. and Meth., A266:234.

Underwood, J. H., Thompson, A. C., Wu, Y., and Giauque, R. D., 1988, X-ray microprobe using multilayer mirrors, Nucl. Instr. and Meth., A266:296.

Vidal, B., and Vincent, P., 1984, Metallic multilayers for X-rays using classical thinfilm theory, Appl. Optics, 23:1794.

Witz, J., 1969, Focusing monochromators, Acta Cryst., A25:30.

Youngman, B. P., 1988, History of thermal/stress analysis methods used at the Stanford Synchrotron Radiation Laboratory, Nucl. Instr. and Meth., A266:525.

Zachariasen, W. H., 1945, "Theory of X-Ray Diffraction in Crystals," Dover, New York.

Ziegler, E., Lepetre, Y., Joksch, St., Saile, V., Mourikis, S., Viccaro, P. J., Rolland, G., and Laugier, F., 1989, Performance of multilayers in intense synchrotron X-rays beams, Proc. SRI 1988 Conf., Tsukuba, Rev. Sci. Instr., (in press).

SAGITTAL FOCUSING OPTICS

J.-L. Staudenmann[1] and W. A. Hendrickson[2]

[1]Howard Hughes Medical Institute
National Synchrotron Light Source
Brookhaven National Laboratory
Upton, NY 11973

and

Department of Electrical Engineering
Columbia University
New York, NY 10027

[2]Howard Hughes Medical Institute
Department of Biochemistry and
  Molecular Biophysics
Columbia University
630 West 168th Street
New York, NY 10032

INTRODUCTION

Ideally, one would like to determine any structure with only one crystal in a "reasonable" time! With present-day detecting, electronic and flux technologies applied to biological substances, this implies neutron diffraction, because the electric field carried by the particle is too weak to cause any damage to the bonding (Hanson and Schoenborn, 1981). The drawbacks with this radiation are mainly twofold: (i) low flux and compounding, and (ii) due to low scattering cross sections, large crystals are required whose qualities may not be as high as small ones (defects and inhomogeneities related to compositional changes within the same crystal).

Over the years, it has been found that X-ray radiation also is convenient (Hendrickson and Teeter, 1981; Hontzatko et al., 1985; Sherriff and Hendrickson, 1987) for biological substances and, for a given flux, much cheaper to produce. However, it is an ionizing radiation which, over the exposure time, gradually destroys such crystals. Thus, most biological structure determinations require several crystals, creating scaling problems and uncertainties about the crystal's true defect structure and chemical compositions. The combination of these effects on the complete data set is to produce a structure in which many of its elements are not known with sufficient accuracy to eliminate any doubt about atomic sequences.

The phasing of reflections through the multi-wavelength anomalous diffraction (MAD) method is possible with neutron diffraction but only in restricted cases; i.e., as in Cd or Sm (Schoenborn, 1975). Unfortunately, it will never become a general method as explained by Hendrickson (these proceedings) for X-ray diffraction where regularly spaced - in energy - absorption edges can be found and used, uniquely identifying atoms. In this sense, MAD experiments are truly (and almost exclusively) designed for synchrotron radiation.

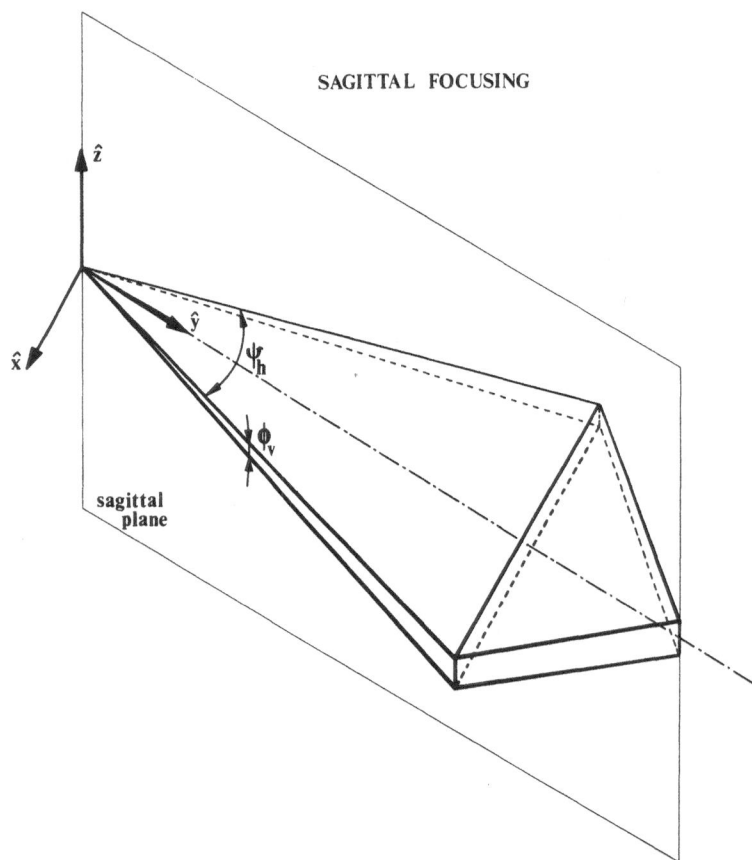

Fig. 1. Illustration of the sagittal focusing. An incident X-ray beam having horizontal, ψ_h, and vertical, ϕ_v, divergences is brought to a line focus by an optical element approximately placed at one-third magnification ratio.

With the introduction of powerful synchrotron radiation sources, such as the National Synchrotron Light Source (NSLS), and adequate optimizations of X-ray fluxes and data collection rates, one hopes to have enough information for solving a given structure with only one crystal. The ideal experimental solution would be to use a complete two-dimensional spherical detector with variable incident energy to increase the data set.

SAGITTAL FOCUSING OPTICS

It is difficult to address the subject to monochromating any X-ray beam without mentioning the pioneer work of DuMond (1937). In 1937, he explained, for example, that the detuning of one monochromator with respect to the other produces a better defined beam, i.e., with a better energy resolution, and suppression of harmonics. Depending upon the detuning level, the incident X-ray flux can be significantly diminished. Some of DuMond's ideas needed about thirty years to be verified as, for instance, the reflection-curve tail suppression by Bonse and Hart (1965). Other aspects required even more time as the "Adjustable Resolution Bragg Reflection System" by Hart et al. (1984).

Various schemes of intercepting large swaths of radiation have been proposed. Some of these projects have been initiated and built, for instance, by Lemonnier et al. (1978) and Sparks et al. (1980); their respective usage are compared and summarized by Deslattes (1980). The reason for implementing "sagittal focusing" in the X-ray optics of any diffraction station is not only the attempt of concentrating as many photons as possible onto the sample but also the easiest mode of focusing allowed for a fixed exit double crystal monochromator: "It is most worthwhile to focus the horizontal divergence which can contribute 100 times more to the intensity than focusing the vertical divergence" (Sparks et al., 1982).

Although sagittal usually refers to a specific mode of focusing X-rays, its common meaning (pertaining or like an arrowhead) does not add to the definition of focus (Dorset and Baber, 1983); thus, it should be avoided. However, by combining explanations found in the Robert (1970) and Oxford English (1979) Dictionaries, one can represent the sagittal focus as in Fig. 1 and describe it as: "a line at the edge of an antero-posterior vertical plane - the sagittal plane or plane of scatter - containing the incident beam propagation axis and which collects the two parts of the beam, equally distributed on each side of this vertical plane, through an optical element." That is, sagittal focusing is not specific about the focusing within that plane (Fig. 1) and thus, it is not restricted to fixed exit double crystal monochromators. Sagittal focusing can readily accomodate mirrors, provided that the beam is equally split between the two sides about its propagation axis. As a consequence, the above definition cannot apply to systems as those described by Lemonnier et al. (1978) or Schildkamp (1988). The above definition truly accounts for the meaning of sagittal while that of Ice and Sparks who wrote "focusing out of the plane of scatter is called sagittal focusing ...." (Ice and Sparks, 1984) is incomplete.

Various bent devices were described to achieve sagittal focusing, either by means of mirrors, monochromators, or combinations of both. Sparks et al. (1980) showed that a cylindrically bent monochromator can collect a larger swath of radiation than that of a mirror, especially for high energy photons where the grazing angle is small, about 1/20 of the monochromator Bragg angle ($\theta \approx 9°$ at 1Å wavelength, or 12.4 keV photons, for Si (111) crystal monochromators). Hence, mirrors become long and wide; their reflectivity is difficult to maintain throughout their area (Takacs, these proceedings). Moreover, Sparks et al. (1980) established that for nearly cylindrically bent crystal monochromators set at a one-third magnification ratio, "the error in the Bragg diffraction angle, θ, for sagittal rays can be eliminated independent of the divergence intercepted. Thus a singly-curved crystal will produce an approximately focused beam at this magnification while maintaining a narrow $\Delta E/E$." The latter aspect of this citation is fundamental to the phase determination through the MAD method as described by Hendrickson (these proceedings). Later, Sparks et al. (1982) demonstrated that conical bending would even be superior to the cylindrical one. In both the cylindrical and conical cases, the anticlastic deformations are strongly reduced by using reinforcing ribs in the back of the crystal monochromators (Sparks et al., 1980; Sparks et al., 1982). The second reference exhibits an outstanding example of a 20-keV focused X-ray beam. Another means of achieving cylindrical bending was proposed by Matsushita et al. (1986) which has the advantage of removing much less material than Sparks' method (Sparks et al., 1980, Sparks et al., 1982). It is not clear, however, if the crystal bending can be as extended as much as in Sparks' method. An alternative bending mode was proposed by Mills et al. (1986) which was acknowledged (Matsushita, 1987) to be better than that proposed earlier (Matsushita et al., 1986). These various bending modes are illustrated in Fig. 2. One common and unfortunate consequence of crystal bending is a net fragility increase of bent monochromators. In any event, it is clear that more work is needed to establish which is the best mean of bending crystal monochromators for achieving the widest swath interception of synchrotron radiation without loosing the reflectivity toward the crystal edges; Fig. 1 of Spark et al., (1980) shows, for instance, a working limit of 20 mrad. It is now acknowledged that this figure is much too large.

THE HOWARD HUGHES MEDICAL INSTITUTE SYNCHROTRON RESOURCE

Fig. 3 illustrates the initial layout of the Howard Hughes Medical Institute (HMMI) Synchrotron Resource facility being built at NSLS, at Brookhaven National Laboratory. Three beamlines are now being developed and will be dedicated to structural biology and to fundamental diffraction physics studies on biological samples, especially to improve the universality of the MAD method. One of the dominant aspects of the HHMI Synchrotron Resource is that each of the three lines will have focusing optics, as described in the preceding section.

X4A, X4B, and X4C will intercept 8, 3, and 6 mrad, respectively, out of a 48-mrad swath of synchrotron radiation. X4B will occupy its center. X4A will be 19 mrad farther from the storage ring, with respect to X4B. X4C will be 18 mrad closer to the ring, with respect to X4B. Specifics are:

(i) X4A, line-designed for MAD experiments, will have a double crystal monochromator built by Kohzu Seiki Company, Ltd. of Tokyo and equipped with sagittal focusing for the second optical element, as used by Matsushita et al., (1986). We decided to choose this kind of monochromator because about a dozen, with various angular ranges, are in action at the "Photon Factory" in Japan; by far the largest "population" of any existing monochromator type. They are sturdily built. They seem very reliable and relatively easy to set. The monochromator will be outside the X4A radiation enclosure, at 18 m from the source and the diffractometer at 24 m; a true three-to-one magnification ratio (Sparks et al., 1980). Typical research on that line will be similar, for instance, to that of Guss et al. (1988) and Hendrickson et al. (1988).

(ii) X4B will be first dedicated, in collaboration with Galayda et al., of NSLS, to statistical studies of vertical beam variations by means of a one-to-one magnification ratio pinhole camera (Yu et al., 1988). It will be transformed into a Laue station to study dynamical processes in proteins at a later date. Synchrotron radiation examples can be found in Helliwell et al. (1986) and Moffat et al. (1986).

(iii) X4C will have focusing optics as described by Schildkamp (1988). This focusing system will be situated inside the X4C hutch, 9.8 m away from the source. This line will be committed to routine protein crystallography structure determinations by using a rotation camera and imaging plates (Miyahara et al., 1986; Amemiya et al., 1988; Bilderback et al., 1988; Whiting et al., 1988).

courtesy CJ Sparks, Jr, GE Ice, J Wong, & BW Batterman
NIM <u>194</u> (1982) 73-78

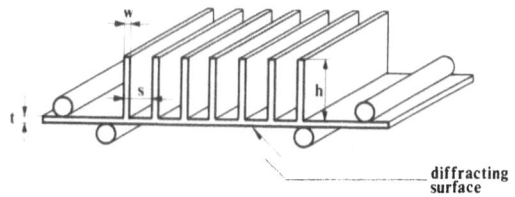

courtesy DM Mills, C Henderson, & BW Batterman
NIM <u>A246</u> (1986) 356-359

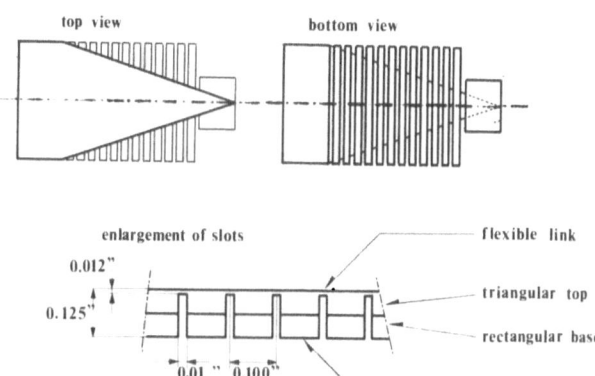

courtesy T Matsushita, T Ishikawa, & H Oyanagi
NIM <u>A 246</u> (1986) 377-379

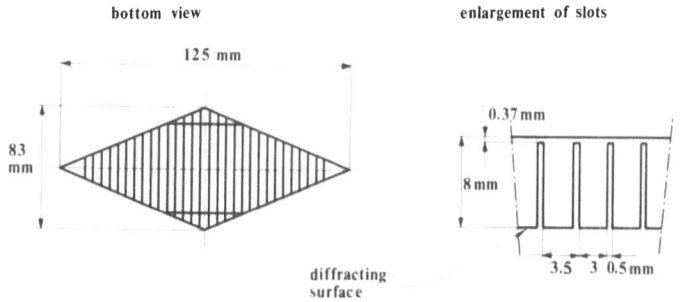

Fig. 2. Illustrations of the three most common bending modes used for sagittal focusing. "NIM" stands for "Nucl. Instr. and Meth." The diffracting surface is continuous in Sparks et al.'s (1982) design
(top sketch) while the diffracting surfaces of the other two modes
are fragmented.

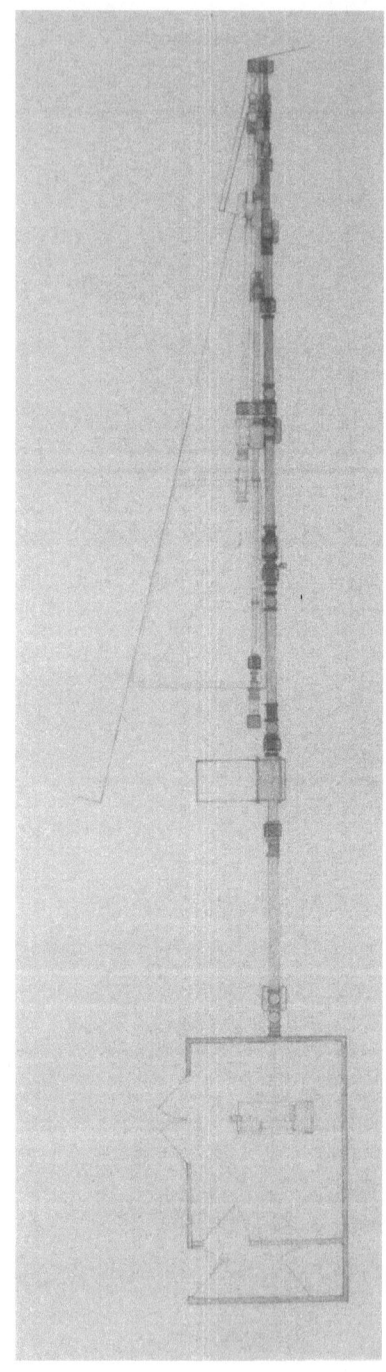

HUGHES SYNCHROTRON RESOURCE

N.S.L.S.   B E A M L I N E   X 4

MAD LINE (A)   ROTATION LINE (B)
Multi-Wavelength Anomalous Diffraction   Routine Diffraction
High Energy Resolution   High X-ray Flux
High Precision   Versatile Cameras

Fig. 3.   Initial layout of the HHMI Synchrotron Resource being built at the National Synchrotron Light
          Source. This facility comprises three lines which will be dedicated to structural biology and
          which are summarized in the text.

Both X4A and X4C radiation enclosures will have the capability of being refrigerated down to 10°C or less and will have a biosafety level #2 rating (HHS Publication No. (CDC) 84-8395, 1984). X4B radiation enclosure cannot physically accomodate the requirements for such biosafety level; thus, toxic biological substances will not be accepted on this line.

Although the primary goal of operation for all three lines is reliability, with performances as close as possible to the peak capacities of the optical elements, each line will be developed in several stages. The optics of the three lines will be constructed to have the best available efficiencies. Synchrotron sources of the "second (or third) generation" will exhibit photon fluxes orders of magnitude higher than today's best optics can handle. From a practical diffraction standpoint, these incredibly high fluxes mean nothing thus far (Freund, these proceedings) because the necessary technologies are not yet developed to take full advantage of them (Takacs, these proceedings).

The first MAD experiments on X4A will be done by using the point to point counting method. The energy spectrum and the treatment of the electronic tail pulses will be sorted out through a multi-channel analyzer. Taking into account the above details, this could be, in many ways, the brief description of any "standard" diffraction line around the X-ray beam port floor of NSLS. Once the line is debugged to satisfaction, the simple scintillation detector will then be replaced by a two-dimensional electronic system whose brand and type are still to be determined. At this point, a pre- and/or post-mirror to the sagittal focusing monochromator may be added subsequently to futher enhance the X4A flux performances.

From a technological point of view, the design and the construction of X4C are more demanding and innovative than those of the X4A initial setup. Thus, the probability of getting results sooner with X4A is much higher than with X4C, even though the latter line has the preference of the HHMI community. X4C will come on-line about one year after X4A.

REFERENCES

Amemiya, Y., Matsushita, T., Nakagawa, A., Satow, Y., Miyahara, J., and Chikawa, J.-I., 1988, Design and performance of an imaging plate system for x-ray diffraction study, Nucl. Instr. and Meth., A266:645.

Bilderback, D., Moffat, K., Owen, J. F., Rubin, B. H., Schildkamp, W., Szebenyi, D., Smith Temple, B., Volz, K., and Whiting, B. R., 1988, Protein crystallographic data acquisition and preliminary analysis using Kodak storage phosphor plate, Nucl. Instr. and Meth., A266:636.

Bonse, U., and Hart, M., 1965, Tailless x-ray single-crystal reflection curves obtained by multiple reflection, Appl. Phys. Lett., 7:238.

Biosafety in Microbiological and Biomedical Laboratories, 1984, "HHS Publication No. (CDC) 84-8395, U.S. Department of Health and Human Services, Public Health Service Centers for Disease Control and National Institutes of Health, U.S. Government Printing Office, Washington, D.C.

Deslattes, R. D., 1980, Primary monochromators using crystal diffraction, Nucl. Instr. and Meth., 177:147.

Dorset and Baber, 1983, Webster's New Universal Unabridged Dictionary.

DuMond, J. W. M., 1937, Theory of the use of more than two successive x-ray crystal reflections to obtain increasing resolving power, Phys. Rev., 52:872.

Freund, A., These proceedings.

Guss, J. M., Merritt, E. A., Phizackerley, R. P., Hedman, B., Murata, M., Hodgson, K. O., and Freeman, H. C., 1988, Phase determination by multiple-wavelength x-ray diffraction: crystal structure of a basic "blue" copper protein from cucumbers, Science, 241:806.

Hanson, J., and Schoenborn, B. P., 1981, Real space refinement of neutron diffraction data from sperm whale carbonmonoxymyoglobin, J. Mol. Biol., 153:117.

Hart, M., Rodrigues, A. R. D., and Siddons, D. P., 1984, Adjustable resolution bragg reflection systems, Acta Cryst., A40:502.

Helliwell, J. R., Papiz, M. Z., Glover, I. D., Habash, J., Thompson, A. W., Moore, P. R., Harris, N., Croft, D., and Pantos, E., 1986, The wiggler protein crystallography workstation at the Daresbury SRS: progress and results, Nucl. Instr. and Meth., A246:617.

Hendrickson, W. A., These proceedings.

Hendrickson, W. A., Anomalous scattering in macromolecular structure analysis, in: "Crystallography in Molecular Biology," Moras, D., Drenth, J., Strandlberg, B., Suck, D., and Wilson, K., eds., Plenum Publ. Corp., New York.

Hendrickson, W. A., Smith, J. L., Phizackerly, R. P., and Merritt, E. A., 1988, Crystallographic structure analysis of lamprey hemoglobin from anomalous dispersion of synchrotron radiation, Proteins, 4:77.

Hendrickson, W. A., and Teeter, M. M., 1981, Structure of the hydrophobic protein crambin determined directly from the anomalous scattering of sulfur, Nature, 290:107.

Hontzatko, R., Hendrickson, W. A., and Love, W., 1985, Refinement of lamprey hemoglobin at 2.0Å resolution, J. Mol. Biol., 184:147.

Ice, G. E., and Sparks, C. J., Jr., 1984, Focusing optics for a synchrotron x-radiation microprobe, Nucl. Instr. and Meth. 222:121.

Lemonnier, M., Fourme, R., Rousseaux, F., and Kahn, R., 1978, X-ray curved-crystal monochromator system at the storage ring DCI, Nucl. Instr. and Methods, 152:173.

Matsushita, T., 1987, personal communication.

Matsushita, T., Ishikawa, T., and Oyanagi, H., 1986, Sagitally focusing double-crystal monochromator with constant exit beam height at the photon factory, Nucl. Instr. and Methods, A246:377.

Mills, D. M., Henderson, C., and Batterman, B. W., 1986, A fixed exit sagittal focusing monochromator utilizing bent single crystals, Nucl. Instr. and Meth., A246:356.

Miyahara, J., Takahashi, K., Amemiya, Y., Kamiya, N., and Satow, Y., 1986, A new type of x-ray area detector utilizing laser stimulated luminescence, Nucl. Instr. and Meth., A246:572.

Moffat, K., Bilderback, D., Schildkamp, W., and Volz, K., 1986, Laue diffraction from biological samples, Nucl. Instr. and Meth., A246:627.

Oxford English Dictionary, 1979, Oxford University Press.

Robert, P., 1970, Le Petit Robert, Societe du Nouvean Littre.

Schildkamp, W., 1988, Design of a tunable and focusing single crystal monochromator for powerful x-ray sources, Nucl. Instr. and Meth., A266:479.

Schoenborn, B. P., 1975, Phasing of neutron protein data by anomalous dispersion, in: "Anomalous Scattering," S. Ramaseshan and S. C. Abrahams, eds., Published for International Union of Crystallography by Munksgaard, Copenhagen.

Sherriff, S., and Hendrickson, W. A., 1987, Location of iron and sulfur atoms in myohemerythrin from anomalous scattering measurements, Acta Cryst., B43:209.

Sparks, C. J., Jr., Borie, B. S., and Hastings, J. B., 1980, X-ray monochromator geometry for focusing synchrotron radiation above 10 keV, Nucl. Instr. and Meth., 172:237.

Sparks, C. J., Jr., Ice, G. E., Wong, J., and Batterman, B. W., 1982, Sagittal focusing of synchrotron x-radiation with curved crystals, Nucl. Instr. and Meth., 195:73.

Takacs, P. Z., These proceedings.

Takacs, P. Z., Hursman, T. L., and Williams, J. T., 1984, Application of silicon carbide to synchrotron radiation mirrors, Nucl. Instr. and Meth., 222:133.

Whiting, B. R., Owen, J. F., and Rubin, B. H., 1988, Storage phosphor x-ray diffraction detectors, <u>Nucl. Instr. and Meth.</u>, A266:628.

Yu, L. H., Galayda, J., and Ma, L., 1988, A preliminary study of the intensity dependence of the split ion chamber beam position monitor, Brookhaven National Laboratory, BNL-41673.

UNDERSTANDING THE PERFORMANCE OF X-RAY MIRRORS

Peter Z. Takacs

Instrumentation Division
Brookhaven National Laboratory
Upton, NY 11973-5000

INTRODUCTION

The manufacture of x-ray mirrors is a rather specialized branch of the optical fabrication industry. As those who have had to deal with the procurement of these components well know, there are only a handful of optical companies who supply most of the grazing incidence optics in use at the synchrotron light source facilities in the United States. There are several reasons for this. Firstly, production of a grazing incidence asphere is usually a one-of-a-kind job. It does not lend itself to mass production methods. Not many optical houses are willing to commit the necessary resources for the fabrication of one-of-a-kind components. Secondly, because it is a one-of-a-kind job, it is labor-intensive, which drives up the cost on each individual item. A great deal of engineering and planning must go into the fixturing and design of the computer control program for numerically-controlled diamond turning machines for production of metal optics. Most end users tend to view the procurement of aspheric x-ray optics as a "catalog" process, i.e., that all one needs to do is choose the desired parameters from the manufacturers' catalog of stock items and then expect delivery of the finished item (after considerable delay beyond the initial delivery date), which will then perform flawlessly as desired. In reality the procurement process is quite complex, starting with the initial trade-off studies to choose between metal optics or glass-ceramic optics, continuing with the selection of a vendor who can best do the job, monitoring the progress of the job and finishing it with the final quality assurance inspection. Many times the component fails to pass the inspection and it is returned for further rework.

Little information is available of practical use to guide the user through any of the above steps. We have been "forced" to develop our own foundation for assessing the performance of various vendors and determining the quality of the components produced by them. Our approach has been to concentrate on the area of metrology of grazing incidence optics and to develop instruments and techniques that can be used to improve the quality of components delivered to us. The major problem hindering the production of grazing incidence optics is the lack of specialized instrumentation of metrology that can be used by the small manufacturing shop to assess the quality of the component under production. We have been engaged over the past several years in developing the theoretical framework and practical measurement techniques to link the metrology to actual performance, providing much-needed feedback to the manufacturer and also educating users and manufacturers in the proper understanding of the language of surface figure and finish metrology.

THEORETICAL FOUNDATION

In order to assess the performance of x-ray mirrors in a meaningful way, we must first decide what it is that we need to measure. Certainly we would like to know what the image quality is for light reflected from the mirror surface at the wavelength of operation -- nominally in the 1 to 10 Å range. To do this test at the operating wavelength would be very time-consuming and would require a very elaborate source and an experimental chamber which could accommodate various types and sizes of mirrors with assorted focal parameters. So we are led to ask: "What is it that we can measure on grazing incidence optics?" We would like to measure mirrors off-line, in the laboratory, and at visible wavelengths. We would like to measure those characteristics of the surface with visible light that affect the performance of the mirror at x-ray wavelengths. The parameters of the surface that meet these criteria are mirror figure and finish. We can measure the surface roughness on a microscopic scale and relate the measured quantities to the amount of light scattered from the surface, and we can measure the gross figure errors and relate them to degradation of the image quality.

The terms "figure" and "finish" are rather vague and do not lend themselves to a good quantitative definition. Historically, the division between figure and finish is based upon the types of instrumentation

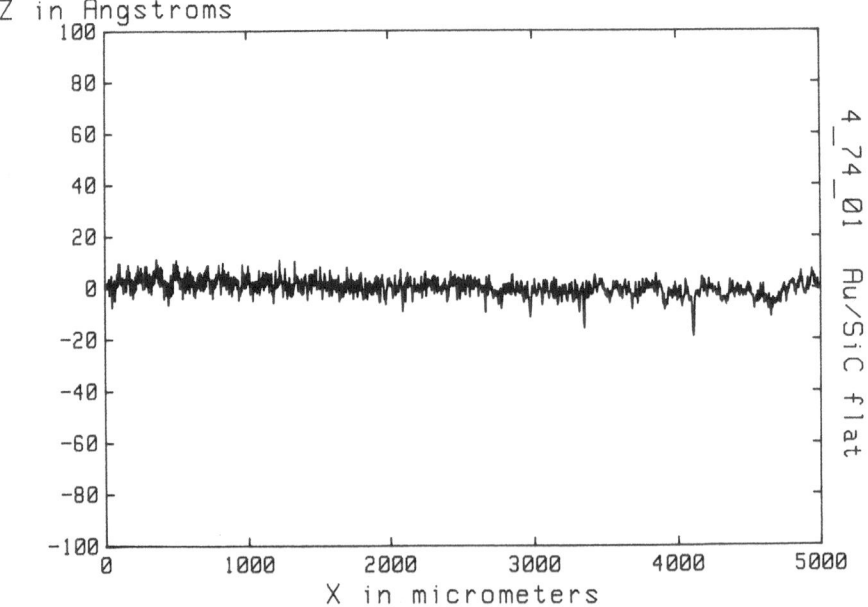

Fig. 1. A typical surface profile for a smooth Au-coated SiC mirror measured with a WYKO NCP-1000 Digital Optical Profiler with a 2.5x magnification objective.

available to make each measurement--conventional interferometry and fringe pattern analysis for figure measurement and visible light-scattering measurements for finish. These methods are totally unsuited for metrology of x-ray optics because conventional interferometry fails at measuring aspheric optics, and visible light-scattering probes the surface at spatial periods that are irrelevant to the cause of scattering at x-ray wavelengths. The region of surface spatial periods that is most important, from an x-ray mirror standpoint, lies between the regions probed by the two conventional methods. Fortunately, over the past few years, commercial instrumentation has been developed that allows one to easily measure the surface finish over the range of spatial periods that is appropriate to near-angle x-ray scattering at grazing incidence (Wyant et al., 1984; Eastman and Zavislan, 1983; Sommargren, 1981; Biegen and Smythe, 1988). These are the spatial periods from several millimeters down to several microns in length. We have employed a microinterferometer system since 1984 to measure surface profiles over a 5 mm trace length with better than 1 Å RMS repeatability (Church and Takacs, 1986). This noncontact method of measuring the surface profile has made it very easy to quantify the quality of full-scale optical components in our laboratory. We have not been so fortunate in the area of figure measurement. No commercial instrument exists that is versatile enough to perform metrology inspections on full-size aspheric optics with a minimum of set-up time. We have been forced to develop our own instrumentation for this task.

But it is not enough to just be able to measure the profile of a surface, one must relate the profile measurements to the desired functional

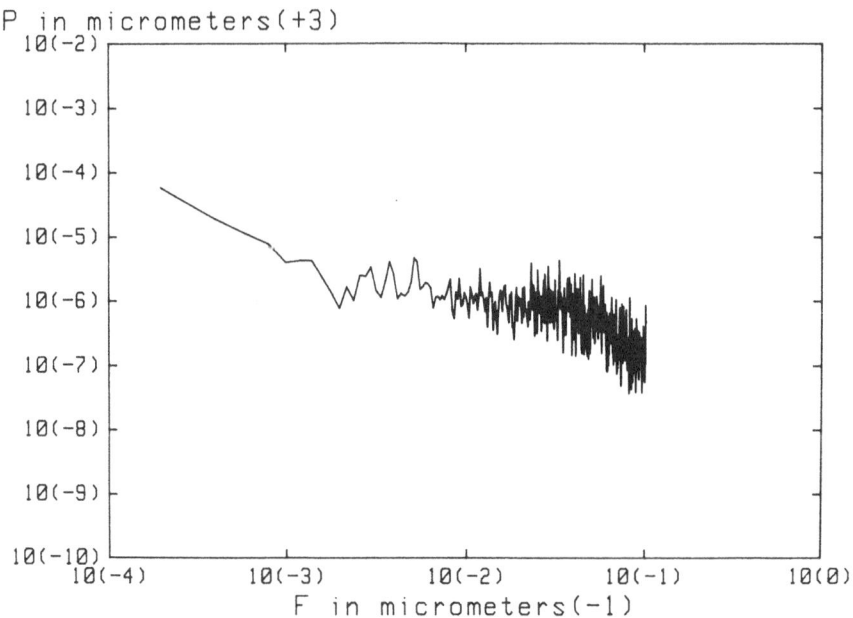

Fig. 2. The power spectral density (PSD) curve for the Au-coated SiC surface, generated by averaging the PSD curves from 10 random locations on the surface. The RMS surface roughness for this surface, computed by integrating the area under the PSD curve, is 3.1 Å over the full measurement width.

parameters of the mirror under actual operating conditions which are quite different from the normal incidence, visible wavelength metrology conditions. We have devoted considerable theoretical effort over the past several years to the development of the tools necessary to make the connection between surface profile measurements and actual performance at x-ray wavelengths (Church et al., 1979; Church and Takacs, 1986). Perhaps the most significant result of our research has been the successful application of the measured surface power spectral density (PSD) function to the prediction of near-angle scattering of x-rays from a smooth mirror surface (unpublished). The surface PSD function is generated by computing the Fourier transform of individual surface profile measurements and then averaging all Fourier transforms for a given surface. We have shown that the Rayleigh-Rice form of the angle-resolved scattering equation at grazing incidence accurately describes the distribution and intensity of x-rays about the specular direction (Church and Takacs, 1986). The scattering equation is:

$$\frac{1}{I_i}\left(\frac{dI}{d\theta}\right)_s = (1-g)\ R(\theta_i)\delta(\theta_s - \theta_i) + 16\pi^2 \left(\frac{\sin\theta_i}{\lambda}\right)^3 \cdot R(\theta_i) \cdot K \cdot S(f) \qquad (1)$$

where $S(f)$ is the PSD curve as a function of surface spatial frequency, f; $R(\theta_i)$ is the Fresnel reflectivity curve at the grazing incidence angle θ_i and K is the materials obliquity factor, which depends on the particular theoretical approach used to generate the grazing incidence approximation. The Rayleigh-Rice formalism appears to provide the best fit to the measured scattering data. The form of K for this case is relatively simple:

$$K = \left(\frac{\theta_s}{\theta_i}\right)^2 \left[\frac{R(\theta_s)}{R(\theta_i)}\right] \qquad (2)$$

The predictions of this theory have been borne out in a set of angle-resolved x-ray scatter measurements at 1.39 Å wavelength from a smooth Au-coated SiC surface at several angles of incidence. Fig. 1 illustrates a typical measured surface profile for this surface. The average PSD curve over ten random locations on the surface is shown in Fig. 2. The RMS roughness for this surface is about 3 Å over the full bandwidth. The PSD curve is multiplied by the appropriate factors in Eq. (1) for each of the three angles of incidence, and the scaled PSD curves are then plotted over the measured x-ray scattering curves in Fig. 3. The fit of the theoretical curves based on the PSD function to the measured x-ray scattering curves is quite good. The Rayleigh-Rice formalism accurately predicts both the shape and magnitude of the scattered light distribution and successfully predicts the asymmetry in the wings of the scattered light distribution, which is especially noticeable at the most grazing angle of incidence.

PRACTICAL CONSIDERATIONS

We demonstrated that we can relate the scattered light distribution at x-ray wavelengths to the surface roughness properties of a mirror surface through the appropriate theoretical framework. The question now is, "How do we apply this information to the procurement of useable optical components for grazing incidence?" The answer is not simple or easy. In the ideal world, if the user knew exactly what he needed in terms of the scattered light distribution from each surface, then it would be a relatively straightforward matter to work backwards through the equation to determine the maximum allowable shape and magnitude for the surface PSD curve. As a rule of thumb one can make a zero-order assumption that the PSD curves for glass-ceramics all have the same power-law shape, only the absolute

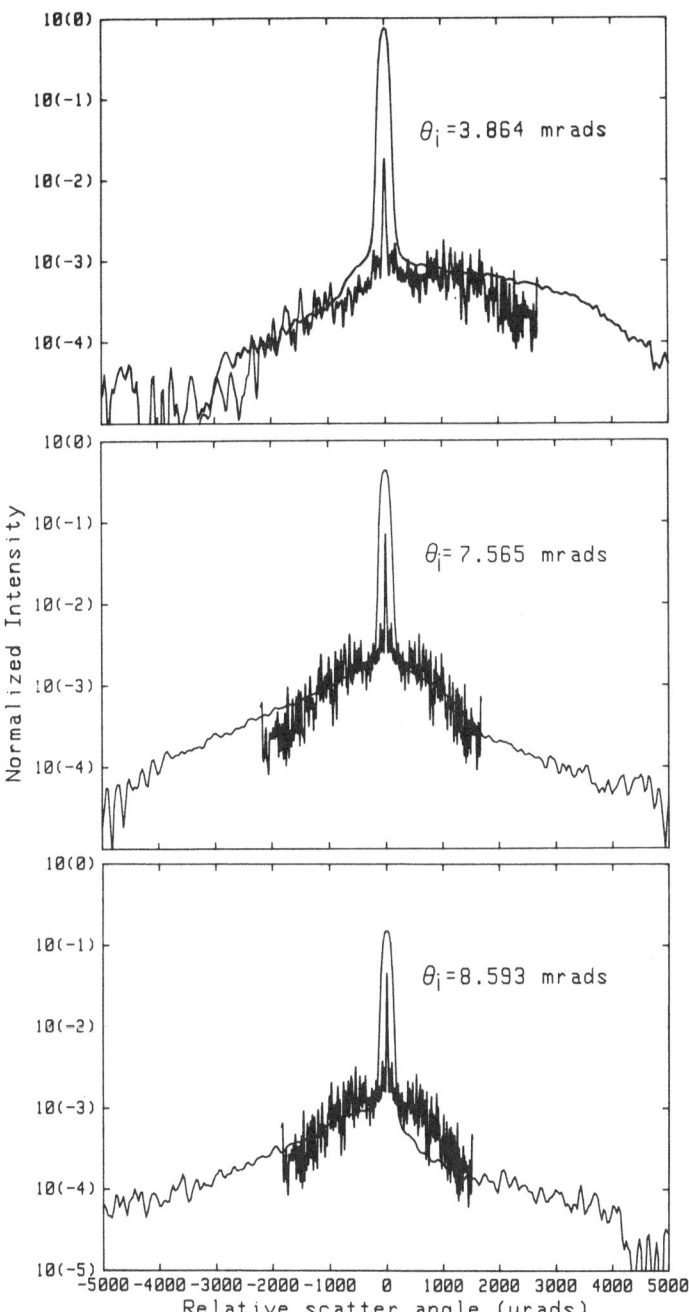

Fig. 3. Predicted and measured scattered light distributions for the Au-coated surface at a wavelength of 1.39 Å at three angles of incidence. The measured scattered light (smooth lines) are normalized to the incident beam intensity: the relative intensity of the specular peak is given by the Fresnel reflectivity of the surface for each particular angle of incidence. The predicted scattered x-ray distributions (rough lines) computed from Eqs. (1) and (2) are overlayed on the appropriate measured curve.

Table I. Suppliers of Synchrotron Radiation Mirrors
Included In This Survey

American Aspheric Co.	P.O. Box 22303 Tucson, AZ 85734
Applied Optics Center Corp.*	Burlington, MA
Continental Optical Corp.	15 Power Drive Hauppauge, NY 11787
Diamond Electro-Optics, Inc.	323 Andover Street Wilmington, MA 01887
Ferranti Astron, LTD.	Unit 1, Aerodrome Way Cranford Lane Hounslow TW5 9QB UNITED KINGDOM
Frank Cooke, Inc.	59 Summer Street North Brookfield, MA 01535
General Optics	554 Finn Avenue Moorpark, CA 93021
Karl Lambrecht Corporation	4204 N. Lincoln Avenue Chicago, IL 60618
Liberty Mirror Libbey-Owens-Ford	851 Third Avenue Brackenridge, PA 15014
Optic-Electronic Corporation	11545 Pagemill Road P.O. Box 740668 Dallas, TX 75374-0668
Research Optics & Development, Inc.*	Waltham, MA
Union Carbide Nuclear Division Y12 Plant (Now under management by Martin Marietta Corporation)	Oak Ridge, TN

*The mirror fabrication facility from this organization
is now a part of Diamond Electro-Optics.

magnitudes are different. The same holds true for the metal surfaces,
although the power-law parameters are usually much different for them.

In practice, the user or optical system designer does not usually have
a complete description of the optical system to allow him to deduce the required surface roughness spectrum. Only the most sophisticated of design
codes include the surface PSD function as a design parameter (Glenn, 1985).
Other factors which need to be considered are the type of detector that will
be used (film, ion chamber, photon counter, etc.), whether or not it is an
imaging system, what is the minimum required resolution, the required spectral purity for a monochromator system, the source size, and the point
spread function caused by "figure" errors. All of these factors tend to
complicate the specification of optical components. Contributing to the

lack of rigor in the design of the grazing incidence system has been that, in the past, it was not possible to perform the required metrology on the mirrors. Mirror quality has been relatively poor and the credibility of the design process has been rather low because the systems have never performed close to their design goals. Designers and users have resorted to simplistic, back-of-the-envelope calculations to specify important surface roughness parameters. Seemingly insignificant changes in these parameters often have a large effect in cost of the final product. Our method of PSD function measurement and performance prediction puts the design process on a much more solid foundation.

MIRROR SUPPLIER DATA BASE

Our measurements of surface roughness began in 1984 after acquisition of a WYKO NCP-1000 Digital Optical Profiler (Wyant et al., 1984). This instrument allows us to measure the surface roughness of full-sized synchrotron mirrors in a noncontact manner, quickly and efficiently. In the several years since, we have compiled a data base of surface roughness as a function of manufacturer, material, and figure type. Recognizing that there is an acute lack of practical information available to the grazing incidence community regarding the above parameters, we have compiled our information in a way that should prove useful for future design considerations.

The mirrors in the database are divided into two distinct classes: flats and aspheres. Conspicuously absent from this database are spherical surfaces. The only spherical surfaces measured to date have been for diffraction gratings and the number of samples is extremely small, so they are not included in this compilation. The aspheric surfaces include cylinders, toroids and ellipsoids. As with the figure types, there are two major material classes: metal substrates and glass or ceramic substrates. The only metal material that is included in this data base is electroless nickel plate (ENP), usually on an aluminum substrate. The reason for this is that, for synchrotron use, no other metal surface can be polished to the smoothness required of an x-ray reflecting surface. The fact that ENP can also be diamond-machined is also an important reason for its use in (SR) Synchrotron Radiation mirror fabrication.

The list of manufacturers providing mirrors to the National Synchrotron Light Source (NSLS) and other synchrotron facilities whose mirrors have been measured in our Metrology Laboratory is presented in Table I. A summary of measured mirrors arranged by figure and material type is given in Table II with the number of mirrors of each type identified by manufacturer. Of all the mirrors in this table, only the General Optics silicon flats are not real SR mirrors. These are single-crystal silicon wafers and are included to demonstrate the high-quality optical finish that can be applied to this material if one is careful. Conventional silicon wafers for semiconductor applications are usually an order of magnitude worse in surface roughness.

Table II also summarizes the number of mirrors of each type that have been measured on the BNL WYKO profiler. Not all of these mirrors are for use at the NSLS: a significant fraction were sent to the NSLS for measurement and then shipped elsewhere for use at other locations, in particular the Stanford Synchrotron Radiation Laboratory (SSRL), Synchrotron Radiation Center (SRC) in Wisconsin and the Cornell High Energy Synchrotron Source (CHESS). Also, some mirrors are entered twice in the statistics, usually a result of measuring before and after a reworking process. If the mirror failed to meet the specifications, it was sent back to the manufacturer and after reworking was returned to the NSLS for a second measurement. The most striking examples of reworking to be noted later are early-fused silica cylinders made by Frank Cooke around 1981-82, prior to the use of the WYKO

Table II. Synchrotron Radiation Mirrors Measured for Surface Roughness

Figure	ASPHERES Material		FLATS Material	
	ENP/Al — Glass, Ceramic		ENP/AI — Glass, Ceramic	
Cylinder	OEC 4	F. Cooke 24	OEC 3	Liberty 5 (Float G)
	ROD 1	Continental 5	AOC 1	F. Cooke 4 (FS)
	Astron 2	Am. Asph. 1	ROD 5	Continental 1 (FS)
			DEO 2	Continental 4 (SiC)
Ellipsoid	DEO 18	Continental 1	Astron 2	Karl L 4 (Float G)
	AOC 1	Unknown 1	Y12 1	Astron 2 (SiC)
	Unknown 2			Genl. Opt. 4 (Si)
	OEC 2			
Toroid	Astron 1	Am. Asph. 1		
	AOC 1	Astron 6		
	OEC 1			
Total Measurements Per Cagegory:	33	39	14	24

profiler for surface metrology. Measurements made on these mirrors in 1985 showed that they had a large amount of low frequency ripple, resulting in RMS roughness numbers that ranged from 25 to 85 Å. The low frequency ripple with spatial periods in the millimeter range is extremely difficult to detect by any other means. Frank Cooke subsequently changed his lapping process and, as a result, new cylinders and reworked old cylinders now routinely measure at or below the 5 Å RMS level.

The RMS roughness of each mirror in each column of Table II is plotted roughly as a function of time in the diagram in Figs. 4 through 7. Each figure and material combination is plotted on a separate graph to better enable visual comparisons between various permutations of the parameters. The horizontal axis labelled "Volume ID Number" refers to the data file identification number in the WYKO profiler log book. It is basically a time series with Volume 1 starting in 1984 and Volume 5 starting in January 1988 covering the first half of 1988. The data points refer to the date on which the mirror was measured in the laboratory, not necessarily to the date the mirror was manufactured. In fact, some mirrors that were manufactured many years ago have only recently been measured. Most measurements in this data base were made with the 2.5X magnification objective on our WYKO instrument. Only a few of the early measurements were made with the 10X objective. The 2.5X objective is now used almost exclusively, since it provides significant information about the roughness in the important spatial period region from 1 to 5 millimeters. This is the region that is most inaccessible to other metrology techniques and is what makes our measurement capability unique. The roughness numbers plotted are for the full 2.5X bandwidth which covers the range from 5.0 millimeters down to 9.8 micrometers.

Figure 4 summarizes the results for glass/ceramic aspheres. The three types of aspheres measured are cylinders, ellipsoids, and toroids. Most of the mirrors are made from fused silica; a few are ZERODUR, two are

single-crystal silicon and two are CVD silicon carbide. Some are coated
with a metal reflecting layer, some are uncoated. We almost never see
evidence that a properly-applied thin metal coating changes the roughness
properties of the surface, although occasionally a bad coating does slip in.

The majority of glass/ceramic aspheres have been right circular
cylinder segments, primarily used for sagittal focusing of a horizontal fan
of synchrotron radiation. The group of 5 cylinders located in a vertical
column at "A" in Fig. 4 were manufactured a few years earlier by Frank
Cooke. As mentioned above, they were only first measured at the time
indicated in the graph. Several other Frank Cooke cylinders with RMS
roughness values above 20 Å were also manufactured in this early time frame,

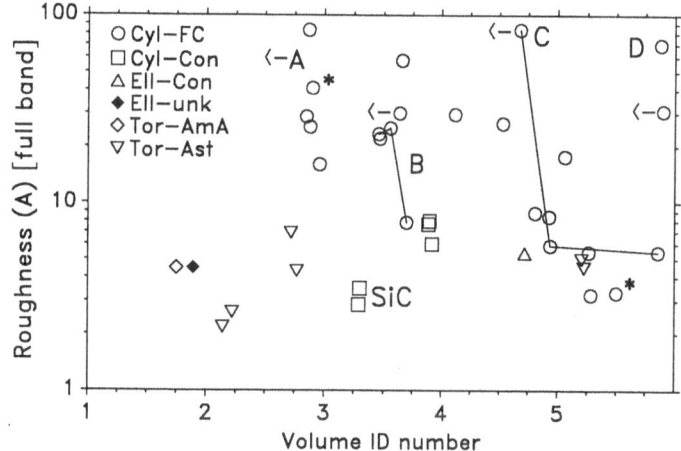

Fig. 4. Measured RMS roughness values plotted as a function of laboratory
logbook volume and page number (roughly a time sequence of when
each mirror was manufactured. Those marked with the left-pointing
arrows were manufactured before 1984. See text for full explana-
tion. The diagram illustrates that fabrication techniques for
glass and ceramic aspheres are good enough to routinely produce
surfaces with roughness values well below 10 Å RMS.

but were only recently measured. Those whose early manufacture date could
be confirmed are indicated with an arrow pointing to earlier times. The
major trend to note is the significant reduction in the average RMS
roughness of the cylinders in more recent times. The early cylinders were
usually always above 20 Å RMS, while now it is unusual to see one that
exceeds 10 Å RMS. A dramatic example of the improvement in the polishing
process based on feedback from the metrology is indicated by the data points
marked with asterisks. An early-fused silica mirror with a roughness of
about 40 Å RMS was reworked and measured in 1988 to have an RMS roughness
slightly more than 3 Å. Another example of improved surface finish is the

series of three measurements on a cylinder at "B" connected by the straight
lines. The initial measurement was 23.3 Å RMS. Additional work on the surface produced a 25 Å surface (not statistically different from the first
measurement), but additional work and a process change produced a surface
with a 7.8 Å RMS finish. The mirror at C was originally made years ago, but
after reworking its roughness level decreased from 83 Å to 5.89 Å RMS. A
subsequent platinum coating was measured at 5.4 Å RMS, again not statistically different from the uncoated surface. Occasionally, however, a new
mirror slips through with a very rough surface, as indicated by the point
labeled "D" at about 70 Å RMS in the upper right corner. This illustrates
the usefulness in being able to make these kinds of measurements on a regular basis. Other products in this class from other vendors all lie below
the 10 Å level, which differs significantly from the metal mirror results.

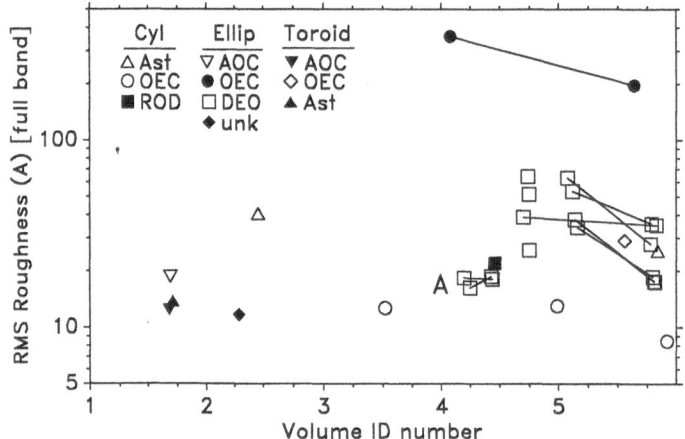

Fig. 5. Measured RMS roughness values for aspheres made from electroless
nickel plate (ENP) on an aluminum substrate. The solid lines
connect measurements made on the same mirror before and after
repolishing. The diagram illustrates the difficulty in achieving
RMS roughness values below 10 Å with this material.

The data for the metal aspheres is plotted in Fig. 5. One can see
almost immediately that there appears to be a lower limit of about 10 Å RMS
to the best that can be achieved on this material, independent of figure.
The best surface to date in the electroless nickel material is the most
recent product from Optic-Electronic Corp., a cylinder with an RMS roughness
of 8.45 Å, but whether or not this low value can be improved upon remains an
open question. All the evidence indicates that it is very difficult to produce a surface below the 10 Å level with ENP, while for glass/ceramic materials it is highly unusual to exceed the 10 Å level. Pairs of data points in
Fig. 5 also indicate the improvement in surface finish of a set of

ellipsoidal mirrors obtained after reworking the mirrors. In each case the rework involved additional polishing on the surfaces. The pair of data points above the 100 Å level are for a rather complex ellipsoid designed to produce an 8 to 1 image size reduction (Jones et al., 1987). The numerical machining control problem was extremely difficult to solve to produce the desired contours with 3 degrees of freedom in the machine motion. Additional polishing was not able to remove the low frequencies left in the surface. The set of open squares labeled "A" indicate before and after measurements on two ENP ellipsoids that were coated with gold-electroplate. Again, no change in the roughness was evident.

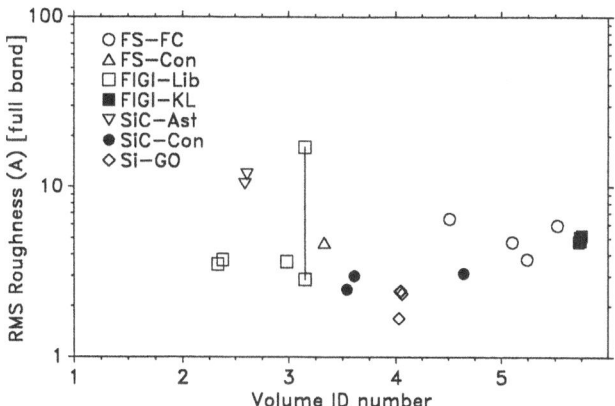

Fig. 6. Measured RMS roughness values for flat surfaces made from glass or ceramic materials. The solid vertical line connects measurements on a piece of float glass that shows an extreme anisotropy in surface roughness between the direction of pull (smooth) and the transverse direction (large amplitude long-period sinusoid). Most measurements lie well below the 10 Å RMS level.

More compelling evidence for the 10 Å roughness barrier can be seen by comparing the glass/ceramic flats in Fig. 6 with the ENP flats in Fig. 7. The two silicon carbide flats above 10 Å are really grating test samples and should not be included in the comparison. The only other flat above 10 Å is a piece of float glass with the unusual property of being very smooth along the pull direction, but with a well-defined, large amplitude sinusoidal low-frequency period in the transverse direction. A vertical line connects the two directions on the graph. Most of these flats are intended to be used in mirror benders to produce focusing in the tangential direction and, in the case of specially-configured cross-section pieces, also focusing in the sagittal direction. The ENP flats in Fig. 7 again show a distinct cutoff at the 10 Å level. There are none better than 8 Å. The only example of degradation in surface roughness after coating is seen at the right-hand side of Fig. 7. A thick platinum coating was evaporated onto a 70 cm long ENP mirror and the surface roughness increased from 15 Å before to 35 Å after coating.

DISCUSSION

The division line at 10 Å between the ENP and glass/ceramic materials appears to be independent of figure type. The same division is seen in the data from the aspheres as from the flats. In fact, the best finishes for each material appear to be independent of figure type. One can achieve the same quality finish on aspheres as on flats, so there should be no need to relax finish performance requirements for nonflat surfaces. Aspheric surfaces are, however, more difficult to produce than are flats, especially when one desires a multiaxis machine cut, as with toroids and ellipsoids.

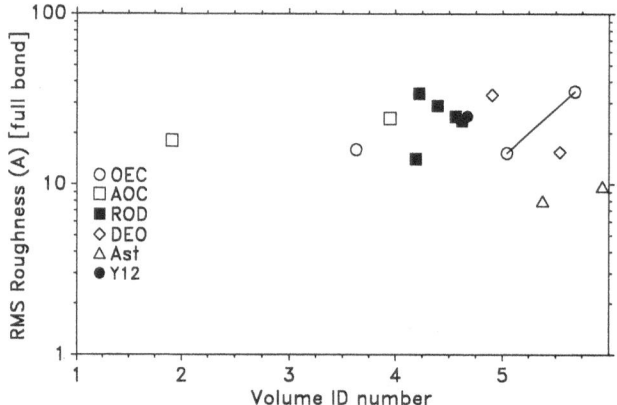

Fig. 7. Measured RMS roughness values for flat surfaces made from ENP/Al material. Most values lie above the 10 Å level. The solid line shows the effect of a poor coating on a previously acceptable surface.

Cylinders can be easily manufactured both in glass/ceramic and metal materials because only one uniform radius needs to be generated in the surface. Higher-order aspheres can be made more easily in metals than in glass, and subsequently for less cost per piece, but the performance may suffer because of the limits on polishing them. The electroless nickel plating process is subject to many factors that are usually beyond the control of the mirror fabricator. There are many variables that go into the ENP process that are not controlled in an ordinary industrial plating operation. Mirror manufacturers often report that they had to have the ENP coating stripped and replated several times before the final diamond-machined and polished surface meet the specifications. These iterations tend to drive the price up on metal mirrors. Very few manufacturers can adequately characterize the quality of the ENP coating before polishing the piece. These considerations are nonexistent for glass/ceramic materials, except for the case of silicon carbide. A layer of chemical vapor deposited (CVD) SiC is deposited on the surface of another substrate material and the CVD layer is then polished to the desired finish. Like the ENP process, the CVD process is also subject to many variables which are often difficult to control. A different set of problems arise in polishing the SiC, but the end result could be the same: the need to redeposit or replate more material and polish again.

Each material has its advantages and its drawbacks. Metal mirrors allow designers more flexibility in terms of optical design by permitting complex surfaces to be machined, but at the present time the price is poor surface finish. Glass/ceramic flats are easily made, but their usefulness in imaging is not very great unless they are bent. Perhaps one will need to spend more effort in engineering the cross sectional shape of a flat mirror and in the design of the bending mechanism to produce the desired tangential and sagittal curvatures from a single piece of float glass, instead of passing on the complexity to the optical fabricator to make an aspheric surface. The bent-mirror solution may not work in all cases, especially those involving actively-cooled mirrors. The trend in optical design for higher-brightness source instrumentation appears to be away from aspheric optics and towards spherical optics used at grazing incidence. The consensus is that spherical surfaces can be made more easily to meet the strict figure tolerances required by newer SR sources than can aspheric surfaces. However, careful consideration must be given to choice of substrate material in order not to compromise the scattered light performance of the system.

ACKNOWLEDGMENT

I would like to acknowledge the assistance of Eugene L. Church in providing the theoretical framework for understanding the relationship between surface roughness and scattered light distribution at grazing incidence. This research was supported by the U. S. Department of Energy: Contract No. DE-AC02-76CH00016.

REFERENCES

Biegen, J. F., and Smythe, R. A., 1988, High resolution phase measuring laser interferometric microscope for engineering surface metrology, in: "Scanning Microscopy Technologies and Applications," E. Clayton Teague, ed., Proc. Society of Photo-Optical Instrumentation Engineers, Bellingham.

Church, E. L., Jenkinson, H. A., and Zavada, J. M., 1979, Relationship between surface scattering and microtopographic features, Opt. Eng., 18:125.

Church, E. L., and Takacs, P. Z., 1986, The Interpretation of glancing incidence scattering measurements, Proc. of the Soc. of Photo-Optical Instrumen. Engs., 640:126.

Church, E. L., and Takacs, P. Z., 1986, Use of an optical-profiling instrument for the measurement of the figure and finish of optical-quality surfaces, WEAR, 109:241.

Eastman, J. M., and Zavislan, J. M., 1983, A new optical surface microprofiling instrument, Proc. of the Soc. of Photo-Optical Instrumen. Engs., 429.

Glenn, P., 1985, Space telescope performance prediction using the OSAC code, in: "Large Optics Technology," Gregory M. Sanger, ed., Proceedings of the Society of Photo-Optical Instrumentation Engineers, Bellingham.

Jones, K. W., Takacs, P. Z., Hastings, J. B., Casstevens, J. M., and Pionke, C., 1987, Fabrication of an 8:1 ellipsoidal mirror for a synchrotron x-ray microprobe, in: "Metrology: Figure and Finish," Bruce Truax, ed., Proc. of the Society Photo-Optical Instrumentation Engineers, Bellingham.

Sommargren, G. E., 1981, Optical heterodyne profilometry, Appl. Optics, 20:610.

Wyant, J. C., Koliopoulos, C. L., Bushan, B., and George, O. E., 1984, An optical profilometer for surface characterization of magnetic media, ASLE Trans., 27:101.

MULTIWAVELENGTH ANOMALOUS DIFFRACTION AS A DIRECT

PHASING VEHICLE IN MACROMOLECULAR CRYSTALLOGRAPHY

Wayne A. Hendrickson, John R. Horton, H. M. Krishna Murthy,
Arno Pahler and Janet L. Smith

Department of Biochemistry and
Molecular Biophysics, Columbia University
New York, NY 10032

INTRODUCTION

The possibility for definitive phase determination from diffraction
measurements made at multiple wavelengths from crystals that contain anomalous scatterers has long been recognized (Okaya and Pepinsky, 1956). This
potential is easy to appreciate since multiwavelength anomalous diffraction
(MAD) experiments can be thought of as in situ multiple isomorphous re-
placements (MIR) arising from the variations in scattering factors that ac-
company changes of wavelengths. These variations, known as anomalous scattering, result from the resonance that occurs between the oscillations of
atomic orbitals and x-ray-induced electronic vibrations. With synchrotron
radiation such experiments are now quite feasible, and with recently developed methods for analyzing MAD measurements, accurate phases can be obtained
for macromolecular crystal structures. Anomalous scattering centers appropriate for MAD experiments can be introduced as for conventional heavy atom
derivatives or they may occur naturally in metalloproteins. In addition,
selenomethionyl proteins produced biologically are suitable for MAD phasing.
Since a single crystalline species (and often a single crystal) suffices for
MAD analysis of such molecules, the MAD method serves as a vehicle for direct structure determination.

MEASUREMENT

In typical experiments we measure data, including Bijvoet mates, at
three to five wavelengths in the vicinity of absorption edges of appropriate
anomalous scatterers. Wavelengths are chosen to optimize the strength of
Bijvoet and dispersive differences (Hendrickson, 1985), and wavelengths for
edge measurements are deduced from x-ray absorption spectra taken on the
very crystals used in diffraction experiments. We take particular care to
reduce systematic errors that can affect the diffraction differences in
which the anomalous scattering signals reside. Thus, all measurements per-
tinent to a particular phase are measured close together in time and with
geometries such that Bijvoet pairs of reflections sustain similar absorption. Many of the residual errors are reduced by parameterized local scaling (Hendrickson and Teeter, 1981). Our experiments were conducted on the
area detector facility at the Stanford Synchrotron Radiation Laboratory
(SSRL) (Phizackerley et al., 1986) and on the diffractometer of beamline 14A

at the Photon Factory (Satow, 1984). We are also developing a beamline for MAD experiments at the Hughes Synchrotron Resource at NSLS.

DATA ANALYSIS

Anomalous scattering factors can be deduced from x-ray absorption spectra. This requires first a scaling of the fluorescence spectra into atomic absorption coefficients on an absolute scale. We devised a procedure for scaling and background correction that matches the observations to theoretical values outside of the edge region. These absorption data are then simply related to the imaginary components of anomalous scattering, f", and, in turn, by Kramers-Kronig transformation to the real components, f'. These procedures, as used for lamprey hemoglobin, are described elsewhere (Hendrickson et al., in preparation) and a typical result is shown in Fig. 1 for selenomethionyl T4 thioredoxin. An alternate approach to scattering factor evaluations is by way of least-squares fitting to the diffraction data. We often note pleiochroism in the x-ray absorption spectra from crystals reflecting an underlying anisotropy in anomalous scattering, as noted before by Templeton and Templeton (1985).

To analyze the diffraction data from MAD experiments, we developed a least-squares procedure (Hendrickson, 1985) based on the algebraic formalism of Karle (1980). The several measurements associated with a specific Bragg reflection are related to terms that separate normal and anomalous scattering contributions into wavelength-invariant and wavelength-dependent factors. Following Karle (1980) we use $|^{\lambda}F(h)|$ to denote the complete

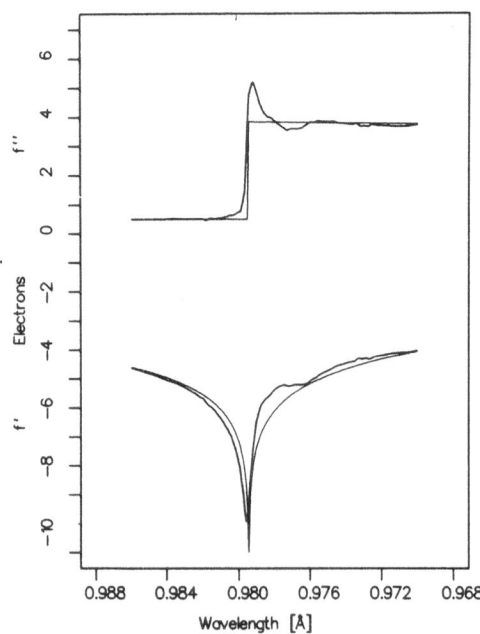

Fig. 1. Experimental scattering factors for the selenium atoms in seleno-methionyl T4 thioredoxin. These data are based on fluorescence measurements of x-ray absorption made at SSRL from a crystal of the thioredoxin. Experimental values are in bold lines and theoretical elemental values are in fine line.

structure factors for reflection h at a particular wavelength λ, but we separate components slightly differently. Then, in the case of a single kind of anomalous scatterer, the diffracted intensity, $I = K|F|^2$, is directly proportional to

$$|^{\lambda}F(\pm h)|^2 = |^{\circ}F_T|^2 \qquad +a(\lambda)|^{\circ}F_A|^2 \qquad\qquad (1)$$
$$+b(\lambda)|^{\circ}F_T||^{\circ}F_A|\cos(^{\circ}\phi_T - ^{\circ}\phi_A)$$
$$\pm c(\lambda)|^{\circ}F_T||^{\circ}F_A|\sin(^{\circ}\phi_T - ^{\circ}\phi_A)$$

The wavelength dependence is encoded in a, b, and c--factors that are ratios of anomalous (f'+if") to normal (f°) scattering components:

$$a(\lambda) = (f'^2 + f''2)/f°^2, \qquad\qquad (2a)$$

$$b(\lambda) = 2(f'/f°), \qquad\qquad (2b)$$

$$c(\lambda) = 2(f''/f°). \qquad\qquad (2c)$$

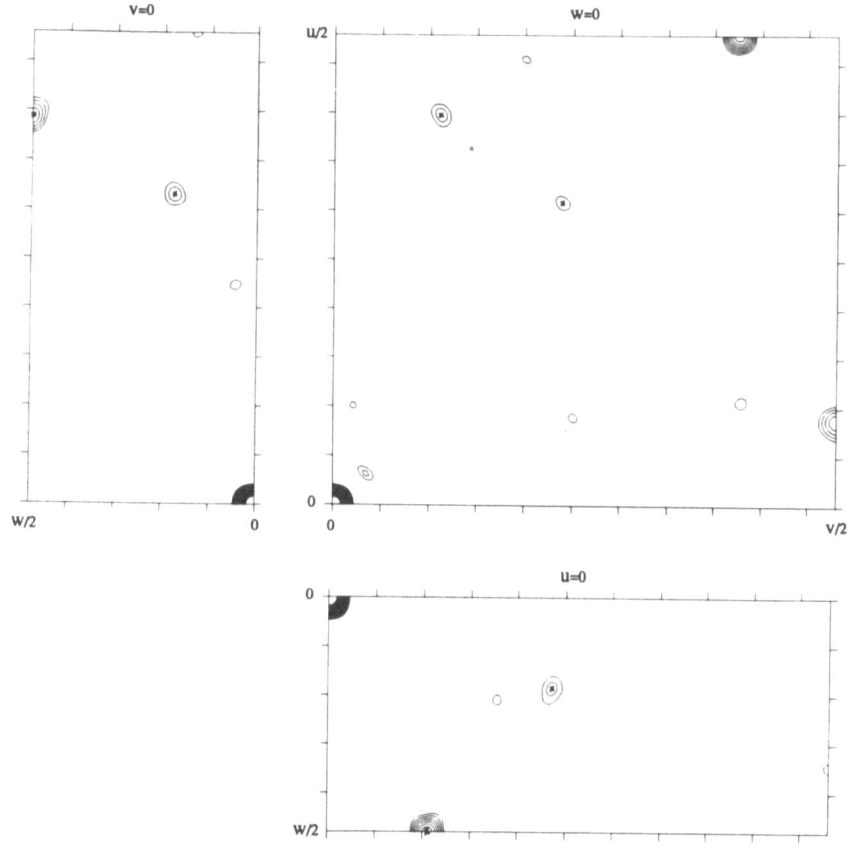

Fig. 2. Harker sections of the selenium Patterson function for seleno-biotinyl streptavidin. This map is based on 1511 of 5050 possible reflections from a 3.0Å data set measured at SSRL. This subset was selected to eliminate reflections having $|^{\circ}F_A| > 100$ or $|^{\circ}F_A| < 3\sigma(|^{\circ}F_A|)$. The map is contoured at equal intervals starting at a minimal contour of three times the intervals.

The wavelength-invariant factors are derived from three quantities of interest for each reflection: the structure factor moduli from the normal scattering contributions of all atoms, $|{}^\circ F_T|$, and from the anomalous scatterers alone, $|{}^\circ F_A|$, and the difference in phase, $({}^\circ\phi_T - {}^\circ\phi_A)$, between these structure factor contributions. Given scattering factors deduced from x-ray absorption spectra, the wavelength-invariant factors can be determined by least squares and then constrained by trigonometric identities to yield the desired structure factor quantities. The $|{}^\circ F_A|$ values obtained from the MAD least-squares analysis can be used to determine the structure of anomalous scattering centers. Since there generally are relatively few such centers, the determination of this substructure is relatively straightforward by Patterson or direct methods. Complications can arise from inaccuracies in the $|{}^\circ F_A|$ values, but standard deviations, fitting residual, and outlier rejections can be used to discriminate against misleading data. An example from the Patterson synthesis for the selenium structure in selenobiotinyl streptavidin is given in Fig. 2. Once a model is found for the anomalous scattering substructure, it can be refined and used to calculate phases, ${}^\circ\phi A$, which can be combined with the MAD phase differences to yield the required phases, ${}^\circ\phi_T$, for the entire structure.

APPLICATIONS

Lamprey Hemoglobin

Our first tests of the MAD phasing procedure were conducted with lamprey hemoglobin using type D2 crystals. These crystals are in space group $P2_12_12_1$ with unit cell dimensions of a=44.57, b=96.62 and c=31.34Å. They contain one globin chain (and one iron atom) per asymmetric unit. This crystal structure was determined previously by the MIR method (Hendrickson et al., 1973) and was subsequently refined by the restrained least-squares method to R=0.14 at 2.0Å resolution (Honzatko et al., 1985). Thus this test afforded us an opportunity to compare MAD phases (a) with MIR phases of high quality (m=0.81 for 5.5Å data) which were used to solve the structure initially, and (b) with phases from the highly refined least-squares model (LSQ).

Data were measured at SSRL on two occasions. In the first experiments, a data set complete to 5.5Å spacings was obtained at four wavelengths (1.800, 1.740, 1.738 and 1.500Å). In a later experiment, a 3.0Å data set was measured again at four wavelengths, but 1.650 was substituted for 1.500Å as the remote high energy point. Results from the 5.5Å experiments showed a phase discrepancy of 47.0° between MAD and MIR phases for the 512 of 528 possible reflections that were measured. An electron density synthesis was similar to the MIR results, but not as clean (Hendrickson et al, 1988). In the 3.0Å experiment, we obtained MAD phases as accurate overall as the MIR phases from which the structure had been interpreted earlier. As shown in Table 1, when a markedly inferior crystal is eliminated (leaving 87% of the set) the agreements with LSQ results are equally good for the MAD (50.5°) and MIR (50.7°) data. When quality tests are imposed that still retain 60% of these reflections, the average MAD vs LSQ discrepancy is reduced to 43.0° whereas the MIR vs LSQ average is unimproved.

Ferredoxin

Our first new structure determination by MAD phasing was for a ferredoxin from the bacterium Clostridium acidi-urici (Murthy et al. 1988). This protein crystallizes in space group $P4_32_12$ with a=b=34.56Å and c=75.27Å, and with one molecule per asymmetric unit. This ferredoxin is clearly homologous with the ferredoxin from Peptococcus aerogenes solved earlier by Adman et al. (1973). It has two 4Fe/4S iron-sulfur clusters and is expected to

Table 1. Phase Comparisons for Lamprey Hemoglobin

Parameter	All MAD Data	Crystal #1 Rejected[a]	Quality Subset[b]
MAD reflections	2898	2586	1552
Fraction of 3Å total	0.97	0.87	0.52
Common reflections[c]	2779	2479	1507
$\langle\Delta\phi\rangle$ (MAD vs MIR)'	59.6	58.0	54.1
$\langle\Delta\phi\rangle$ (MAD vs LSQ)	52.2	50.5	43.0
$\langle\Delta\phi\rangle$ (MIR vs LSQ)	50.6	50.7	52.8

[a]Data obtained from one of the three crystals were markedly inferior and these contributions are eliminated.
[b]Quality indicators obtained from the MAD analysis have been applied:
$|F_T|>10\sigma(F)$, $|{}^{\circ}F_A|>2\sigma(F)$, $|{}^{\circ}F_A|<200, \sigma(\Delta\phi)<50$, and Q<20 where Q is the average least-squares residual.
[c]Common reflections refer to the presence of phase information from each set (MAD, MIR and LSQ).

have very strong anomalous signals. Data were collected out to 2.5Å spacings at five wavelengths (1.9000, 1.8000, 1.7419, 1.7390 and 1.5000Å) but, unfortunately, due to the small size of the crystal and a rather weak

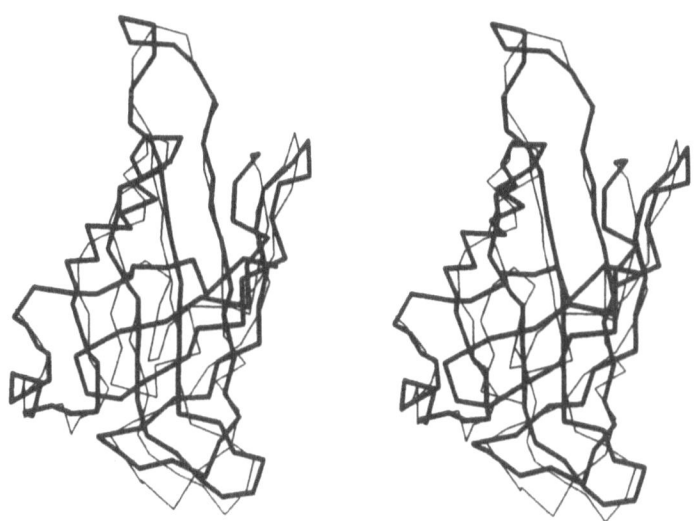

Fig.3. A stereo view of the alpha-carbon backbones for two independent tracings of the streptavidin structure. The model in bold lines agrees well with the refined structure and the one in light lines, which was traced without reference to the amino-acid sequence, differs only in one excursion.

beam, the high angle data proved to be unusable for phasing. However, with references to the P. aerogenes model, it was possible to solve the structure at 5Å resolution.

The iron-sulfur cluster positions were first located by direct methods (MULTAN), and were then refined to obtain phases. Two maps were computed, corresponding to the two enantiomorphous space-group possibilities, and the correct choice of enantiomer was found by observing an expected intramolecular pseudodiad symmetry. Then the P. aerogenes structure was placed onto the iron centers with coincident pseudodiad axes. This placement presented two ambiguities that were resolved by a density correlation function, rigid-body least squares refinements, and packing contact calculations. The structure determination to high resolution can now proceed by molecular replacement extension.

Streptavidin

The first application of MAD phasing to a molecule from a family without known three-dimensional structures was to streptavidin. Core streptavidin has been crystallized in three different modifications (Pahler et al., 1987) and one type was used for structure determination by the MAD method. These crystals are in space group I222 with a=95.07, b=105.45 and c=47.74Å, and they have two subunits of the 54,000 dalton tetramer per asymmetric unit. Streptavidin binds biotin with high affinity, and we found that the seleno analog of biotin would bind as well. Thus, we proceeded to analyze the crystal structure of selenobiotinyl streptavidin from MAD data.

An initial experiment conducted at SSRL involved a rather large and badly cracked crystal. A complete 3.0Å data set was measured at four wavelengths (0.9000, 0.9789, 0.9792 and 1.1000Å). Data processing required rather complicated scaling procedures to extract the anomalous signals and the results were quite noisy. Nevertheless, it was possible to interpret the Patterson map shown in Fig. 2 in terms of two selenium positions in space group I222 (not $I2_12_12_1$) that agreed with self-rotation function orientations for the tetrameric structure. Unfortunately, the resulting electron-density maps were uninterpretable.

In a second experiment carried out at the Photon Factory, we succeeded in obtaining a data set, complete to minimal Bragg spacings of 3.1Å, at three wavelengths (0.9000, 0.9795 and 0.9809Å). The outer data suffered considerable radiation damage, so an electron-density map was computed at 3.3Å resolution. This map proved to be excellent and was readily interpreted. One of us traced out the alpha-carbon backbone for one of the subunits in the asymmetric unit and another independently traced the backbone for the other subunit. These two tracings are compared in Fig. 3, and they agree very well with the refined structure. A complete atomic model was fitted to the 3.1Å map after molecular averaging refinement, and this model has now been refined using PROLSQ to R=0.17 against a 2Å data set collected with CuKα radiation.

Thioredoxin

Our crystallographic analysis of thioredoxin by MAD phasing methods is still in progress, but in a sense the major objective has already been accomplished. This study was undertaken to test the feasibility of incorporating selenomethionine into proteins in place of methionine. That this should be possible was known from work of Cowie and Cohen (1957) who successfully grew a met⁻ E. coli strain on selenomethionine. In light of modern molecular biology, arbitrary genes introduced into bacterial expression systems should then be induced to express selenomethionyl proteins. The abundance of methionine (1 in 60 residues on average) is such that it provides adequate MAD phasing strength (Hendrickson, 1985; 1987). As a first test of this approach, we collaborated with David Le Master at Yale who produced an appropriate methionine auxotrophic host for an expression plasmid for E. coli and bacteriophage T4 thioredoxin (Le Master and Richards, 1985). E. coli thioredoxin has one methionine in 108 residues and T4 thioredoxin has three in 87 residues. Composition analyses show essentially quantitative replacement with selenomethionine in the recombinant proteins produced in the selenomethionine incorporation system.

Crystals have been grown for both T4 and E. coli selenomethionyl thioredoxins, and MAD experiments have been performed with the E. coli protein. Crystals were grown under conditions similar to those used in the structural analysis by Holmgren et al. (1975), but PEG was used instead of MPD and the resulting crystals are not isomorphous (e. g., β changes from 113° to 100°) although they are clearly related. As with the Uppsala crystallization, copper is incorporated into the crystal lattice. We have measured 3Å data at SSRL from these crystals at five wavelengths about the selenium and copper edges. Thus, in addition to testing the applicability of the MAD phasing to selenomethionyl proteins, we will test our extension of the MAD algebraic analysis to multiple types of anomalous scatterers.

PROSPECTS

Absorption edges for elements of atomic numbers in the ranges from 20(Ca) to 47(Ag) and from 50(Sn) to 92(U), K and L respectively, are accessible to x-ray wavelengths in the range from 0.5 to 3.0Å. These elements include a number of metals that are integral components of metalloproteins, and a number of such applications have already been made (Harada et al., 1986; Korzun, 1987; lamprey hemoglobin and ferredoxin work described here; cucumber basic protein by M. Guss, E. Merritt and coworkers, personal communication). Most of the compounds commonly used as heavy-atom derivatives for MIR phasing are also useful in the MAD context, and, if well occupied and immobilized, these L-edge scatterers will produce impressive anomalous signals. One such study using a lanthanide derivative has been reported (Kahn et al., 1985). In addition to conventional heavy atoms, the MAD approach is applicable to elements that are normally not heavy enough for MIR phasing. As we have seen with streptavidin, selenium is such an example. This opens up what is perhaps the most exciting prospect for MAD structural analysis, namely, the possibility to generate proteins in which selenomethionine systematically replaces methionine. Brominated nucleic acids offer a similarly general approach for this class of macromolecular structure. Finally, one can conceive experiments based on the sulfur or phosphorous atoms that naturally occur in biological macromolecules. Such experiments are technically difficult due to the high absorption that occurs at the long wavelength (ca. 5Å) corresponding to these edges, and the large scattering angles that are required ($2\theta_{max}$ for 3Å data of approximately 120 Å). Nevertheless, the prospect is appealing.

ACKNOWLEDGMENTS

We thank Paul Phizackerley and Ethan Merritt for participating in the experiments at SSRL, Yoshinori Satow for his involvement in experiments at the Photon Factory, and Warner Love, Bill Orme-Johnson, Dave Le Master, and Apcel, Ltd for supplying protein samples. This work was supported in part by a grant, GM34102, from the National Institutes of Health. The facilities used at SSRL are supported by the Department of Energy and the National Institutes of Health.

REFERENCES

Adman, E. T., Sieker, L. C., and Jensen, L. H., 1973, The structure of a
bacterial ferredoxin, J. Biol. Chem., 248:3987.

Cowie, D. B., and Cohen, G. N., 1957, Biosynthesis by Escherichia coli of
active altered proteins containing selenium instead of sulfur,
Biochem. Biophys. Acta, 26: 252-261.

Harada, S., Yasui, M., Murakawa, K., Kasai, N., and Satow, Y., 1986, Crystal
structure analysis of cytochrome c' by the multiwavelength anomalous
diffraction method using synchrotron radiation. J. Appl. Cryst.
19:448.

Hendrickson, W. A., 1985, Analysis of protein structure from diffraction
measurements at multiple wavelengths. Trans. Amer. Cryst. Assn.
21:11-21.

Hendrickson, W. A., 1987, Anomalous scattering in macromolecular structure
analysis, in: "Crystallography in Molecular Biology," D. Moras, J.
Drenth, B. Strandberg, D. Suck and K. Wilson, eds., Plenum Press, New
York.

Hendrickson, W. A., and Teeter, M. M., 1981, Structure of the hydrophobic
protein crambin determined directly from the anomalous scattering of
sulphur, Nature, 290:107.

Hendrickson, W. A., Love, W. E., and Karle, J., 1973, Crystal structure
analysis of sea lamprey hemoglobin at 2Å resolution, J. Mol. Biol.
74:331.

Holmgren, A., Soderberg, B.-O., Eklund, H., and Branden, C.-I., 1975,
Three-dimensional structure of Escherichia coli thioredoxin-S2 to 2.8
Ångstroms resolution, Proc. Natl. Acad. Sci., USA 72:2305.

Honzatko, R. B., Hendrickson, W. A., and Love, W. E., 1985, Refinement of a
molecular model for lamprey hemoglobin from Petromyzon marinus, J.
Mol. Biol., 184:147.

Kahn, R., Fourme, R., Bosshard, R. Chaimdi, M. Risler, J. L., Dideberg, O.,
and Wery, J. P., 1985, Crystal structure study of Opsanus tau
parvalbumin by multiwavelength anomalous diffraction, FEBS Lett. 179:
133.

Karle, J., 1980, Some developments in anomalous dispersion for the structure
investigation of macromolecular systems in biology, Int. J. Quant.
Chem., 7:357.

Korzun, Z. R, 1987, The tertiary structure of azurin from Pseudomonas
denitrificans as determined by Cu resonant diffraction using
synchrotron radiation, J. Mol. Biol., 196:413.

Le Master, D. M., and Richards, F. M., 1985, 1H-15N heteronuclear NMR
studies of Escherichia coli thioredoxin in samples isotropically
labelled by residue type, Biochemistry, 24:7263.

Okaya, Y., and Pepkinsky, R., 1956, New formulation and solution of the
phase problem in x-ray analysis of noncentric crystals containing
anomalous scatterers, Phys. Review, 103:1645.

Pahler, A., Hendrickson, W. A., Gawinowicz-Kolks, M. A., Argarana, C. E..,
and Cantor C. R., 1987, Characterization and crystallization of core
streptavidin, J. Biol. Chem. 262:13933.

Phizackerley, R. P, Cork, C. W., and Merritt, E. A., 1986, An area detector
data acquisition system for protein crystallography using
multiple-energy anomalous dispersion techniques, Nucl. Instr. and
Meth., A246:579.

Satow, Y., 1984, Photon factory data collection systems for x-ray crystallography using synchrotron radiation, in: "Methods and Applications
in Crystallographic Computing," S. R. Hall and T. Ashida, eds.
Oxford Univ. Press, Oxford.

Templeton, D. H., and Templeton, L. K., 1985, Tensor x-ray optical
properties of the bromate ion, Acta Cryst., A41:133.

LAUE PHOTOGRAPHY FROM PROTEIN CRYSTALS

Keith Moffat, Donald Bilderback+, Wilfried Schildkamp,
Doletha Szebenyi and Tsu-yi Teng

Section of Biochemistry, Molecular and
Cell Biology, and +CHESS
Cornell University
Ithaca, NY 14853

INTRODUCTION

The recent advent of intense, polychromatic, pulsed synchrotron x-ray sources has prompted a re-examination of the Laue diffraction technique, particularly as applied to crystals of proteins and other macromolecules. This article reviews briefly the main aspects of the Laue technique, and how it may be applied to the general area of time-resolved crystallography. Applications have as their goal the elucidation of the structure of short-lived intermediates in such processes as enzymatic catalysis, ligand binding and release, and protein folding and unfolding. Knowledge of the structure of such intermediates is critical to a full understanding of molecular mechanisms of action, yet they are inaccessible to conventional x-ray techniques since their lifetimes are typically very much less than one second.

LAUE PHOTOGRAPHY

In Laue photography (Amoros et al., 1975), a polychromatic beam of x-rays from a synchrotron bending magnet, wiggler, or undulator source falls on a stationary crystal. A set of planes of spacing d then selects the wavelength λ from the polychromatic beam spanning λ_{min} to λ_{max} that will satisfy Bragg's Law, and gives rise to a Laue reflection in which integrated intensities are obtained automatically by integration over wavelength. Such a Laue diffraction pattern is characterized by several features.

First, The Laue pattern is a mapping of the distribution of central lines in reciprocal space, denoted rays, rather than of reciprocal lattice points as in a conventional, monochromatic oscillation or precession pattern.

A second feature is the problem of overlapping orders that arises from the simultaneous solution of Bragg's Law by the values (d,λ), $(d/2,\lambda/2)$, $(d/3,\lambda/3)$.... at the same scattering angle 2θ. This affects only a small fraction of the Laue pattern, never greater than 17% (Cruickshank et al., 1987). Thus, contrary to earlier belief (Wyckoff, 1924; Bragg, 1975), the Laue technique is indeed suitable for extraction of precise structure factors.

Third, for typical values of λ_{min} and λ_{max}, say 0.5 and 2.0 Å, many reciprocal lattice points are stimulated simultaneously (Bilderback et al., 1988) and give rise to a dense Laue pattern in which the reflections are closely spaced. This generates a spatial overlap problem, which is nevertheless tractable both experimentally (Hajdu et al., 1988b; Helliwell et al., unpublished results) and theoretically (Cruickshank et al., unpublished results). By suitable collimation of the incident beam and location of the detector, only a small fraction of the data will be spatially overlapped, and this fraction can be minimized further by more sophisticated profile-fitting algorithms (Hajdu et al., 1988b).

Fourth, by suitable orientation of the crystal, one can obtain a large fraction of the crystallographically unique data in a single Laue photograph. The higher the symmetry of the Laue group, the higher is this fraction (Chen and Moffat, unpublished results).

Fifth, exposure times for a Laue photograph are substantially shorter than for an oscillation photograph (Moffat et al., 1987) and lie in the 10-500 ms range for strongly scattering crystals such as those of lysozyme (Moffat et al., 1986) and phosphorylase b (Hajdu et al., 1987b) when a CHESS bending magnet or Daresbury wavelength-shifter wiggler source is used. If the time resolution Δt_R is defined as that exposure time which will permit the measurement of one-half of all intensities with a precision better than or equal to P%, then it may be shown (Moffat, 1989) that to a good approximation

$$\Delta t_R = \frac{2}{j_e i I_{Max}} \cdot \frac{V^2}{V_c} \cdot \frac{0.33 D^{*2}}{M_n \lambda_m^2} \cdot \frac{10^4 K^2}{P^2}$$

where I_{max} is the maximum of the spectral intensity distribution falling on the crystal that depends on the properties of both the source and the optics that deliver the photons emitted by the source to the crystal; i is the current circulating in the synchrotron; j_e is the (radius of the electron)^2; V is the unit cell volume; V_c is the crystal volume illuminated; D^* is the limiting resolution of the crystal in reciprocal space; M_n is the total molecular weight in the unit cell; λ_m is the mean wavelength, namely $0.5(\lambda_{min} + \lambda_{max})$; and K is a numerical factor > 1 which expresses the fact that a realistic, imperfect detector must accumulate $10^4 K^2/P^2$ photons to achieve a precision of measurement of P%. When experimental values are inserted into this equation, exposure times in good agreement with experiment are obtained (Table 1 of Moffat, 1989).

At CHESS, a realistic value of $i I_{max}/2$ is 5×10^9 photons (s.mm^2.10^{-3} Å)$^{-1}$ at 1Å, for a 200 μm aperture located 12 m from a bending magnet source with no focussing and a synchrotron current of 30 mA. Since a single bunch of electrons requires 2.56 μs for one revolution in CESR, this corresponds to an intensity per bunch of 30 mA of 1.3×10^4 photons (mm^2.10^{-3} Å)$^{-1}$. To obtain a Laue diffraction pattern from the x-rays emitted by a single bunch of electrons lasting for 120 ps (Moffat et al., 1987) intensities at least three orders of magnitude larger are required. This necessitates either much higher synchrotron currents plus focussing white beam optics with wiggler sources, or the use of undulators such as those planned for the Advanced Photon Source (APS) at Argonne*.

---

*Note added. Single bunch Laue patterns were obtained during the weeks following this Symposium, with an exposure time of 120 ps, using a prototype APS undulator installed at CHESS. Crystals of an indole alkaloid and the protein lysozyme were examined, and data were recorded on the Kodak storage phosphor detector (Szebenyi et al., 1988; Moffat et al., unpublished results).

A sixth feature is that when the first Laue diffraction patterns were obtained from protein crystals (Moffat et al., 1984; Bilderback et al., 1984; Helliwell, 1984, 1985), it was immediately clear that the signal-to-noise ratio was good: Laue reflections stood out distinctly from a low, uniform background. At first sight this is surprising, as the effect of a progressively wider bandpass is to stimulate more reflections, to increase the noise (background) around each reflection, but to leave the signal (integrated intensity) of each reflection unaltered. That is, the wider the bandpass, the worse is the signal-to-noise ratio, S_{Laue}, for each reflection. There is a corresponding effect in the monochromatic oscillation case: the larger the oscillation range, the worse is the signal-to-noise ratio, S_{mono}, for each reflection.

Since monochromatic oscillation data are deemed to have good signal-to-noise characteristics, how do such data compare quantitatively with Laue data? This may be answered by an extension of the treatment of integrated intensities in the Laue and monochromatic cases given by Kalman (1979). The background scattering is, to a good approximation, spherically symmetrical and thus given by $b^2(d*) = b^2(2\sin\theta \cdot k)$. Let the incident spectral intensity distribution be $I(k)$, corresponding to $I(\lambda)$ in the fifth feature above, where $k = 1/\lambda$; and let $I(k)b^2(2\sin\theta \cdot k) = B(k,\theta)$. Then it may be shown that (Moffat et al., unpublished data), under certain reasonable assumptions, the ratio of the S_{Laue} to S_{mono} for the Laue and monochromatic cases is given by

$$S_{Laue}/S_{mono} = \frac{B(k,\theta)}{\int_{k_{min}}^{k_{max}} B(k,\theta)dk} \cdot k\cot\theta \cdot \phi$$

where ϕ is the monochromatic oscillation range and k is the magnitude of the wave vector for a reflection stimulated at Bragg angle θ. As the wavelength range and bandpass diminish, k_{min} approaches k_{max} and the first term approaches $1/(k_{max} - k_{min}) = 1/\Delta k$. Hence

$$S_{Laue}/S_{mono} = (k/\Delta k) \cdot \cot\theta \cdot \phi = (\lambda/\Delta\lambda) \cot\theta \cdot \phi.$$

As expected, S_{Laue}/S_{mono} increases as ϕ increases, and decreases as the bandpass $\Delta\lambda/\lambda$ increases. Although its exact value depends on the form of $B(k,\theta)$ and on k_{min} and k_{max}, it is typically a number of order unity. That is, the signal-to-noise is comparable in Laue and monochromatic oscillation experiments. For equal numbers of stimulated reflections, $\phi = 0.6 D*(\lambda_{max} - \lambda_{min})$ where $D*$ is the diffraction limit of the crystal in reciprocal space (Moffat and Helliwell, unpublished), and then

$$S_{Laue}/S_{mono} = 0.6 D*\lambda_m \cot\theta.$$

For $D* = 0.5 \text{ Å}^{-1}$, $\lambda_m = 1 \text{ Å}$, $\theta = 10°$, this yields a value of 3.4 : the signal-to-noise is superior in the Laue experiment under these assumptions.

A final feature is that despite the necessity for applying wave-length-dependent corrections to each Laue intensity to obtain structure amplitudes, accurate values can indeed be obtained (Smith Temple and Moffat, 1987, unpublished results; Helliwell et al., 1988; Hajdu et al., 1988b).

TIME-RESOLVED CRYSTALLOGRAPHY

By virtue of the short exposure times, large numbers of stimulated reflections, good signal-to-noise, and production of integrated intensities from a stationary crystal, the Laue technique is well suited to time-resolved crystallography, in which the x-ray intensities change with time in response to a structural perturbation of the crystal (Wood et al., 1983; Bilderback et al., 1984).

Time-resolved crystallography is founded on the independence of the conformations of protein molecules in the crystal lattice. This independence comes from the weakness of the intermolecular interactions that stabilize the lattice. The principles are discussed in detail elsewhere (Moffat and Helliwell, 1989; Moffat, 1989). Briefly, it must be possible to initiate a structural reaction in a crystal uniformly, rapidly, and in a manner that does not damage the crystal, to monitor the evolution of the Laue intensities in time (with retention of isomorphism), and to analyze the time-dependent changes in structural terms. This might be accomplished either as a time-dependent difference Fourier (Bilderback et al., 1984; Moffat et al., 1986), or if circumstances permit, as a series of static difference Fouriers where each pertains to a particular structural intermediate (Moffat, 1989).

As yet, few results have been obtained on single crystals in this challenging area (Moffat, 1989). Time-resolved diffraction from less ordered systems such as protein solutions, or fiber and membrane arrays, is more readily studied and was reviewed recently (Gruner, 1987). The largest problems are likely to arise in reaction initiation. Except in a few specialized cases (Hajdu et al., 1987a, 1987b, 1988b) diffusion is too slow, and rapid photochemical techniques that either exploit the natural photosensitivity of certain crystals, or confer it on other crystals through the incorporation of photoactivable compounds, seem likely to be more generally applicable (Moffat et al., 1986; Hajdu et al., 1988a; Moffat and Helliwell, 1989; Moffat, 1989). An excellent example of the type of results that may be obtained, albeit so far on a very much slower time scale, is provided by studies that employed both monochromatic (Hajdu et al., 1987a) and Laue (Hajdu et al., 1987b) techniques to study catalysis in crystals of glycogen phosphorylase b. These revealed chemically interpretable differences at the active site, as the conversion of the poor substrate heptenitol to heptulose-2-phosphate proceeded, on a time scale of tens to hundreds of minutes.

CONCLUSIONS

The Laue technique is maturing: the ability to obtain substantial data sets very rapidly, and to extract accurate structure amplitudes, is now established. The major difficulties to be encountered in its application to time-resolved studies seem likely to be enzymological rather than crystallographic in nature. Can suitable reaction initiation schemes be devised? Will the crystals maintain their diffraction pattern throughout the reaction or will they rupture? What time resolutions are accessible? How may data be collected in a time-dependent mode with integrating or photon-counting detectors? What are the best strategies for data analysis? These and other questions must now be addressed, if the goal of "molecular movies" is to be attained.

ACKNOWLEDGMENTS

Supported by NIH grants GM29044, GM36452 and RR01646. We thank Boris Batterman, Ian Brooks, Howard Chen, Ying Chen, Alan LeGrand and Brenda Smith Temple for helpful discussions, and Durward Cruickshank and John Helliwell for a stimulating collaboration.

REFERENCES

Amoros, J. L., Buerger, M. J., and Canut de Amoros, M., 1975, "The Laue Method," Academic Press, New York.

Bilderback, D. H., Moffat, K., and Szebenyi, D. M. E., 1984, Time-resolved Laue diffraction from protein crystals: Instrumental considerations, Nucl. Instr. and Meth., 222:245.

Bilderback, D., Moffat, K., Owen, J., Rubin, B., Schildkamp, W., Szebenyi, D., Smith Temple, B., Volz, K., and Whiting, B., 1988, Protein crystallographic data acquisition and preliminary analysis using Kodak storage phosphor plates, Nucl. Instr. and Meth., A266:636.

Bragg, W. L., 1975, "The Development of X-ray Analysis," G. Bell and Sons LTD, London.

Cruickshank, D. W. J., Helliwell, J. R., and Moffat, K., 1987, Multiplicity distribution of reflections in Laue diffraction, Acta Cryst., A43:656.

Gruner, S. M., 1987, Time-resolved x-ray diffraction of biological materials, Science, 238:305.

Hajdu, J., Acharya, K. R., Stuart, D. I., McLaughlin, P. J., Barford, D., Oikonomakos, N. G., Klein, H., and Johnson, L. N., 1987a, Catalysis in the crystal: synchrotron radiation studies with glycogen phosphorylase b, EMBO J., 6:539.

Hajdu, J., Machin, P. A., Campbell, J. W., Greenhough, T. J., Clifton, I. J., Zurek, S., Gover, S., Johnson, L. N., and Elder, M., 1987b, Millisecond x-ray diffraction and the first electron density map from Laue photographs of a protein crystal, Nature, 329:178.

Hajdu, J., Acharya, K. R., Stuart, D. I., Barford, D., and Johnson, L. N., 1988a, Catalysis in enzyme crystals, TIBS, 13:104.

Hajdu, J., Greenhough, T. J., Clifton, I. J., Campbell, J. W., Shrive, A. K., Harrison, S. C., and Liddington, R. C., 1988b, "Brookhaven Symposium on Quantitative Biology," this volume.

Helliwell, J. R., 1984, Synchrotron x-radiation protein crystallography: instrumentation, methods and applications, Rept. Prog. Phys., 47:1403.

Helliwell, J. R., 1985, Protein crystallography with synchrotron radiation, J. Mol. Struct., 130:63.

Helliwell, J. R., Habash, J., Cruickshank, D. W. J., Harding, M. M., Greenhough, T. J., Campbell, J. W., Clifton, I. J., Elder, M., Machin, P. A., Papiz, M. Z., and Zurek, S., 1988, The recording and analysis of synchrotron X-radiation laue diffraction photographs from the protein pea lectin and small molecule crystals using a broad wavelength bandpass, 0.2 Å $< \lambda < 2.5$ Å, J. Appl. Cryst., in press.

Kalman, Z. H., 1979, On the derivation of integrated reflected energy formulae, Acta Cryst., A35:634.

Moffat, K., 1989, Time-resolved macromolecular crystallography, Ann. Rev. Biophys. Biophys. Chem., 18:309.

Moffat, K., Szebenyi, D., and Bilderback, D., 1984, X-ray Laue diffraction from protein crystals, Science, 223:1423.

Moffat, K., Bilderback, D., Schildkamp, W., and Volz, K., 1986, Laue diffraction from biological samples, Nucl. Instr. and Meth., A246:627.

Moffat, K., Bilderback, D., and Schildkamp, W., 1987, Protein crystallography with Laue geometry on wigglers and undulators, in: "Workshop on PEP as a Synchrotron Radiation Source," R. Coisson and H. Winick, eds., SSRL, Stanford, California.

Moffat, K., Cruickshank, D. W. J., and Helliwell, J., 1987, Laue diffraction from protein crystals theoretical aspects, in: "Biophysics and Synchrotron Radiation Research," A. Bianconi and A. Congiu Castellano, eds., Springer-Verlag, New York.

Moffat, K., and Helliwell, J. R., 1989, The laue method and its use in time-resolved crystallography, in: "Applications of Synchrotron Radiation," E. Mandelkow, ed., Springer-Verlag, New York (in press)

Smith Temple, B. R., and Moffat, K., 1987, Computational aspects of protein crystal data analysis, in: "Proceedings of a Daresbury Study Weekend," J. R. Helliwell, P. A. Machin, and M. Z. Papiz, eds., DL/SCI/R25, SERC, Daresbury.

Szebenyi, D. M. E., Bilderback, D., LeGrand, A., Moffat, K., Schildkamp, W., and Teng, T.-Y., 1988, <u>Trans. Amer. Cryst. Assoc</u>., in press.

Wood, I. G., Thompson, P., and Mathewman, J. C., 1983, A crystal structure refinement from Laue photographs taken with synchrotron radiation, <u>Acta Cryst</u>., B39:543.

Wyckoff, R. W. G., 1924, "The Structure of Crystals," Chemical Catalog Co., New York.

LAUE CRYSTALLOGRAPHY: APPLICATION TO VIRUS CRYSTALS

Janos Hajdu[1], Trevor J. Greenhough[2,3], Ian J. Clifton[2], John
W. Campbell[2], Annette K. Shrive[3], Stephen C. Harrison[4] and
Robert C. Liddington[4]

[1]Laboratory of Molecular Biophysics, Oxford University,
U. K.; [2]SERC Daresbury Laboratory, Daresbury, U. K.
[3]Department of Physics, University of Keele, U. K.; [4]Howard
Hughes Medical Institute and Department of Biochemistry and
Molecular Biology, Harvard University, Cambridge, MA, USA

INTRODUCTION

Seventy-six years ago, Friedrich, Knipping and von Laue (1912) demonstrated the diffraction of X-rays on a crystal of copper sulphate using
white X-radiation. With a stationary crystal and white X-radiation, a
large number of lattice planes diffract simultaneously as the Bragg condition is satisfied for each of these planes by at least one wavelength of
the spectrum. The wider the wavelength range the more lattice planes
become accessible to the Laue geometry (Amoros et al., 1975; Cruickshank et
al., 1987). Moreover, with crystals of high symmetry, a full data set may
be recorded on a single photograph. The Laue technique did not become a
method of data collection because conventional X-ray sources do not have a
satisfactory spectrum and because of the difficulties in unravelling the
complicated diffraction patterns.

There has been a recent revival of the method at synchrotron radiation
sources (Helliwell, 1984; Moffat et al., 1984). The prime driving force
behind this revival was the possibility of collecting full diffraction data
sets in very short times. This led to substantial developments in both the
necessary experimental and computational techniques. With existing synchrotron radiation facilities, data acquisition on the millisecond-second
scale has already been achieved. At high energy storage rings (e.g., the
PEP in Stanford), the total time needed to collect a full data set from a
protein or virus crystal may be as short as 10-100 picoseconds. With data
acquisition times well below the NMR time scale, X-ray crystallography may
enter a new era.

The first successful use of the Laue technique with synchrotron radiation was in the determination of a small molecule structure by Wood,
Thompson and Matthewman (1983). The diffraction image was recorded with
radiation generated on a dipole bending magnet of SRS Daresbury. Later,
the full white radiation produced on the superconducting wiggler magnet of
the SRS was used in experiments that produced the first electron density
maps from Laue diffraction photographs of a protein crystal, glycogen phos-
phorylase-b (Hajdu et al., 1985, 1987). The same source was used in a

study to find heavy atom positions in crystals of xylose isomerase (Farber et al., 1988). Electron density maps calculated from these data compare favourably with maps calculated from monochromatic data sets.

This paper discusses the possibility of using the Laue method for studying virus crystals and gives a preliminary account of a flow cell experiment with a crystal of tomato bushy stunt virus. The experiment described here indicates that molecular processes like those associated with the disassembly of a virion, drug binding, or infectivity may be studied this way.

KINETIC LAUE EXPERIMENTS WITH TOMATO BUSHY STUNT VIRUS CRYSTALS

The structure of tomato bushy stunt virus is known to 2.9Å resolution (Harrison et al., 1978). The virus undergoes a conformational change when two calcium ions are removed from the subunit interfaces. This is followed by a swelling if the pH is raised above 7.0 (Robinson and Harrison, 1982). Icosahedral viruses crystallize in high symmetry crystal forms so that a single Laue photograph may contain an almost complete data set.

The Synchrotron Radiation Source

The full unfiltered, unfocussed white radiation from the wiggler magnet (Greaves et al., 1983) of beamline 9 (station 9.7) at SRS Daresbury was used in the experiment. The synchrotron was operating at 2.0 GeV energy with a circulating electron current of 150-200 mA in multi-bunch mode. The effective wavelength range was 0.25 Å - 2.1 Å. The very dense nature of the virus Laue patterns made it necessary to use a narrow and well-collimated beam. Considering optimal geometric conditions for the data collection (Arndt, 1968), a tungsten apperture with a 100 µm opening was placed in the beam pipe upstream from the camera (and about 9 m from the source) and a tungsten collimator with an opening of 120 µm was used in the standard Arndt-Wonacott camera at a distance of 1 m from the 100 µm aperture. This produced an average spot size of 100-110 µm on the film. Beam "lifetime" was 30 hours. The experiments were performed in January and April 1988 and benefitted from the update of the ring with the high brightness lattice.

Experimental Set-up

A crystal of tomato bushy stunt virus (I23, a=380Å, size=0.4mm x 0.4mm x 0.25mm) was mounted by carefully wedging it into a slightly tapered quartz capillary. The capillary was subsequently cut below and above the crystal. Fine silicon rubber tubing was attached to both ends of the capillary tube and the assembly was transferred to a flow cell holder at 21C. The flow rate of the solutions was set to 75 µl/hour allowing for a complete exchange of the liquid around the crystal in every second. With this experimental set-up we hoped to study the time course of the conformational change induced by the removal of calcium and the first steps in the process of swelling once the pH had been raised.

Calculations suggest that a crystal orientation about 10 degrees off the 110 axis allows the recording of more than 90% of the unique set of reflections to 3Å resolution (Table 1). The crystal was alligned optically to fit this orientation and Laue photographs were taken before, during, and after the removal of calcium and after the pH was raised to 6.5. Exposure times of 4x6 s were used. Each film pack consisted of 6 CEA Reflex 25 films (5 inch x 5 inch) without any interleaving foils.

Table 1. Theoretical Distribution of the Unique Set of Reflections
Accessible for the Laue Geometry on the Photograph of Fig. 2.

Resolution shell (in Angstroms)	% unique recorded
infinity - 10	46.3
infinity - 8	58.2
infinity - 6	72.7
infinity - 4	85.9
infinity - 3	92.7
infinity - 2	97.4

The first photograph, with crystal in its mother liquor (1M ammonium sulphate, pH 5.0), showed sharp spots. Ten minutes later, the mother liquor around the crystal was exchanged with a solution containing 2 mM EDTA, 1.0 M ammonium sulphate, pH 5.0 in order to remove the structurally bound calcium ions. The photograph, taken 30 seconds after the introduction of this solution, had slightly streaked spots. These spots could be due to a transient order-disorder transition (the next photograph had well-ordered spots), or it could be due to radiation damage since the crystal

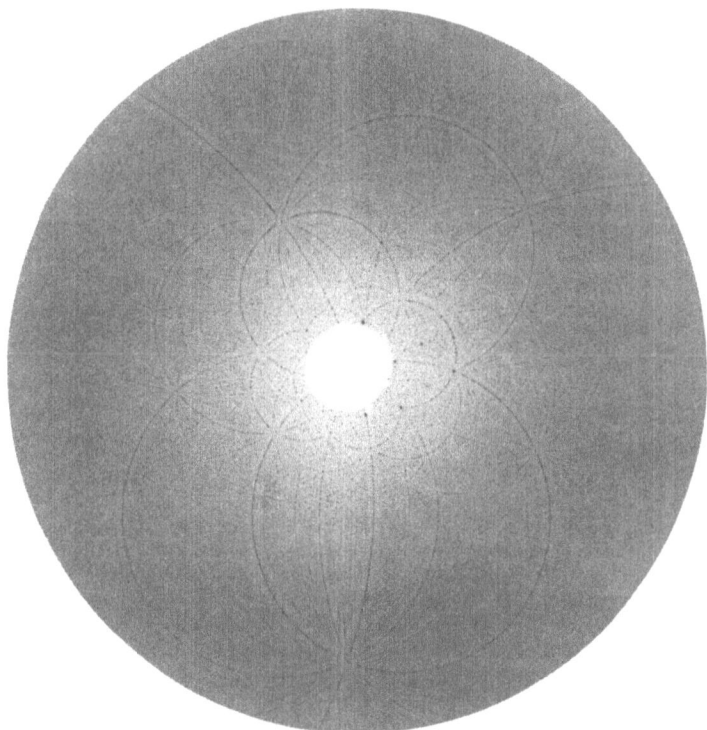

Fig. 1. Laue diffraction photograph of tomato bushy stunt virus in 1.0 M ammonium sulphate, 2 mM EDTA, pH 5.0. Crystal to film distance: 223.78 mm, refined misorientation angles: PHIX=135.22, PHIY=37.39, PHIZ=-5.79, effective wavelength range: 0.25-2.1Å.

was not translated between the first 2 exposures. On the following photograph, some 4 hours later, a fresh part of the crystal was exposed. The picture showed a perfectly ordered diffraction pattern (Fig. 1). Data processing statistics described in this paper relate to this picture. After 5 minutes, the crystal was translated again and the pH of the EDTA - ammonium sulphate solution was raised from 5.0 to 6.5. Another photograph was taken 30 seconds after raising the pH. This produced streaked spots indicating yet another order-disorder transition, but the crystal did not recover from this one.

DATA PROCESSING

A report on data processing from the photograph corresponding to the calcium-free form of the virus (Fig. 1) is given below. The Daresbury set of Laue programs (Machin, 1987; Campbell et al., 1987) was used after extensive modifications to cope with the extremely dense virus Laue pattern.

Determination of Crystal Orientation and Prediction of the Laue Pattern

An autoindexing routine, based on the location of nodal spots on the photograph (Elder, 1986), was used to determine a starting orientation of the crystal. The mismatch between the predicted and observed pattern was minimized by refinement of the missetting angles (PHIX, PHIY, PHIZ), crystal to film distance, and camera constants to produce a final predicted pattern that fit the observation with an RMS displacement of 0.029 mm (about one raster unit). Unit cell parameters were kept constant throughout. Input parameters included unit cell dimensions, crystal to film distance (224 mm), wavelength range (0.25-2.1 Å) nominal resolution limit of the data (3.0 Å) and spatial overlap cut off (0.12 mm). The wavelength range and the resolution limit used to predict the Laue pattern were deliberately overestimated (i.e., wider range, higher resolution). This overestimation is a safety measure to prevent prediction of multiplets as singlets. With the refined parameters, a set of 203,618 reflections was generated out of which 71,217 reflections were closer than 0.12 mm (spatial overlaps). The remaining reflections made up 104,812 separated spots on the film out of which 15,810 spots were multiplets leaving 89,002 singlets. Because of the uncertainities in deconvolution of multiplets, only the singlets were used after the integration of the films.

Integration With Profile Fitting

Films were scanned on a Scandig 3 rotating drum microdensitometer at 25 μm raster size. The standard profile for each of 17 integration bins distributed over the area of the film was determined by a preliminary pass through the data. The profile determined with this method (Fig. 2) is not the average profile as defined by Rossmann (1979) but the profile of the average defined by Greenhough and Suddath (1986). All reflections were integrated with profile fitting and the profiles were weighted based on the position of the reflection in the bin. The integration bins were arranged radially in two concentric circles. This fit the most common variation in spot profiles on a Laue photograph: radial streaking. Table 2 gives a summary of the integration.

Deconvolution of Overlapping Reflections

An outlier rejection scheme to determine an improved least squares background plane for the reflections was developed and tested in the integration stage. This procedure enables a revised version of the profile-fitting technique to be used for determining the total integrated intensity of spatially overlapped reflections. Many more good spots can then be

Table 2. Summary of Integration

SPOTS (1) PREDICTED: 160747

SINGLETS	139871	MULTIPLETS	20876
Separated	89002	Separated	15810
Overlapped	50869	Overlapped	5066

SEPARATED SPOTS MEASURED

	A Film	B Film	C Film	D Film	E Film	F Film
TOTAL	104812	46525	31222	23405	17359	12528
SINGLETS	89002	32840	19583	13695	9529	6334
REJECTED (OVERLOADS)	2219	1030	649	362	219	141
REJECTED (ZERO BACKGROUND)	11	4	5	1	2	2
REJECTED (BGRATIO, I/SIGI)	7110	597	137	83	42	24

(1) A distinction is made between predicted reflections and actual spots
    since harmonic overlaps produce a single spot containing more than one
    reflection.

integrated (in this case, 55,935 more spots in addition to the previously
measured 104,812 spots on the A-film), the majority of which are singlets
(50,869) thus making most of the overlapped data available. In this paper
however, results without this overlap deconvolution are described.

```
0   0   0   0   0   0   0   0   0   0   0

0   1   3   1   3   4   3   4   3   2   0

0   2   6   6  14  21  21  15   7   2   0

0   3   9  22  44  61  58  39  18   3   0

0   4  13  39  76  93  86  62  26   4   0

1   6  15  43  86 100  89  63  23   4   0

0   4  12  34  68  80  69  45  13   4   0

0   3   8  17  33  40  34  18   5   2   0

0   2   7   8  11  12  11   5   3   2   0

0   1   3   2   3   4   3   2   1   0   0

0   0   0   0   0   0   0   0   0   0   0
```

Fig. 2. Average spot profile in one of the 17 integration bins.
         1411 strong spots were used to form this profile.

Interfilm Scaling

Two strategies were applied: either the standard AFSCALE program of the Daresbury Package was used to derive wavelength-dependent interfilm scale factors for the six films in the pack, or each film in the pack was put through an individual wavelength normalisation procedure which resulted in 6 normalised data sets. The data sets were then scaled and merged using an anisotropic temperature factor. Results with data from the first approach are described here. After merging, a set of 54,827 merged reflection was obtained with I > 0, 42,303 had I > 1 sigma, 30,715 had I > 2 sigma and 19,514 had I > 3 sigma.

Wavelength Normalisation Using Symmetry-Related Reflections

This step serves to compensate for wavelength-dependent factors such as incident intensity, absorbtion, Lorentz and polarisation factors, obliquity and detector response. There are several ways for the determination of the normalisation curve: for example, the curve can either be derived externally, by using a monochromatic reference set (Machin, 1985), or internally, using the Laue data only (Campbell et al., 1986). In the first case, the Laue reflections are divided into wavelength bins and scaled, bin by bin, to a monochromatic reference set (LAUESCALE). In the second case, intensity differences of symmetry-related Laue reflections (that come up at different wavelengths because of a misorientation of the crystal) can be used to determine the normalisation curve internally (LAUENORM, Campbell et al., 1986).

Since reflections of longer wavelengths are extremely sensitive to small variations in parameters, rejection criteria for the modified LAUENORM program were set to exclude reflections with wavelength > 1.7Å (only 1701 positive reflections were lost). Before wavelength normalisation, the merging R-factor [defined as R = SUM(ABS(I(I) - (I(J)))/(I(I) + I(J))] was 39.8% for data with intensities greater than 1 sigma. After normalisation, this value dropped to 16.1% for all data and to 11.5% for data with I > 3 sigma. These values compare well with corresponding monochromatic measurements (Hogle et al., 1983). The normalisation was based on the analysis of 5975 symmetry-related reflections. A unique set of 34,855 reflections resulted from all measurements with I > 1 sigma, 25,800 measurements with I > 2 sigma and 16,565 from measurements with I > 3 sigma. Normalised reflections with intensities greater than 2 sigma values were scaled to the native monochromatic data set using an isotropic temperature factor with the program RSTATS. This gave a mean isomorphous difference of 20%, similar to the difference found with the monochromatic data only.

Calculation of Electron Density Maps

Data to 3.5Å resolution were included. A further filter was applied to the differences between the monochromatic native and the Laue derivative data. This excluded reflection pairs where the difference was less than one sigma of the difference (Hogle et al., 1983). A difference map (Fig. 3) was calculated with coefficients F(Laue)-F(mono) and native monochromatic phases, and then was icosahedrally averaged (Bricogne, 1976). The quality of the map was as good as the quality of the monochromatic difference map of Hogle et al. (1983). The map clearly shows the locations of two Ca^{+2} ions at each of the trimer interfaces (Fig. 3). In addition, we observed features corresponding to an apparent change in one of the liganding loops at the A/C interface. This change appears to involve some disorder in residues 223-226 that occurs on removal of the bound ions. No comparable feature was seen at the other interfaces, probably reflecting the small but significant differences among these "quasi-equivalent" positions.

LIMITATIONS

One of the limitations of the Laue technique is that crystals with increased mosaicity produce streaked reflections instead of small, well defined spots on the film. Streaking makes the dense diffraction patterns particularly difficult to analyse. Two of the photographs in the sequence with tomato bushy stunt virus show streaking and have therefore been excluded from the current analysis. Although data processing from the other films is not yet complete, it is clear from the preliminary results that the Laue method can be used on systems with very large unit cell dimensions. The data compare well with other Laue data sets collected on smaller proteins like phosphorylase (Hajdu et al., 1987), xylose isomerase (Farber et al., 1988), gamma-chymotrypsin (Almo et al., in preparation) and turkey egg white lysozyme (Howell et al., in preparation). Additional work is required to analyse the possible order-disorder-order transition observed upon the removal of bound calcium.

OUTLOOK

Time-resolved X-ray studies on macromolecules and viruses may be carried out on systems with intermediate half-life periods ranging from hours, to a few seconds, and perhaps down to picoseconds (stroboscopic experiments using the time structure of the storage ring). These methods have great potential in the three-dimensional study of a variety of phenomena, not only in protein crystallography but in other areas of solid state physics too.

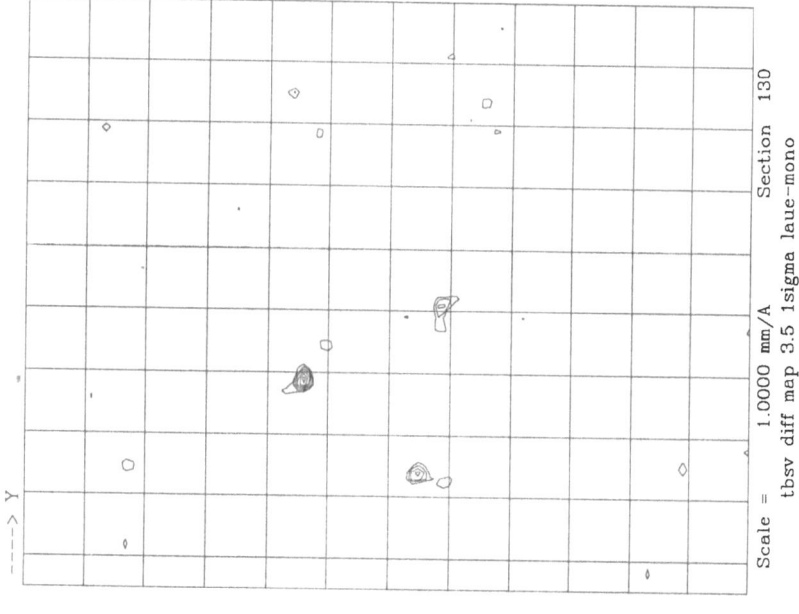

Fig. 3. Difference Fourier map between monochromatic native and Laue derivative data showing the calcium sites in tomato bushy stunt virus. The map shows a section normal to z=130A. The grid has 10 Å divisions.

With existing technology several methods are available for the three-dimensional study of transient structures (Hajdu et al., 1988): (1) collection of a sequence of complete monochromatic data sets if the lifetime of the intermediate is longer than about 30 minutes; (2) collection of a sequence of Laue photographs if the lifetime of the intermediate is longer than about 3 secs; (3) collection of streaked Laue photographs if the lifetime is of the order of 0.1 secs (The film is translated so that each diffraction spot is drawn into a streak in which the intensity at a particular position corresponds to a particular time. This saves the time needed to change the film.) and (4) stroboscopic experiments utilising the pulsed nature of the synchrotron radiation for intermediates with lifetimes of a few hundred picoseconds.

In a stroboscopic experiment at the SRS, a crystal will have to survive 500 million excitations during a 1 second long exposure. High energy storage rings might allow the recording of a full Laue pattern by a single passage of a single electron bunch resulting in an overall exposure time of about 10-60 ps and an increase of 10-15 orders of magnitude in data rates. Elementary molecular processes could be studied this way. The combination of the crystallographic experiments with spectroscopic techniques could provide a deeper understanding of dynamic processes involved in protein function. The results may have implications, among others, in understanding thermal stability, the role of solitons in function, structrues of excited molecules, factors important in molecular dynamic calculations, or details of catalytic processes. The short data collection times could allow for better synchronization since processes could be studied shortly after the trigger. This will result in a reduced Boltzmann effect, i.e., the population distribution will be sharper. Moreover, with exposure times in the ps range, radiation damage will have practically no effect on data quality since 'excised atoms' will not have time to leave their original position far behind (the distance light travels in a ps is 0.3 mm). Developments in fast data collection methods are of great importance offering a possible route to the dissection and detailed understanding of atomic and molecular events in crystals. Equally important developments are needed in computer software to enable mixtures of structural states to be analysed and in the biochemistry and chemistry for the design of suitable triggers, substrates, and ligands.

ACKNOWLEDGMENTS

Work was supported by grants from the SERC and MRC (to J.H.) and by HHMI and NIH (grant CA-13202 to S.C.H.). We are grateful to Professor Leslie L. Green, the Director of SRS Daresbury for making available spare beam time at station 9.7 to J.H.. We wish to thank Simon Clarke, Ian Glover, Lynne Howell, Colin Jackson, Gordon Locke-Scobie, Phil More, Colin Nave, Alf Neild, Miroslav Papiz, and the numerous shift fitters for their help during this work.

REFERENCES

Amoros, J. L., Buerger, M. J., and Canut de Amoros, M., 1975, "The Laue Method," Academic Press, New York.
Arndt, U. W., 1968, The optimum strategy in measuring structure factors, Acta Cryst., B24:1355.
Bricogne, C., 1976, Methods and programs for direct-space exploitation of geometric redundancies, Acta Cryst., A32:832.
Campbell, J. W., Moffat, K., and Helliwell, 1986, Information Quarterly for Protein Crystallography, Daresbury Laboratory, 18:23.

Campbell, J. W., Clifton, I. J., Elder, M., Machin, P. A., Zurek, S., Helliwell, J. R., Habash, J., Hajdu, J., and Harding, M. M., 1987, in: "Springer Series in Biophysics, Biophysics and Synchrotron Radiation," A. Bianconi, and A. Congiu Castellano, eds., Springer Verlag, Berlin, Heidelberg, New York, London, Paris, Tokyo, 2:52.

Cruickshank, D. W. J., Helliwell, J. R., and Moffat, K., 1987, Multiplicity distribution of reflections in laue diffraction, Acta Cryst., A43:656.

Elder, M., 1986, Information Quarterly for Protein Crystallography, Daresbury Laboratory, 19:31.

Farber, G. K., Machin, P. A., Almo, S., Petsko, G. A., and Hajdu, J., 1988, X-ray Laue diffraction from crystals of xylose isomerase, Proc. Nat. Acad. Sci. USA, 85:112.

Friedrich, W., Knipping, P., and Laue, M. von, 1912, Sitzungsberichte der Math. Phys. Klasse, (Kgl.) Bayerische Akademie der Wissenschaften, Munchen, 303-322.

Greaves, G. N, Bennett, R., Duke, P. J., Holt, R., and Suller, V. P., 1983, X-ray optics and spectral brightness of the superconducting SRS wiggler, Nucl. Inst. Meth., 208:139.

Greenough, T. J., and Suddath, F. L., 1986, Oscillation camera data processing. 4. Results and recommendations for the processing of synchrotron radiation data in macromolecular crystallography, J. Appl. Cryst., 19:400.

Hajdu, J., and Stuart, D. I, 1985, Information Quarterly for Protein Crystallography, Daresbury Laboratory, 15:17.

Hajdu, J., Machin, P., Campbell, J. W., Greenhough, T. J., Clifton, I., Zurek, S., Gover, S., Johnson, L. N., and Elder, M., 1987, Millisecond X-ray diffraction and the first electron density map from laue photographs of a protein crystal, Nature, 329:178.

Hajdu, J., Acharya, K. R., Stuart, D. I., Barford, D., Johnson, L. N., 1988, Catalysis in enzyme crystals, TIBS, 13:104.

Harrison, S. C., Olson, A. J., Schutt, C. E., Winkler, F. K., and Bricogne, G., 1978, Tomato bushy stunt virus at 2.9Å resolution, Nature, 276:368.

Helliwell, J. R., 1984, Synchrotron x-radiation protein crystallography: instrumentation, methods and applications, Rep. Prog. Phys., 47:1403.

Hogle, J., Kirchhausen, T., and Harrison, S. C., 1983, Divalent cation sites in tomato bushy stunt virus. Difference maps at 2.9Å resolution, J. Mol. Biol., 171:95.

Machin, P. A., 1985, Information Quarterly for Protein Crystallography, Daresbury Laboratory, 15:1.

Machin, P. A., 1987, Computational aspects of protein crystal data analysis, J. R. Helliwell, P. A. Machin, M. Z. Papiz, eds., Daresbury Laboratory, DL/SCI/R25, 75-83.

Moffat, K., Szebenyi, D. M. E., and Bilderback, D. H., 1984, X-ray Laue diffraction from protein crystals, Science, 223:1423.

Robinson, I. K., and Harrison, S. C., 1982, Structure of the expanded state of tomato bushy stunt virus, Nature, 297:563.

Rossmann, M. G., 1979, Processing oscillation diffraction data for very large unit cells with an automatic convolution technique and profile fitting, J. Appl. Cryst., 12:225.

Wood, I. G., Thompson, P., and Matthewman, J. C., 1983, A crystal structure refinement from laue photographs taken with synchrotron radiation, Acta Cryst., B39:543.

PREPARATION OF MAGNETICALLY ORIENTED SPECIMENS

FOR DIFFRACTION EXPERIMENTS

Lee Makowski

Department of Physics
Boston University
Boston, MA 02215

INTRODUCTION

Macromolecular assemblies that are exceedingly anisotropic in shape present substantial problems for structural analysis using diffraction techniques. The subunits that make up these structures are frequently difficult to crystallize, and when they do, may take on conformations in the crystal that are distinct from their conformation in the assembly. Structural analysis of the intact assemblies suffers from the fact that the assemblies cannot, in general, be crystallized. Consequently, noncrystalline specimens must be used for structural studies.

Analysis of diffraction data from noncrystalline specimens is frequently limited by the degree of orientation that can be obtained from the specimens (Makowski, 1978). Orientation of an anisotropically shaped object can be obtained by any method that applies a torque to the object. For instance, oriented specimens have been obtained by the use of flow, centrifugation, electric and magnetic fields. Spontaneous orientation may occur on precipitation or the formation of a liquid crystalline phase.

Magnetic orientation has been used on a wide variety of macromolecular assemblies from whole retinal rods (Chalazonitis et al., 1970) to filamentous bacterial viruses (Torbet and Maret, 1979). Double-stranded DNA and sickle cell hemoglobin fibers orient perpendicularly to magnetic fields. Polymerization of fibrinogen in a magnetic field leads to the formation of highly oriented specimens of fibrin (Freyssinet et al., 1983). Even icosahedral viruses exhibit magnetically induced birefringence in a strong magnetic field (Torbet, 1986; Torbet et al., 1986).

In most biological specimens, the origin of the torque causing orientation in a magnetic field is diamagnetic anisotropy. When an external magnetic field is applied to an electron associated with a nucleus, its motion is changed so as to induce a magnetic moment in the direction opposite to the applied field. This induced moment generates the well known weak repulsion of diamagnetic materials from magnetic fields. When a molecule or assembly has anisotropic diamagnetic properties, the same effect tends to orient the molecule in the magnetic field.

In this paper, the physical basis of diamagnetic anisotropy in biological molecules is reviewed; the application of magnetic fields to the

Table 1. Diamagnetic Anisotropy of Chemical Groups

	Molar $\Delta\chi_m(\times10^6 cm^3/mol)$	Per Molecule $\Delta\chi_0(\times10^{29}erg/Gauss^2)$
Benzene	-59.7	- 9.91
Trp	-95.3	-15.83
Tyr, phe	-60.0	- 9.97
DNA(bp)	-60.0	-10.0
Peptide bond	- 5.4	- 0.89

orientation of filamentous bacteriophages is discussed; and an analysis is made of the size of the cooperative unit required to obtain a high degree of orientation in a magnetic field.

PHYSICAL BASIS OF DIAMAGNETIC ANISOTROPY IN BIOLOGICAL MOLECULES

Substantial diamagnetic anisotropy occurs in molecular groups where there are delocalized electrons. For instance, when a benzene ring is inserted into a magnetic field with its plane perpendicular to the magnetic field, a strong ring current is induced, resulting in a strong induced magnetic moment opposing the applied field. The induced moment tends to be repelled from the field. When the benzene ring is inserted with its plane parallel to the applied field, the induced moment is much smaller, and the associated repulsive force also is smaller. Consequently, a benzene ring tends to orient with its plane parallel to a magnetic field. Observed diamagnetic anisotropy in proteins may be attributed to induced ring currents in aromatic side chains (Pauling, 1936) and the diamagnetic properties of the peptide bond (Worcester, 1978; Pauling, 1979). DNA base pairs appear to have a diamagnetic anisotropy very similar to that of the benzene ring. Table 1 lists diamagnetic anisotropies for molecular groups commonly found in biological assemblies.

Diamagnetic anisotropy occurs because the motion of bonding electrons in a molecule is anisotropic. In general, the diamagnetic susceptibility is a tensor. However, the treatment can be simplified by assuming that the particles under consideration have axes of rotational symmetry. In that case, the diamagnetic anisotropy, χ, is defined to be the difference between the susceptibilities parallel and perpendicular to the axis of symmetry; $\Delta\chi = \chi_{||} - \chi_\perp$. Various definitions and units of diamagnetic susceptibility are used in the literature and their relationships to one another are reviewed by Maret and Dransfield (1985).

For anisotropic structures with axial symmetry, the magnetic energy as given by Worcester (1978) is

$$E(\theta,B) = -B^2[(\chi_{||} - \chi_\perp)cos^2\theta +\chi_\perp]/2. \qquad (1)$$

Expanding the cosine and keeping only quadratic terms, for relatively highly oriented specimens, the angular dependence of the magnetic energy is

$$E(\theta,B) \; \alpha \; -B^2\Delta\chi\theta^2/2. \qquad (2)$$

Consequently, the probability of an assembly being· oriented at an angle, θ, to the applied field is proportional to

$$exp(B^2\Delta\chi\theta^2/2kT). \qquad (3)$$

X-ray diffraction measurements of the orientation of macromolecular assemblies in a magnetic field indicate that the angular distribution often closely approximates that of a Gaussian;

$$\exp(-\theta^2/2\sigma^2).\qquad\qquad(4)$$

Equating the exponents in equations (3) and (4), we can relate the anisotropic diamagnetic susceptibility to the observed angular distribution

$$\Delta\chi \frac{-kT}{B^2\sigma^2}.\qquad\qquad(5)$$

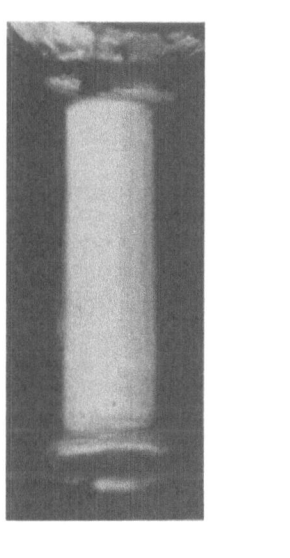 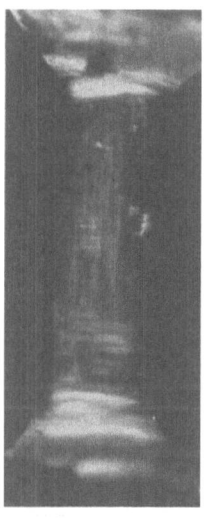

1a   1b

Fig. 1.   Optical micrographs of a fiber of filamentous bacteriophage Pf1 between crossed polarizers. (a) Axis of the fiber tilted 45° to the polarizer axes exhibiting strong birefringence. (b) Axis of the fiber parallel to polarizer axes showing nearly complete extinction. The fiber is approximately 300 microns in diameter.

Observed values of σ result in derived values for $\Delta\chi$ much larger than would be expected for a single molecular group, or even for a single assembly. Consequently, the anisotropy derived using equation (5) must relate to a group of assemblies orienting cooperatively. An estimate of the maximum possible diamagnetic anisotropy for a single assembly may be made by using the values in Table 1 and the chemical composition of the assembly. From this number and the derived anisotropic susceptibility, (5), an estimate for the minimum number of assemblies orienting cooperatively can be derived. This number provides us with a measure of the importance of liquid-crystalline cooperative behavior in the formation of oriented specimens for diffraction studies.

APPLICATION TO FILAMENTOUS BACTERIOPHAGES

We used two types of specimens for diffraction studies of filamentous bacterial viruses: fibers and oriented solutions. The fibers were formed by placing 8-10 µlitres of virus solution at a concentration of 30-50 mg/ml between two glass rods separated by about a millimeter (Nave et al., 1981). The solution was then allowed to dry for several days in a magnetic field of 2.5 or 5.8 Tesla. Optical micrographs of a fiber of filamentous bacteriophage Pf1 are shown in Figs. 1a and 1b. Using x-ray beams 0.2mm in diameter, diffraction patterns were obtained with Gaussian orientation distributions with σ of order 1.0-1.5°. A diffraction pattern from a fiber of Pf1 is shown in Fig. 2a.

Oriented solutions are produced by placing a 30-50 mg/ml solution of filamentous bacteriophage in a closed container in a magnetic field and allowing it to orient. Diffraction patterns may then be taken of the solution while it is in the magnetic field (Glucksman et al., 1986). Alternatively, the solution may contain acrylamide, which is then polymerized using ultraviolet light while the solution is still in the magnetic field (Stark et al., 1988). The acrylamide polymerizes around the virus particles, resulting in a specimen in which the virus particles remain stably oriented for many months. Fig. 2b is a diffraction pattern from an oriented solution of Pf1 that was embedded in acrylamide while in a 5.8 Tesla magnetic field. Diffraction patterns from these specimens have orientational distributions with σ of 6-8°.

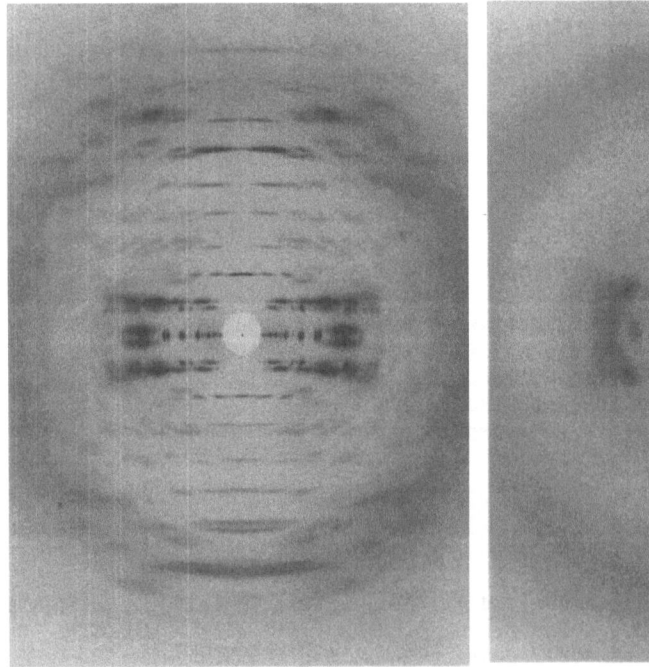 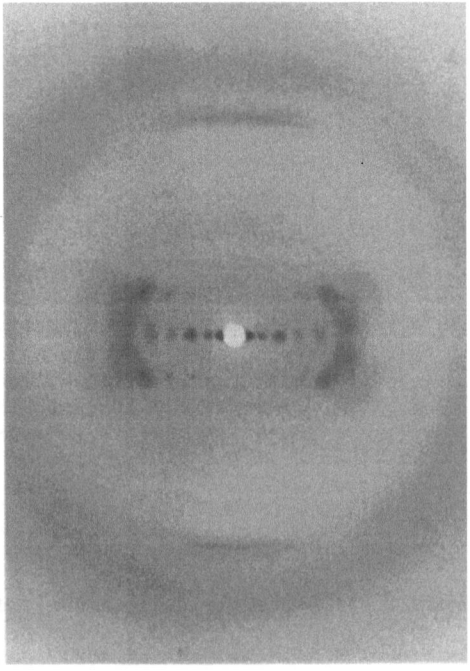

2a   2b

Fig. 2.   (a) Diffraction pattern from a fiber of filamentous bacteriophage
          Pf1 with a half-width of disorientation, σ of about 1.5°. (b)
          Diffraction pattern from an oriented solution of Pf1 embedded in
          acrylamide. The half-width of disorientation, σ, of this specimen
          is about 6-8°.

	σ	B (Tesla)	$\Delta\chi(\times10^{20}erg/G^2)$	Number of Particles Pf1	M13
Fibers	1.5°	5.8	-1.793	3735	5022
		2.5	-9.649	20100	27028
Solution	6.0°	5.8	-0.112	233	314
		1.6	-1.475	3073	4123

COOPERATIVITY IN THE FORMATION OF ORIENTED SPECIMENS

The known chemical compositions of the filamentous bacteriophages allow us to make estimates of their maximum possible diamagnetic anisotropy. Assuming that the anisotropy comes entirely from the peptide bonds, aromatic side chains, and associated DNA bases, the maximum diamagnetic anisotropy for a single Pf1 particle is approximately $-4.8 \times 10^{-24}erg/G^2$, and for a single M13 particle it is $-3.6 \times 10^{-24}erg/G^2$. Given the observed distribution of orientations, the minimum size of the cooperative unit orienting in these specimens varies from about 200 to over 20,000 virus particles. These values are given in Table 2.

The diamagnetic anisotropy listed is the minimum that a cooperative unit must have in order to produce the observed distributions of particle orientations. Also tabulated are the minimum number of virus particles required to produce cooperative units with these anisotropies.

The size of these numbers indicates that liquid crystalline behavior is essential to obtain orientation in these specimens. Evidently, a single particle will exhibit little or no orientation in a magnetic field at room temperature. However, liquid crystalline domains that orient as single units and hold several thousand particles are likely to orient in a magnetic field to an extent useful for diffraction experiments.

DISTRIBUTION OF PARTICLE ORIENTATIONS IN SINGLE LIQUID CRYSTAL DOMAINS

These considerations do not, however, completely explain the experimental observations. Equation (5) indicates that σ should be proportional to the magnetic field. We found that under the best possible conditions, fibers tend to give rise to diffraction patterns with σ not substantially smaller than 1.2°, and oriented solutions give rise to patterns with observed σ of 5-8° independent of magnetic field over variations of field strength of a factor of 2-3. These observations have included many fibers grown at 2.5 and 5.8 Tesla, and oriented solutions formed at 1.6 and 5.8 Tesla. Qualitatively, there was little difference in the degree of orientations observed at the higher fields. The cause of this observation may be as follows: The distribution of orientations obtained by diffraction patterns includes a contribution from disorientation of liquid crystalline domains plus the disorientation of individual particles within the domains. The fact that the observed distribution of particle orientations is independent of magnetic field suggests that the orientation of domains has become saturated, and the variation observed is due largely to single particle variations within the domains. Consequently, the sizes of liquid crystalline domains in filamentous bacteriophage specimens are likely to be much larger than the minimum sizes listed in Table 2.

In oriented solutions of filamentous bacteriophages, the typical interparticle distance is of the order of 300Å, about 5 times the diameter of a single virus particle. The distribution of orientations within a liquid crystalline domain of this kind will depend critically on the interparticle potential which in turn depends on the detailed structure of the particle.

In fibers, the phage particles are virtually in contact with one another, forming a two-dimensional hexagonal lattice with lattice constant equal to or slightly less than the expected maximum diameter of the virus particles. Disorientation in these domains may originate during the final drying when small variations in orientation become frozen into the fiber, which eventually dries to a glass-like consistency. These appear to depend quite sensitively on the details of the procedure for fiber formation. (It is quite easy to produce poorly oriented fibers by small changes in the protocol.) The presence of aggregates, defects and other imperfections will contribute to the observed degree of disorientation.

SUMMARY

The diamagnetic anisotropy of molecular groups in macromolecules has been used as a basis for predicting the minimum size of a cooperative unit required to obtain useful orientation for a macromolecular assembly in a magnetic field. In the case of filamentous bacteriophages, to explain the observed orientation, at least several thousand virus particles must be orienting cooperatively. A single particle will exhibit little or no orientation at room temperature even in a relatively strong magnetic field. The observation that there is little variation in the quality of specimens produced over moderate ranges of magnetic field strength, suggests that the orientation of the cooperative units is saturated at relatively modest magnetic fields; that the cooperative units (liquid crystalline domains) are substantially larger than the minimum size calculated, and that the observed orientational distribution is almost entirely representative of the distribution of orientations within single liquid crystalline domains. When using magnetic fields for orientation, liquid crystalline cooperative behavior appears to be a requirement to obtain the kind of orientation needed for detailed structural studies.

REFERENCES

Chalazonitis, N., Changeuz, R., and Arvanitaki, A., 1970, Rotation des segments externes des photorecepteurs dans le champ magnetique constant, C. R. Hebd. Seances Acad. Sci. Ser. D, 271:130.
Freyssinet, J. M., Torbet, J., Hudry-Clergeon, G., and Maret, G., 1983, Fibrinogen and fibrin structure and fibrin formation measured by using magnetic orientation, Proc. Natl. Acad. Sci. USA, 80:1616.
Glucksman, M. J., Hay, R. D., and Makowski, L., 1986, X-ray diffraction from magnetically oriented solutions of macromolecular assemblies, Science, 231:1273.
Makowski, L., 1978, Processing of x-ray diffraction data from partially oriented specimens, J. Appl. Cryst. 11:273.
Maret, G., and Dransfeld, K., 1985, Topics in applied physics, in: "Strong and Ultrastrong Magnetic Fields and Their Applications," F. Herlach, ed., Springer-Verlag, New York.

Nave, C., Brown, R. S., Fowler, A. G., Ladner, J. E., Marvin, D. A.,
     Provencher, S. W., Tsugita, A., Armstrong, J., and Perham, R. N.,
     1981, Pfl filamentous bacterial virus. X-ray fibre diffraction
     analysis of two heavy atom derivatives, J. Mol. Biol., 149:675.
Pauling, L., 1936, Diamagnetic anisotropy of aromatic groups, J. Chem.
     Phys., 4:673.
Pauling, L., 1979, Diamagnetic anisotropy of the peptide group, Proc. Natl.
     Acad. Sci. USA, 76:2293.
Stark, W., Gluckman, M. J., and Makowski, L., 1988, Conformation of the coat
     protein of filamentous bacteriophage Pfl determined by neutron
     diffraction from magnetically oriented gels of specifically
     deuterated virions, J. Mol. Biol., 199:171.
Torbet, J., and Maret, G., 1979, Fibres of highly oriented Pfl bacteriophage
     produced in a strong magnetic field, J. Mol. Biol., 134:843.
Torbet, J., 1986, Using magnetic orientation to study structure and
     assembly, TIBS, 12:327.
Torbet, J., Timmins, P. A., and Y. Lvov, 1986, Internal structural aniso-
     tropy of spherical viruses studied with magnetic birefringence,
     Virology, 155:721.
Worcester, D. L., 1978, Structural origins of diamagnetic anisotropy in
     proteins, PNAS, 75:5475.

PARTICIPANTS

ALMO, Steven
 Dept. of Chemistry, Room 2-212,
Massachusetts Inst. of Technology
77 Massachusetts Avenue
Cambridge, MA 02139

ARMIJO, Michael C.
 Biology Dept.
Brookhaven National Lab.
Upton, NY 11973

BELLAMY, Henry
 Stanford Synchrotron
    Radiation Lab.
Stanford Linear Accelerator Cntr.
Stanford U.
P.O. Box 4349, Bin 69,
Stanford, CA 94305

BERNSTEIN, Frances C.
Chemistry Dept.
Brookhaven National Lab.
Upton, NY 11973

BJORKMAN, Pamela J.
 Dept. of Medical Microbiology
Stanford U.
Stanford, CA 94305

BLASIE, J. Kent
 Depts. of Chemistry and
    Biochemistry/Biophysics
U. of Pennsylvania
Philadelphia, PA 19104

BOMMANNAVAR, Arun
 National Synchrotron Light Source
Brookhaven National Lab.
Upton, NY 11973

BORDAS, Joan
 Science and Engineering
    Research Council
Daresbury Lab., Warrington
Cheshire, England WA4 4AD

CAPEL, Malcolm S.
 Biology Dept.
Brookhaven National Lab.
Upton, NY 11973

CAVARELLI, Jean
 Dept. of Biological Sciences
Lilly Hall of Life Sciences
Purdue U.
West Lafayette, IN 47907

CHEN, Chun-Zhang
 Biology Dept.
Brookhaven National Lab.
Upton, NY 11973

CHENG, Xiaodong
 Biology Dept.
Brookhaven National Lab.
Upton, NY 11973

CRUICKSHANK, Durward W. J.
 Chemistry Dept.
U. of Manchester Institute of
    Science and Technology
105 Moss Lane, Alberley Edge
Cheshire, SK9 7HW, England

CYR, Donna
 Biology Dept.
Brookhaven National Lab.
Upton, NY 11973

DE CROMBRUGGHE, Marie A.
 Biology Dept.
Brookhaven National Lab.
Upton, NY 11973

DEWAN, John C.
 Chemistry Dept.
New York U.
4 Washington Place
New York, NY 10003

DONG, Baozhong
 National Synchrotron Light Source
 Brookhaven National Lab.
 Upton, NY 11973

DUBENDORFF, John
 Biology Dept.
 Brookhaven National Lab.
 Upton, NY 11973

EALICK, Steven E.
 Center for Macromolecular
   Crystallography
 U. of Alabama at Birmingham
 BHS 268, THT 79
 Birmingham, AL 35294

ENEMARK, John H.
 Dept. of Chemistry
 U. of Arizona
 Tucson, AZ 85721

FREUND, Andreas K.
 European Synchrotron Radiation
   Facility
 B.P. 220
 F-38043 Grenoble Cedex, France

GARNER, Christopher David
 Chemistry Dept.
 U. of Manchester
 Manchester M13 9PL, England

GEACINTOV, Nicholas E.
 Chemistry Dept.
 New York U.
 New York, NY 10003

GLINKA, Charles J.
 Reactor Radiation Div.
 National Bureau of Standards
 Gaithersburg, MD 20899

HAJDU, Janos
 Lab. of Molecular Biophysics
 Rex Richards Building
 Oxford U.
 South Parks Road
 Oxford OX1 3QU, England

HENDRICKSON, Wayne A.
 Dept. of Biochemistry, BB 523
 Columbia U.
 630 West 168th Street
 New York, NY 10035

HIND, Geoffrey
 Biology Dept.
 Brookhaven National Lab.
 Upton, NY 11973

HONG, Yuqun
 Biology Dept.
 Brookhaven National Lab.
 Upton, NY 11973

HUXLEY, Hugh E.
 The Rosenstiel Basic Medical
   Sciences Research Center
 Brandeis U.
 Waltham, MA 02254

ITO, Takashi
 Institute of Physics
 College of Arts and Sciences
 U. of Tokyo
 Meguroku, Komaba 3-8-1
 Tokyo 153, Japan

JANIN, Joel
 U. Paris-Sud, Orsay
 Box 433, 91405 Orsay
 France

JOHNSON, John E.
 Dept. of Biological Sciences
 Lilly Hall of Life Sciences
 Purdue U.
 West Lafayette, IN 47907

KAHN, Richard
 LURE, Batiment 209
 91405 Orsay
 France

KATAOKA, Mikio
 Dept. of Molecular Biology
   and Biophysics
 Yale U.
 260 Whitney Avenue
 New Haven, CT 06511

KNIGHT, Stefan
 Dept. of Molecular Biology
 Swedish U. of
   Agricultural Sciences
 Uppsala Biomedical Center
 Box 590, S-751 24
 Uppsala, Sweden

KNOTEK, Michael L.
 National Synchrotron Light
   Source
 Brookhaven National Lab.
 Upton, NY 11973

LANGE, Gudrun
Max-Planck-Unit for Structural
  Molecular Biology
Ohnhorststrasse 18, D-2000
Hamburg 52, West Germany

LAWS, William R.
Dept. of Biochemistry
Mt. Sinai School of Medicine
One Gustave L. Levy Place
New York, NY 10029

LEVIN, Paul
Biology Dept.
Brookhaven National Lab.
Upton, NY 11973

MAKOWSKI, Lee
Dept. of Physics
Boston U.
590 Commonwealth Avenue
Boston, MA 02215

MANDELKOW, Eckhard
Max-Planck-Unit for Structural
Molecular Biology
Ohnhorststrasse 18, D-2000
Hamburg 52, West Germany

MATSUDAIRA, Paul
Whitehead Institute
9 Cambridge Center
Cambridge, MA 02142

MC CRAY, James A.
Dept. of Physics
Drexel U.
32nd and Market
Philadelphia, PA 19104

MC MILLAN, Martin
Analytical Technology Div.
Eastman Kodak, Bldg. 49
2nd Floor, Kodak Park
Rochester, NY 14650

MOFFAT, J. Keith
Section of Biochemistry
Molecular and Cell Biology
252 Clark Hall
Cornell U.
Ithaca, NY 14853

ORME-JOHNSON, William H.
Dept. of Chemistry
Bldg. 18, Room 023
Massachusetts Institute of
  Technology
Cambridge, MA 02139

PENNER-HAHN, James E.
Dept. of Chemistry
U. of Michigan
930 N. University
Ann Arbor, MI 48109

PHIZACKERLEY, R. Paul
Stanford Synchrotron
  Radiation Lab.
Stanford Linear Accelerator Cntr.
Stanford U.
P.O. Box 4349, Bin 69
Stanford, CA 94305

POLEWSKI, Krzysztof
Biology Dept.
Brookhaven National Lab
Upton, NY 11973

PRATER, James
Dept. of Chemistry
Massachusetts Inst. of Technology
77 Massachusetts Avenue
Cambridge, MA 02139

RADEKA, Veljko
Instrumentation Div.
Brookhaven National Lab.
Upton, NY 11973

RAMAKRISHNAN, Venki
Biology Dept.
Brookhaven National Lab.
Upton, NY 11973

ROBINSON, Ian
National Synchrotron Light Source
Brookhaven National Lab.
Upton, NY 11973

SAXENA, Anand M.
Biology Dept.
Brookhaven National Lab.
Upton, NY 11973

SAYERS, Zehra
European Molecular Biology Lab.
Hamburg Outstation
Notkestrasse 85
2000 Hamburg 52, West Germany

SCARROW, Robert C.
Dept. of Chemistry
U. of Minnesota
207 Pleasant Street, S.E.
Minneapolis, MN 55455

SCHILDKAMP, Wilfried
 Section of Biochemistry,
   Molecular and Cell Biology
 281 Wilson Laboratory
 Cornell U.
 Ithaca, NY 14853

SCHMID, Michael F.
 Dept. of Biochemistry
 U. of Arizona
 Tucson, AZ 85721

SCHNEIDER, Dieter K.
 Biology Dept.
 Brookhaven National Lab.
 Upton, NY 11973

SCHOENBORN, Benno P.
 Biology Dept.
 Brookhaven National Lab.
 Upton, NY 11973

SETLOW, Richard B.
 Biology Dept.
 Brookhaven National Lab.
 Upton, NY 11973

SINGER, Paul
 Dept. of Pharmacological Sciences
 State U. of New York
 Stony Brook, NY 11794

SOLTIS, Michael
 Stanford Synchrotron
   Radiation Lab.
 Stanford Linear Accelerator Cntr.
 Stanford U., P.O. Box 4349
 Bin 69
 Stanford, CA 94305

STAUDENMANN, Jean-Louis
 Howard Hughes Medical Inst.
 Columbia U.
 c/o National Synchrotron Light
   Source
 Brookhaven National Lab.
 Upton, NY 11973

STEVENS, Eugene S.
 Dept. of Chemistry
 State U. of New York
   at Binghamton
 Binghamton, NY 13901

STUHRMANN, Heinrich
 U. of Mainz
 c/o Abteilung Makromolekulare
   Chemie
 Forschungszentrum Geesthacht GMBH
 Max-Plank-Strasse Postfach 1160
 2054 Geesthacht, West Germany

SUTHERLAND, John C.
 Biology Dept.
 Brookhaven National Lab.
 Upton, NY 11973

SWEET, Robert M.
 Biology Dept.
 Brookhaven National Lab.
 Upton, NY 11973

TAKACS, Peter Z.
 Instrumentation Div.
 Brookhaven National Lab.
 Upton, NY 11973

THEIL, Elizabeth C.
 Dept. of Biochemistry
 North Carolina State U.
 Raleigh, NC 27695

WANG, Dewu
 National Synchrotron Light Source
 Brookhaven National Lab.
 Upton, NY 11973

WESTBROOK, Edwin M.
 Biological, Environmental,
   and Medical Research Div.
 Argonne National Lab.
 9700 South Cass Avenue
 Argonne, IL 60439

WHITE, Stephen W.
 Biology Dept.
 Brookhaven National Lab.
 Upton, NY 11973

WILLIAMS, Grahame J. B.
 Enraf-Nonius Service Corp.
 390 Central Avenue
 Bohemia, NY 11716

WILSON, Keith S.
 European Molecular Biology Lab.
 Hamburg Outstation, c/o DESY
 Notkestrasse 85, 2000 Hamburg 52
 West Germany

WOODHEAD, Avril D.
 Biology Dept.
 Brookhaven National Lab.
 Upton, NY 11973

WU, Bomu
 Dept. of Biological Sciences
 Lilly Hall of Life Sciences
 Purdue U.
 West Lafayette, IN 47907

Pyrex-glass mirror (continued)
reflection curves, 262-263

Quadrant and linear x-ray detectors
comparison of output of, 68-69
Quaternary structures
of isometric RNA viruses, 141
of viruses, 156

Ray tracing methods, 258
Real space focusing, 275
Receptor mediated endocytosis, 67,69
Reciprocal space diagrams, 257,
259-269
symmetric reflections, 270
Resolution ellipsoid
in reciprocal space, 274-275
Resonant scattering, 85
Resonant x-ray scattering, 83
Rh. rubrum enzyme, 113
Rieske clusters
EXAFS spectra, 180
nitrogen coordination, 181
XANES structure, 182
Rieske-like clusters, 178
curve-fitting results, 181
EPR spectra, 177
redox potentials, 177
Rieske-like site
EPR spectrum, 179
EXAFS spectra, 179
Fe-S cluster, 177
RNA
in comoviruses, 155
in picornaviruses, 155
RNA replication, 157
in plant viruses, 156
proteins, 156
subunit role in, 156
x-ray crystallography, 141
Rotating anode
as X-ray source, 5, 124
Rotation camera facility, 26
Rubisco
active site of, 118, 120
active site residues, 111-112, 118
photorespiration, 111
photosynthesis, 111
types of subunits, 111
Rubisco subunit
schematic of, 114-115

Sagittal focusing, 271, 274
bending modes, 296-297
illustration of, 294
mirrors, 295
monochromators, 295, 299
optics, 293-295
Sarcoplasmic reticulum
Ca^{2+}ATPase of, 77

Sarcoplasmic reticulum (continued)
functional studies, 77-78
structural studies, 77-78
Sarcoplasmic reticulum ATPase
large-scale structural changes,
77, 82
Sarcoplasmic reticulum membrane
calcium active transport, 77
structure of, 80
SAXS camera
vertical linear position sensitive
detector (LPSD), 30
Scattering curves
of microtubules, 95-96
of protofilament fragments, 95-96
of rings, 95-96
of tubulin dimers, 95-96
Screened precession photography, 130
Signal-to-noise ratio, 327
Silicon pad detectors, 251-252
Silicon pixel detectors, 250
Simulated fluorescence decay curve
analysis of, 211
Single bunch Laue patterns, 326
Single-crystal Bragg-Fresnel
elements, 285
Single crystal monochromators, 87,
265-280
Single isomorphous replacement (SIR)
map, 127-128
Small angle x-ray scattering
for dynamic studies of muscle, 246
Small RNA viruses
capsids of, 151
evolution, 156
polypeptide chains of, 154
Small spherical RNA virus subunits
polypeptide chains of, 154
Soft x-ray microscopy, 65
SPEAR storage ring
electron bunches, 19
protein dynamics, 216
pulse instruments, 216
Stanford Linear Accelerator Center
(SLAC), 19-20, 23
synchrotron radiation research,
19, 22
time-resolved experiments, 19, 21
SPEAR beamlines
schematic layout, 21
Spin-trapping technique, 227, 238
Spinach Rubisco
active site, 116-117
crystals, 112
large subunit of, 113-114
lysine residues, 116
metal binding site, 117-118
quaternary structure, 114-115
Spot integration
HLA films, 126